POPULATION GENETICS OF FOREST TREES

FORESTRY SCIENCES

Volume 42

The titles published in this series are listed at the end of this volume.

Population Genetics of Forest Trees

Proceedings of the International Symposium on Population Genetics of Forest Trees Corvallis, Oregon, U.S.A., July 31–August 2, 1990

Edited by

W. T. ADAMS

Oregon State University

STEVEN H. STRAUSS

Oregon State University

DONALD L. COPES

USDA Forest Service, Pacific Northwest Research Station, Oregon

and

A. R. GRIFFIN

Shell International Petroleum Co., London

Reprinted from New Forest 6 (1–4)

KLUWER ACADEMIC PUBLISHERS

DORDRECHT / BOSTON / LONDON

Library of Congress Cataloging-in-Publication Data

International Symposium on Population Genetics of Forest Trees (1990 : Corvallis, Or.)
Population genetics of forest trees : proceedings of the International Symposium on Population Genetics of Forest Trees, Corvallis, Oregon, U.S.A., July 31-August 2, 1990 / edited by W.T. Adams ... [et al.].
p. cm. -- (Forestry sciences ; 42)
ISBN 0-7923-1857-9 (alk. paper)
1. Forest genetics--Congresses. 2. Plant population genetics--Congresses. 3. Biochemical markers--Congresses. I. Adams, W. T. II. Title. III. Series.
SD399.5.I56 1990
634.9'56--dc20 92-18860

ISBN 0-7923-1857-9

Published by Kluwer Academic Publishers,
P.O. Box 17, 3300 AA Dordrecht, The Netherlands.

Kluwer Academic Publishers incorporates
the publishing programmes of
D. Reidel, Martinus Nijhoff, Dr W. Junk and MTP Press.

Sold and distributed in the U.S.A. and Canada
by Kluwer Academic Publishers,
101 Philip Drive, Norwell, MA 02061, U.S.A.

In all other countries, sold and distributed
by Kluwer Academic Publishers Group,
P.O. Box 322, 3300 AH Dordrecht, The Netherlands.

Printed on acid-free paper

Printed in the Netherlands

Contents

New Forests **6**: 1, 1992.

Introduction

This book contains 19 invited papers from the International Symposium on Population Genetics of Forest Trees, held in Corvallis, Oregon, USA, July 31 – August 2, 1990. The theme of this joint meeting of the International Union of Forestry Research Organizations (IUFRO) Working Parties on Biochemical Genetics (S2.04.05) and Population and Ecological Genetics (S2.04.01) was "the contribution of biochemical markers to the understanding of population genetics of forest trees". Over 140 individuals from 15 countries attended. In addition to IUFRO, sponsors included the Forest Science Department of Oregon State University, and the USDA Forest Service, Pacific Northwest Research Station.

The increased availability of biochemical markers has resulted in extensive research directed to the population genetics of forest trees during the past 15 years. These studies have added greatly to our understanding of the genetics of tree species, especially in the areas of population genetic structure, evolutionary biology, and mating systems. The goals of this symposium were to summarize and review the results of these efforts and to identify fruitful areas for future research.

The papers in this volume address all of the major applications of biochemical markers to forest genetics research. These include description of the patterns of genetic diversity within and between species, and assessment of their underlying causes (8 papers); investigation of mating systems, gene dispersal, and genetic structure within populations (4 papers); and, application of biochemical markers to problems in forest management (4 papers).

While the emphasis in the above papers is on studies employing isozymes or terpenes, two additional papers specifically address the potential of DNA as a biochemical marker for forest genetics research. The final paper was written by invitation as a commentary on the symposium. This paper does not attempt, however, to summarize the major conclusions of the individual papers, but rather offers personal critiques and insights on current approaches to population genetics research in forest trees.

We thank the symposium sponsors for helping to make this symposium a success. We are also grateful to Dr. Mary Duryea, Editor-in-Chief of *New Forests*, and to Mr. Ad C. Plaizier of Kluwer Academic Publishers, for their encouragement in this project.

Special Editors

W. T. Adams
S. H. Strauss
D. L. Copes
A. R. Griffin

New Forests **6**: 3, 1992.

Dedication

We dedicate these proceedings to our colleague and friend, Dr. James Hanover, who passed away on February 21, 1992. In a distinguished career of more than 35 years as teacher and scholar, Jim contributed greatly to the fields of forest genetics and tree physiology.

New Forests **6**: 5–22, 1992.

Review paper

Genetic diversity—seeing the forest through the trees

M. THOMPSON CONKLE

Institute of Forest Genetics, Pacific Southwest Forest and Range Experiment Station, USDA Forest Service, Box 245, Berkeley, California, 94701, USA

Received 7 December 1990; accepted 13 October 1991

Key words: allozymes, biochemical markers, heterozygosity

Application. Processing biochemical markers from individual trees provides a means to resolve and measure genetic diversity. Identifying that variation and interpreting that rich source of genotypic information leads to effective means for managing populations to maintain forest health. Looking ahead, research in molecular forestry will extend toward discovering how the biology of forest species is effected by variation in individual genes.

Abstract. Forest trees, populations, races, species, and taxonomic groups above the species level display rich variation in biochemical markers. The variation stems from inherited modifications that trace back in time, through converging ancestries, towards common progenitors. Past movements of continents, mountain building events, and climate changes isolated forest populations one from another and provided critical challenges to the lineages that survived to the present day. A wealth of molecular variants in forest trees characterize these widely-distributed, large, long-lived, outbreeding, organisms. Forest trees have an abundance of rare variants and over one third of all the alleles (different forms of one gene) occur only rarely ($<2\%$ frequency) in a few trees of a species sample. Those rare alleles may either represent new variation or persistent forms of genes that have low adaptive value under present conditions. From another perspective, however, the largest share of genetic variation in forest trees is due to the presence of multiple alleles found at intermediate frequencies for only a small percentage of all the genes, and those alleles are commonly widespread throughout species areas. These common alleles may mark genes that track historical events in lineages or mark genes with adaptive significance in present populations. Evidence from enzyme studies supports the conclusion that highly comparable functional genes are common to different forest taxa. Future research will be toward understanding the phenotypic expression of particular genes and revealing the relative importance of genetic variants to adaptation and growth.

Introduction

"You can't see the forest for the trees!" That familiar phrase is often intended as light hearted criticism of work on narrow problems. It carries a double meaning for forest geneticists who gathered at a symposium in

Corvallis, Oregon, to reflect, report, and speculate on a broad spectrum of information about forest diversity. This opening contribution to the symposium addressed a mix of three subjects ranging from the origins of forest taxa, through examination of diversity in a sample of species, to concluding remarks about the construct of forest species genomes. It covered general subject matter and intentionally avoided the specific themes in molecular forestry that were addressed by the other participants.

The timing of the symposium was appropriate because a sizable volume of findings had become available since July, 1979, when a similar group convened for discussions on isozymes of North American forest trees and insects (Conkle 1981). The presentations in 1979 demonstrated the high value of biochemical markers (primarily mobility variants for numerous, specific enzymes coded by nuclear genes) for evaluating variation in forest tree populations. Since then, the necessity to conserve and manage genetic variation has become a worldwide priority. And with few efficient means available to evaluate diversity, the rapid acquisition of precise genetic information using biochemical markers plays a central and critical roll in assessing populations and monitoring species in native forest ecosystems.

A complete consideration of forest diversity would include mention of variation in natural morphology, progeny growth in plantation trials, and survival responses in stress tests. Those measurements, trials, and tests are essential for evaluating whole plant responses but the differences observed in those trials are controlled by variation in unidentified genes. Here, consideration focuses on knowledge about variation in alleles of identified genes. Currently, these alleles and variation in these genes appear to have only minor predictive value of specific growth responses in whole plants.

Clues about diversity from the past

Skilled analysis of morphological characteristics has provided reasonably dependable and time proven methods for classifying organisms. Phylogenetic analyses of extant and fossil taxa when combined with events in paleohistory gives a basis for appreciating the ancestral relationships among modern species and speculating about the origins of their various adaptations. It is not trivial to observe that the relationships among forest species are best pictured as a tree with branches and twigs (clades) that trace back to a single root, a progenitor common to all plants.

Forest seed-plant origins are in the Coniferopsida (a subdivision of the gymnosperms which includes orders of ginkgos, conifers, and yews) and date from the Carboniferous, roughly 300 million years ago (Ma). Constance Millar (Millar 1991; Millar and Kinloch 1991) evaluated information about key events in the history of pines, a genus with about 100 species, wide distribution in the northern hemisphere, and an informative example of forest tree species evolution. She used evidence from the oldest pine fossil (dated about 130 Ma) and from the wide abundance of other fossil pines shortly thereafter to estimate the derivation of the genus *Pinus* at about 200 Ma. The first pines evolved in the middle latitudes of the northern hemisphere at a time when continents formed a single great land mass (Pangaea) and ice blanketed extensive areas (Fig. 1a) of the south polar masses (Harrington 1987; Parrish 1987). Continental collisions during formation of that land mass raised mountains near points of impact (Appalachian and Atlas Mountains, and elevated topography in Greenland and Norway) that gave variety to habitats for the evolving lineages.

First angiosperm fossils appeared at about 120 Ma and differentiation

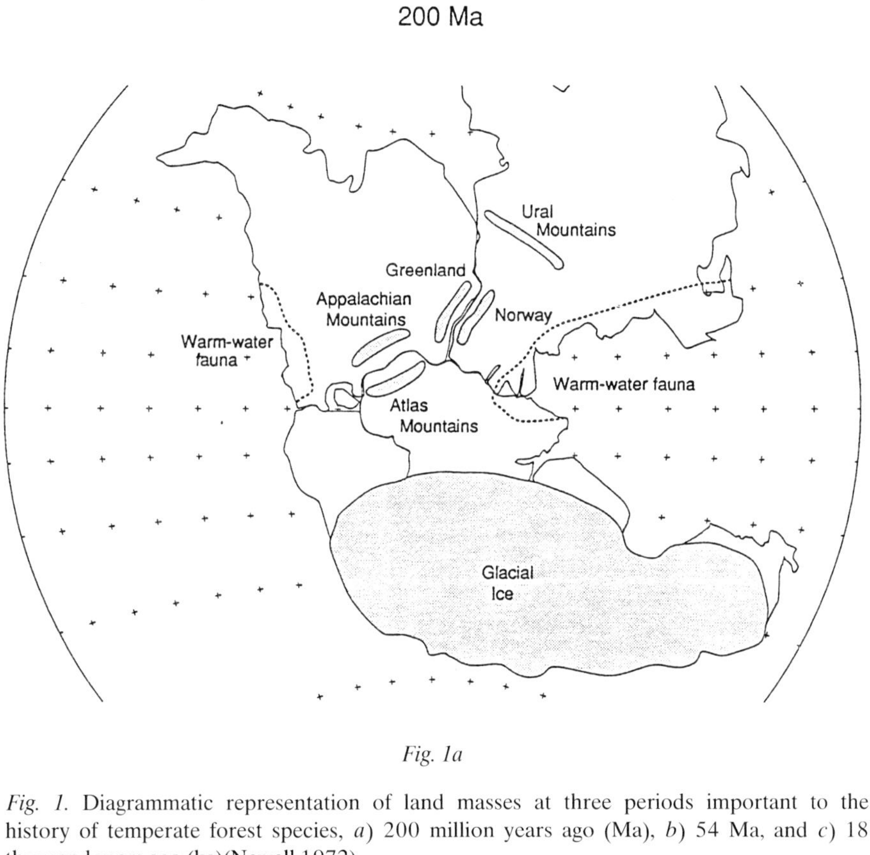

Fig. 1a

Fig. 1. Diagrammatic representation of land masses at three periods important to the history of temperate forest species, *a*) 200 million years ago (Ma), *b*) 54 Ma, and *c*) 18 thousand years ago (ka)(Newell 1972).

into dicotyledon and monocotyledon divisions virtually coincided with their first appearance (Friis et al. 1987). An explosive radiation of angiosperms from low latitudes established them world wide by 100 Ma as the overwhelming dominant land plant group (Crane 1987). Conifers, by that time, had spread east and west throughout North America and Eurasia (Laurasian land mass) and they equaled all the other land plants combined in absolute species diversity (Nichols et al. 1985).

At about 100 Ma a warm sea separated the northern from the southern land masses. The North American and Eurasian masses were drifting apart then, but they were still joined in their far northern areas. A group of continents (South America, Africa, India, and Australia joined to Antarctica) were drifting apart from the southern conglomerate (Gondwana). North—south interior seaways bisected North America (from the Gulf of Mexico to the Arctic) and Eurasia (from the eastern Mediterranean to the Arctic). Waters of those interior seaways isolated substantial regions of both of the northern land masses.

Events beginning about 54 Ma (Fig. 1b) profoundly influenced the

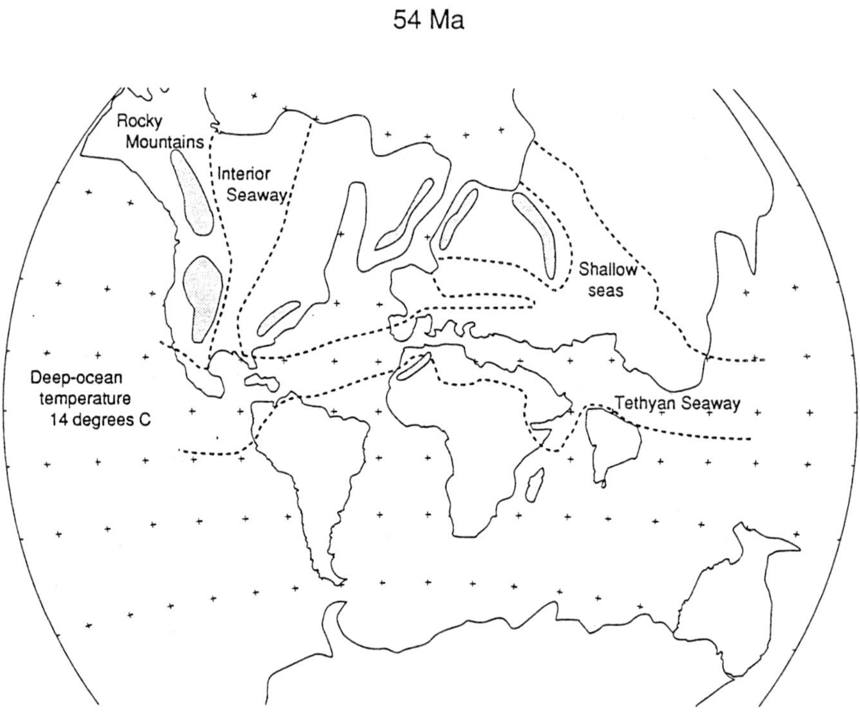

Fig. 1b

evolution of conifers (evidence from pines, Millar 1991) and deciduous forest genera (Critchfield 1980). During a period of warm humid climates, sub-tropical floristic associations became established in relatively high latitudes (centered 75 °N), fracturing and displacing the temperate species into either more northern, more southern, or more upland areas. Pine species adaptations and geographic patterns associated with modern subsections of *Pinus* largely reflect evolution in refugial regions for the period from about 54 Ma to 37 Ma (Millar 1991). Appalachian forests displayed a mix of moist tropical and deciduous hardwood vegetation suggesting that some seasonally dry and warm-temperate climates prevailed there during the same period (Critchfield 1980).

The continued westward drift of North and South Americas further isolated them from the eastern continents and their lineages from their eastern vegetation counterparts. The genetic differentiation of hard pines continued on separate continents and now genetic barriers to crossing precludes hybridizations between species native to the Eastern and Western hemispheres (Critchfield 1986).

Dramatic topographic and climatic events took place between 54 and 37 Ma. Rocky Mountain topography (with a 100 million year period of

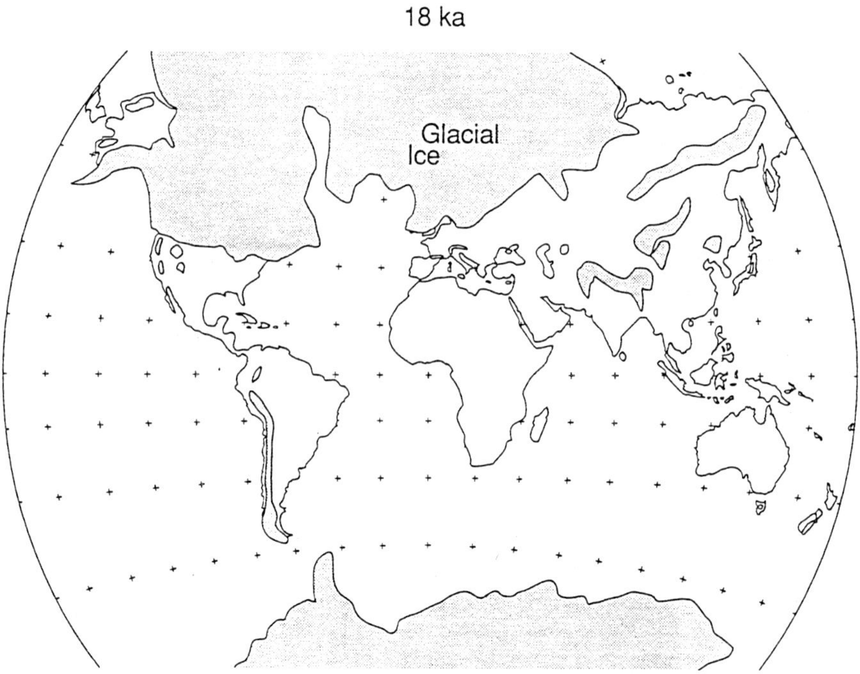

Fig. 1c.

uplift) emerged in full development. India collided with Asia and raised the Himalayas. Australia separated from Antarctica. Ocean currents, then shunted through cold polar seas in a new seaway south of Australia, brought about significant ocean cooling (Kennett 1977). A corresponding severe drop in temperatures on land (along with increased seasonality) brought the warm periods to a close. Tropical and subtropical plants were lost from mid and upper latitudes, and the previously displaced taxa reclaimed the temperate mid latitudes. The present differentiation among subsections of *Pinus* points to their evolution in the isolated refugial regions they occupied during the warm periods of the Eocene (Millar 1991). Further differentiation of lineages radiating from those refugia may have established many modern day species.

During a period between 37 and 2 Ma., the collision of Africa and Europe resulted in complete desiccation of the Mediterranean Sea (Hsu 1972) and development of the Pyrenees, Alps, and Carpathian Mountains. Continued uplift accompanied by volcanic activity along western margins of North, Central, and South America raised the Cascade-Sierra Nevada, elevated the Sierra Madre Occidental and the Transverse volcanic belt in Mexico, and formed the Andes in South America. Deep ocean cooling continued unabated. Deep ocean temperatures dropped from 14 °C, about 75 Ma, to 3 °C at the onset of the ice ages (Newell 1972).

Repeated cycles of the advance and retreat of glacial ice, beginning about 2.5 Ma, forced mid and upper latitude forest species into large scale migrations. The current ice ages began with about 40 full climatic cycles in 41 thousand-year rhythms. They then shifted to about 8 full cycles with long periods of glaciation, of 80 to 100 thousand years each, followed by short interglacial intervals, of 10 to 15 thousand years each (Ruddiman and Wright 1987). During each cycle, glacial ice advanced over polar and upper latitude areas and in high mountain terrain. Masses of ice at the culmination of the recent cold cycle, 18 thousand years ago, were the most extensive of all masses during the ice ages (Fig. 1c). Those great ice volumes lowered global seas about 130 m below the present levels (Cronin et al. 1981).

Forest species in mid and high latitudes shifted north and south, and higher and lower in elevation, in response to shifting habitats during the cold and warm periods of glacial cycles (Barnosky et al. 1987; Jacobson et al. 1987). The magnitude of these migrations can be appreciated by noting that northern boreal species of eastern North America were centered in southeastern United States about 18 thousand years ago (Webb et al. 1987). Vegetation in more stable environments (locations moderated by oceanic influences, arid rain-shadow locations, and equatorial regions) was effected less drastically. The migration of forest species revealed the limits of their respective adaptations. The near universal survival of forest species plus their general lack of morphological change during extreme stress imposed by the ice age climates are evidence of their persistence and the well buffered genetic stability of their traits.

Critchfield (1984) used data from genetic markers to summarize several consequences of recent glacial events for North American conifers. He noted populations in previously-glaciated northern areas to be less variable than their southern counterparts; the migrating founders of some species carried forward only a limited sample of progenitor genomes. Critchfield identified extant species that lack or have distinctly low levels of variation, resulting presumably from being driven to near extinction. He identified geographic races that formed as populations radiated from different refugia. Some races had transient existence owing to introgression or fusion upon encountering compatible taxa. Other races, evolved during periods of isolation to become varieties, subspecies, or species. Critchfield also provided evidence that some species increased variation by gene exchange during periods of contact with compatible species.

Given the past histories of boreal and temperate lineages, taxa differ widely in their adaptations to different climatic conditions. An expected response of some successful species to repeated glacial cycles might be the accumulation of wide climatic tolerances. Since temperate forest species have long been mobile geographic units responding to climate change, they might be expected to have accumulated adaptations for colonizing new habitats along their leading fronts and for resisting elimination along their trailing boarders. Each climatic cycle would result in shifts, from advance to retreat and from leading to trailing margins. Viewed today, some noteworthy species have extremely wide adaptations, they span significantly large ranges of elevations and latitudinal distances. Other species, presumably those for whom migration and survival were severe challenges, have narrow distributions. Some narrowly distributed taxa possess moderate variation, but others are known to be genetically depauperate and have little or no capacity remaining for significant evolutionary change. Biochemical markers, whether adaptive or neutral in present day environments, reveal amounts of allelic variation, identify genetic relationships, and track patterns of differentiation within and among populations.

Summary items for past evolution

- Forest species evolved from common ancestors;
- Conifers have about a 200 million year history that began when all continents were one land mass;
- Forest angiosperms evolved about 100 million years ago;

- Major differentiation of boreal and temperate forest groups occurred about 54 to 37 million years ago when warm climates favored development of subtropical vegetation in mid and upper latitudes, forcing temperate vegetation into cooler or drier refugia;
- Climatic cooling beginning about 37 Ma favored speciation of boreal and temperate forest taxa during periods of migration into new areas;
- Ice ages of recent history caused mass migration of species and contributed to further differentiation of lineages as populations repeatedly advanced and retreated, some into and out of restricted refugia, and others across latitudes and elevations that may have resulted in the development of wide environmental tolerances;
- Amounts and distribution patterns of variation in modern taxa reflect the molecular history of lineages as well as the disequilibrium inherent in mobile groups.

Profiles of species diversity

Diversity data from a number of species in California are available (Fig. 2). Species that are included all had genetic analyses performed using similar analytical and interpretative techniques and many grow together in natural mixtures (sympatry strengthens the comparisons among them). Wide spread species that span a notable range of elevations are: ponderosa (*Pinus ponderosa* Dougl. ex Laws.), Jeffrey (*P. jeffreyi* Grev. & Balf.) , and sugar pines (*P. lambertiana* Dougl.), Douglas-fir (*Pseudotsuga menziesii* [Mirb.] Franco), incense cedar (*Calocedris decurrens* Torr.) and California white fir (*Abies concolor* var. *lowiana* [Grod.] Lemm.). Others — Bishop (*Pinus muricata* D. Don), knobcone (*P. attenuata* Lemm.), Monterey (*P. radiata* D. Don), Coulter (*P. coulteri* D. Don), and Digger (*P. sabiniana* Dougl.) pines — are restricted to relatively narrow elevational bands and occupy well defined habitats. A third group consists of Monterey cypress (*Cupressus macrocarpa* Hartw.), Torrey (*Pinus torreyana* Parry ex Carr.) and Washoe (*P. washoensis* Mason & Stockwell) pines that seem near extinction because they grow in only a few diminutive native populations.

Three pines in the California sample — ponderosa, Washoe, and Jeffrey — are members of the *Ponderosae* subsection and stem from a common lineage. All members of their subsection inhabit western North America, and differentiation of the lineage traces to the period between 54 and 37 million years ago. These three species have northern distributions and northern ecological affinities but they are close relatives with a large group of western yellow pines endemic to Mexico and Central America (Conkle and Critchfield 1988). Jeffrey pine crosses to a moderate extent with

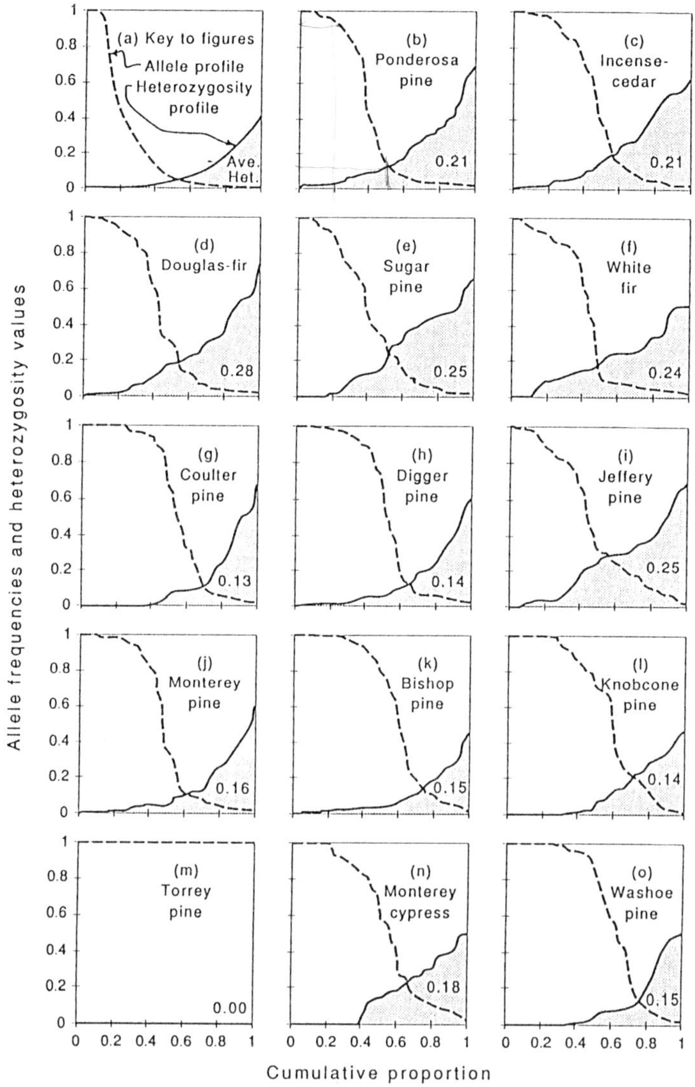

Fig. 2. Diversity profiles for 14 conifer species displaying cumulative distributions for: 1) alleles of different frequencies, irrespective of locus of origin, plotted in descending order, 2) expected heterozygosities by locus given in ascending order, and 3) mean expected heterozygosities (shaded areas under curves) whose species values are given: *a*) key to figures, *b*) ponderosa pine (*Pinus ponderosa*), *c*) incense-cedar (*Calocedris decurrens*, Harry 1984), *d*) Douglas-fir (*Pseudotsuga menziesii*), *e*) sugar pines (*Pinus lambertiana*), *f*) white fir (*Abies concolor*), *g*) Coulter pine (*Pinus coulteri*), *h*) Digger pine (*Pinus sabiniana*), *i*) Jeffrey (*Pinus jeffreyi*, Furnier 1984), *j*) Monterey pine (*Pinus radiata*, Millar et al. 1988), *k*) bishop pine (*Pinus muricata*, Millar et al. 1988), *l*) knobcone pine (*Pinus attenuata*, Millar et al. 1988), *m*) Torrey pine (*Pinus torreyana*, Ledig and Conkle 1983), *n*) Monterey cypress (*Cupressus macrocarpa* Conkle 1987), and *o*) Washoe Pine (*Pinus washoensis*, Niebling and Conkle 1990).

Coulter pine, and Jeffrey pine appears to be a species linking the *Ponderosa* and *Sabinianae* subsections (Conkle and Critchfield 1988). Coulter, Digger, and Torrey pines comprise the *Sabinianae* subsection. Their differentiation from yellow pines in Mexico and Central America traces to about 65 million years ago (Millar 1991). California closed cone pines — Monterey, Bishop, and knobcone — have historical differentiation times and geographic origins similar to the *Sabinianae.* They are the northern most members of the *Oocarpae* and are related to a group of extant species in Mexico and Central America. All pine species, just mentioned, are hard pines (subgenus *Pinus*).

Sugar pine is the only white pine (subgenus *Strobus*) in this sample. Its origins are obscure, but other members of subsection *Strobi* are distributed throughout North America and Eurasia. Douglas-fir, white fir, and Monterey cypress likewise are related to species distributed throughout North America and Eurasia. Northern hemisphere distributions for *Pinus*, *Pseudotsuga*, *Abies*, and *Cupressus* support the conclusion that these conifers evolved on northern land masses after separation from the southern land masses. Of the species sampled, only incense-cedar with its range centered in Oregon and California appears to have differentiated to a high degree prior to the break up of the universal land mass (Pangaea). Other Northern hemisphere incense-cedars (*Calocedrus*) are in China and Taiwan. Southern hemisphere incense-cedars (*Libocedrus*) are native to Chile in South America (also included under *Austrocedrus*) and to New Zealand and New Caledonia in the South Pacific.

Protein samples from forest trees, whether from seeds or vegetative tissues, when separated during electrophoresis and stained, reveal the presence of bands for specific enzymes. The majority of mobility variants found in enzyme studies of genetic diversity segregate in equal proportions with 50% of germ cells having one form and 50% having the alternative. Equal segregation of different alleles for genes tracing to egg or pollen cells provides strong evidence that they are alleles coded by nuclear genes (allozymes). Data from laboratory studies of allelic mobility variants for particular enzymes in species are often summarized in the form of allele frequencies and expected heterozygosities for numerous genes.

Species diversity characteristics can be generated from allozyme data by ignoring gene identities and by considering enzyme genes to be random samples of a large class of functional genes. Cumulative distribution frequencies characterize allele frequencies (frequency of occurrence) and provide a visual description of species heterozygosity (average proportion of loci with different alleles per gene in an individual, and also the average proportion of individuals in populations with different alleles per gene). Figure 2 pictures three comparisons for each species:

1. allele profiles based on a descending sorted sequence of allele frequencies for all genes;
2. heterozygosity profiles based on ascending sorted sequences of expected heterozygosities for all genes; and
3. average heterozygosity values (Ave. Het.).

The horizontal axis, marked in unit scale, is the same size as the vertical axis (allele frequencies and heterozygosity values) to make the figure a unit of area (the shaded area below the heterozygosity profile is equal to the heterozygosity for each species). Since alleles with frequencies less than 0.02 only have minor effects on heterozygosities, they were excluded from the allele frequency distributions (Fig. 2). For ponderosa pine, 20% of alleles have frequencies greater that 0.90, 50% of loci have expected values less than 0.10, and the average expected heterozygosity is 0.21.

Tree species commonly have minor proportions of fixed alleles (frequencies equal 1.00), and heterozygosities, when sorted, depict a linear to exponential increase in diversity from fixation (one allele, heterozygosity equals 0.00) to the most variable loci (heterozygosities >0.6). The exception is Torrey pine populations (Fig. 2 [m]) which have all genes with fixed alleles (average heterozygosity equals 0.00).

The wide spread species — ponderosa pine, Fig. 2 (b), incense-cedar (c), Douglas-fir (d), sugar pine (e), white fir (f), and Jeffrey pine (i) — have a range of average heterozygosity values from 0.21 to 0.28 (mean 0.24). Species with somewhat limited distributions — Coulter pine (g), Digger pine (h), Monterey pine (j), Bishop pine (k), and knobcone pine (l) — on examination have comparatively more fixed alleles in their allele profiles and somewhat more concave than linear heterozygosity profiles. The latter species range from 0.13 to 0.16 (mean 0.14) for average heterozygosities.

The rare endemic species — Torrey pine Fig. 2 (m), Monterey cypress (n), and Washoe pine (o) — contrast with the previous species. Torrey pine has no measurable diversity, while Monterey cypress and Washoe pine have profiles and heterozygosities that are nearly comparable with species having large distributions. A rare event in Torrey pine's history, perhaps the present species establishment by one or only a few founders, or a near extinction drop in population size, presumably caused fixation in its genes (Ledig and Conkle 1983).

Genes in the sampled species differ in the numbers of alleles per locus (from 1 to about 10). Since rare alleles occur almost exclusively in the heterozygous condition and since alleles with intermediate frequencies are major contributors to population heterozygosities, it is informative to examine the percent of alleles at different frequencies for genes with differing number of alleles (Table 1). Table 1 was developed from allele

Table 1. Proportions of alleles at various frequencies grouped by the number of alleles that are detected for genes in a sample of 13 conifer species (see Fig. 2, and excluding Torrey pine)

Allele frequency	Number of alleles per gene					Total(%)
	1	2	3	4	>4	
	Number of alleles (% of total)					
	65 (6)	163 (15)	283 (26)	250 (23)	326 (30)	1,087 (100)
	(%) in each class					
>0.95	100	31	14	6	3	17
>0.90	0	38	19	12	5	21
>0.80	0	44	26	17	11	27
>0.50	0	51	34	24	16	33
>0.20	0	57	42	32	21	39
>0.10	0	63	50	39	26	45
>0.05	0	69	57	49	36	53
>0.02	0	75	69	61	51	64
0.02 and lower	0	25	31	39	49	36

frequency data (total of 1,087 alleles) for all the sampled species except Torrey pine.

Overall, 6% of the genes had only one allele, 15% had 2 alleles, and close to 80% had 3 or more alleles (Table 1). About one third of all alleles had frequencies greater than 0.50. Thirty one % of genes with 2 alleles had frequencies in excess of 0.95 (>0.95), but genes with more the 4 alleles had only 3% of alleles with frequencies in excess of 0.95. Comparisons across other classes with different numbers of alleles per gene also show declining percentages; only about 20% of the genes with greater than 4 alleles have frequencies exceeding 0.20. Rare alleles, here noted as ones in frequencies equal to 0.02 and lower, by tally are 25% of the 2 allele per gene cases, but increase to near 50% for genes with more than 4 alleles. Rare alleles are abundant when expressed as a percent of all alleles, their percentages exceed one third of all alleles.

Heterozygosity values were also tabulated by the number of alleles per gene for all genes in the sample species (Table 2). Of a total of 360 loci, 82% were polymorphic. Genes with 2 and 3 alleles made up 52% of the total. Note in the right hand total column that 50% of the genes had heterozygosity values greater than 0.10. There is a positive relationship between the number of alleles per gene and the percent values in classes for each row of cumulative heterozygosity values. Only 19% of genes with two alleles have heterozygosities greater than 0.20, while 70% of genes with greater than 4 alleles exceed 0.20 heterozygosity. It is noteworthy that the genes (including monomorphics) in these species had an average of 3 alleles each (1,087 alleles / 360 genes; Tables 1 and 2).

Table 2. Proportions of genes with various heterozygosity values grouped by the number of alleles that are detected for genes in a sample of 13 conifer species (see Fig. 2, and excluding Torrey pine)

Heterozygosity value	Number of alleles per gene: 1	2	3	4	>4	Total (%)
	Number of genes (% of total)					
	65 (18)	93 (26)	93 (26)	56 (16)	53 (15)	360 (100)
	(%) in each class					
>0.50	0	0	13	14	19	8
>0.20	0	19	36	50	70	32
>0.10	0	41	62	71	85	50
>0.05	0	52	78	84	98	67
>0.02	0	63	89	96	100	69
0.02 and lower	100	37	11	4	0	31

Summary items for profiles of forest diversity

- A number of forest species in California, commonly exhibit high levels of diversity in biochemical markers;
- Wide spread species that span a significant range of latitudes and elevations have lower proportions of fixed loci (having just one allele) and higher average heterozygosities than more geographically restricted species;
- Rare endemic species vary from complete fixation and no detectable biochemical variation to relatively high levels of variation characteristic of more wide spread taxa;
- As the number of alleles per gene increases, alleles in high frequencies decrease;
- Rare (less than 0.02 frequency) alleles are abundant;
- About 85 % of all loci have 4 or fewer alleles;

- Diversity measured by allozyme heterozygosity traces to relatively few alleles with intermediate frequencies;
- Genes overall averaged 3 alleles each.

Biochemical markers and genomes

After noting long histories for forest lineages and monitoring some levels of diversity in allozymes, it is appropriate to briefly move somewhat closer to individual genes by considering their characteristics in cells and whole organisms.

The basic elements of gene action are well understood. The order of four different bases in DNA of chromosomes is transcribed and translated into the variety of molecules that sustain life processes. Higher plant cells carry three genomes in separate organelles: the nucleus contains several to many pairs of large homologous chromosomes (diploid with the alleles for similar gene regions on separate homologous chromosomes, and each chromosome consisting of one long strand of DNA) — chloroplasts and mitochondria (each having unique and numerous copies of small circular chromosomes, haploid in number).

Individual genes are regions of DNA with segments that regulate transcription and segments with the codes for functional molecules. Listing the many variations of DNA are beyond this discussion, but enzymes are a particularly important class of secondary DNA products having catalytic and metabolic functions needed for the vast majority of all cellular reactions. The enzymes included in most forest species studies average about 425 amino acid residues in length (range from about 325 to 700, Ledig and Conkle 1983), their corresponding DNA codes may approach several thousand bases in length. The relationship between enzymes and DNA is so close that enzyme genes, and alleles of enzyme genes are terms in common usage.

The complement of enzyme genes that are analyzed in different forest species appears common to them all. Most workers report that different species, even those from different forest genera, have similarities for the number of genes per enzyme system. Each set of bands stained for a particular enzyme have distinct phenotypes. Groupings of bands for each gene and overall patterns when several genes express on the same electrophoresis gel are similar in different taxa. The size, staining intensities, and occasional multiple patterns of bands of genes indicate their close homology in different taxa. And different genes have characteristic phenotypes common across species when developmental patterns are compared and when different plant tissues are examined.

Several conifer enzyme systems may provide further evidence of wide interspecies gene homology through gene product compartmentalization. It is known that specific nuclear gene products associate with and are active only in specific cellular compartments: in chloroplasts, mitochondria, or cytoplasm. Reports on plants consistently validate that three genes are active in an enzyme system that different workers refer to as either GOT or AAT (glutamate oxaloacetate transaminase or aspartate aminotransferase). In corn, a fast mobility gene, stained for GOT on gels from electrophoresis, is expressed in mitochondria, another fast mobility GOT gene is expressed in plastids, and a third GOT gene with slow mobility is expressed in the cytoplasm (Goodman and Stuber 1983). The products of all three genes are independent, all only form intragene molecular associations (no intergene heterodimers are noted.)

GOT enzyme bands from dry and imbibed seeds of conifers similarly have an intensely staining high mobility gene with a single band in haploid tissues, a light staining high mobility gene (also single banded in haploids), and a consistently stained lower mobility gene (or perhaps duplicated genes), the latter commonly encodes for a characteristic stack of two or more bands on gels. The light staining high mobility bands increase in stain intensity as embryos enlarge and germination progresses.

Observations after separation and staining of enzymes from chloroplast enriched, mitochondria enriched, and cytoplasmic (mitochondria and chloroplasts removed) fractions of germinating conifer seeds in our laboratory support the conclusion that bands of the dark-staining fast-mobility GOT gene are associated with the mitochondrial fraction. Their dark stain reaction correlates with high levels of mitochondrial respiration expected in nutrient tissues during germination. Bands of the second GOT gene (often with mobilities slightly slower that the darker bands) are associated with the chloroplast enriched fraction. Their light stain reaction in early germination stages and darker stain in latter stages correlates with the onset of photosynthesis. Slow mobility GOT bands, commonly having a two banded haploid phenotypes, are associated with the cytoplasmic fraction. In conifers as in corn, the significant implication is that individual genes are unique in form, function, and site of action; many genes are homologous throughout all plants and must trace back through plant lineages to common progenitor codes.

While several lines of evidence support the hypothesis that forest trees in wide taxonomic groups share the same complement of genes, substantial numbers of reports on gene linkage (many based on recombination ratios of segregating alleles in haploid tissues in conifer seeds) further indicate there is significant conservation of gene order on chromosomes of different species (Cheliak and Pitel 1985). Conserved linkage relationships extend across conifer genera in the case of closely linked gene pairs.

Summary items for organization of the genome

- Perhaps the majority of different functional genes are similar in different plant taxa, homologous gene products have specific sites of action in cells;
- Gene linkages on chromosomes appear to be moderately conserved even across diverse conifer genera;
- Individual gene similarities trace far back through branching lineages to common ancestors in all plants.

In conclusion, it is appropriate to note that new research on biochemical markers is unlimited in scope and importance due to techniques now available for decoding and manipulating DNA. All hereditary codes are open to investigation using enzymes that cut DNA strands at specific base sequences and use molecular tags that identify strands carrying codes of interest. Nuclear, plastid, and mitochondrial genomes can be tagged with numerous new markers. Is seems safe to predict that a similar group of forest researchers gathering ten years from now will report on more precise phylogenetic relationships among taxa, provide more accurate information about the evolution of specific gene families, and discuss single gene regulation and expression in individuals and taxa. Reports may even provide information about phenotypic modifications of tree growth, pest resistance, and environmental tolerances induced by genetic manipulation. The unifying nature of biochemical and molecular research, that results from common phylogeny, means that findings on one plant, one family, a population, or taxon can be rapidly applied to a wide variety of other organisms.

References

Barnosky, C. W., Anderson, P. M. and Bartlein, P. J. 1987. The northwestern U.S. during deglaciation; Vegetational history and plaeoclimatic implications, pp. 289—312. In: Ruddiman, W. F. and Wright, J. E., Jr. (Eds) North America and adjacent oceans during the last deglaciation. Boulder, Colorado, Geological Society of America, The Geology of North America, V. K—3.

Cheliak, W. M. and Pitel, J. A. 1985. Inheritance and linkage of allozymes in Larix laricina. Silvae Genetica 34: 142—148.

Conkle, M. T. 1981. Proceedings of the symposium on isozymes of North American forest trees and forest insects, July 27, 1979, Berkeley, California. Gen. Tech. Rep. PSW—48, Pacific Southwest Forest and Range Exp. Stn., Forest Serv., U.S. Dep. Agric., Berkeley, California, 64 pp.

Conkle, M. T. 1987. Electrophoretic analysis of variation in native Monterey cypress (*Cupressus macrocarpa* Hartw.), pp. 249—256. In: Elias, T. S. (Ed) Conservation and Management of Rare and Endangered Plants. Proc. of a Conference held 5—8 Nov., 1986, Sacramento, California. California Native Plant Soc., Sacramento, California.

Conkle, M. T. and Critchfield, W. B. 1988. Genetic variation and hybridization of ponderosa pine, pp. 27—43. In: Ponderosa Pine: The Species and its Management. Washington State Univ. Coop. Ext.

Crane, P. R. 1987. Vegetational consequences of the angiosperm diversification, pp. 107—144. In: Friis, F. M., Chaloner, W. G. and Crane, P. R. (Eds) The Origin of Angiosperms and their Biological Consequences. Cambride University Press.

Critchfield, W. B. 1980. Origins of the eastern deciduous forest, pp. 1—14. In: Proceed. Dendrology in the Eastern Deciduous Forest Biome, September 11—13, 1979. Va. Polytech. Inst. and State Univ.

Critchfield, W. B. 1984. Impact of the Pleistocene on the genetic structure of North American conifers, pp. 70—118. In: Lanner, R. M. (ed.), Proceed. 8th North America Forest Bio. Workshop. Logan, Utah.

Critchfield, W. B. 1986. Hybridization and classification of the white pines (*Pinus* section *Strobus*). Taxon 35: 647—656.

Cronin, T. M., Szabo, B. J., Ager, T. A., Hazel, J. E and Owens J. P. 1981. Quaternary climates and sea levels on the U.S. Atlantic coastal plain. Science 211: 579—584.

Friis, E. M., Chaloner, W. G. and Crane, P. R. 1987. Introduction to angiosperms, pp. 1—15. In: Friis, F. M., Chaloner, W. G. and Crane, P. R. (Eds) The Origin of Angiosperms and their Biological Consequences. Cambridge University Press.

Furnier, G. R. 1984. Population genetic structure of Jeffery pine. Dissertation, Doctor of Philosophy, Oregon State U., Corvallis.

Goodman, M. M. and Stuber, C. W. 1983. Maize, pp. 1—33. In: Tanksley, S. D. and Orton, T. J. (Eds), Isozymes in Plant Genetics and Breeding, Part B. Elsevier Science Publishers B. V., Amsterdam.

Harrington, J. B. 1987. Climatic change: a review of causes. Can. J. For. Res. 17: 1313—1339.

Harry, D. E. 1984. Genetic structure of incense-cedar (Calocedrus decurrens) populations. Dissertation, Doctor of Philosophy, Un. of California, Berkeley.

Hsu, K. J. 1972. When the Mediterranean dried up. Sci. Amer. 227: 26—36.

Jacobson, G. L., Jr., Webb. T., III and Brimm, E. C. 1987. Patterns and rates of vegetation change during the deglaciation of eastern North America, pp. 277—288. In: Ruddiman, W. F. and Wright, J. E., Jr. (Eds) North America and Adjacent Oceans During the Last Deglaciation. Boulder, Colorado, Geological Society of America, The Geology of North America, V. K—3.

Kennett, J. P. 1977. Cenozoic evolution of Antarctic glaciation, the circum-Atarctic Ocean, and their impact on global paleoceanography. J. Geophy. Res. 82: 3843—3860.

Ledig, F. T. and Conkle, M. T. 1983. Gene diversity and genetic structure in a narrow endemic, Torrey pine (*Pinus torreyana* Parry ex Carr.). Evolution 37: 79—85.

Millar, C. I. 1991. Impact of the Eocene on the evolution of pines. Annals of Missouri Bot. Garden, in press.

Millar, C. I., Strauss, S. H., Conkle, M. T. and Westfall, R. D. 1988. Allozyme differentiation and biosystematics of the Californian closed-cone pines (Pinus subsect. Oocarpae). Syst. Bot. 13: 351—370.

Millar, C. I. and Kinloch, B. B. 1992. Taxonomy, phylogeny, and coevolution of pines and their stem rusts. In: Hiratsuka, Y. (Ed) Proceedings of the Third IUFRO Rusts of Pine Working Party Conference. Northern Forestry Centre, Forestry Canada.

Newell, N. D. 1972. The evolution of reefs. Sci. Am. 226: 54—65.

Niebling, C. R. and Conkle, M. T. 1990. Diversity of Washoe pine and comparisons with allozymes of ponderosa pine races. Can. J. For. Res. 20: 298—308.

Niklas, K. J., Tiffney, B. H. and Knoll, A. H. 1984. Patterns in vascular land plant diversification: a factor analysis at the species level, pp. 97—128. In: Valentine, J. W. (Ed) Phanerozoic Diversity Patterns: Profiles in Macroevolution. Princeton, Princeton University Press.

Parrish, J. T. 1987. Global palaeogeography and palaeoclimate of the late Cretaceous and early Tertiary, pp. 51—73. In: Friis, F. M., Chaloner, W. G. and Crane, P. R (eds), The origins of angiosperms and their biological consequences. Cambridge Univ. Press.

Ruddiman, W. F. and Wright, H. E., Jr. 1987. Introduction. pp. 1—12. In: Ruddiman, W. F. and Wright, J. E., Jr. (Eds) North America and Adjacent Oceans During the Last Deglaciation. Boulder, Colorado, Geological Society of America, The Geology of North America, V. K—3.

Webb, T., III, Bartlein, P. J. and Kutzbach, J. E. 1987. Climate change in eastern North America during the past 18,000 years; Comparisons of pollen data with model results, pp. 447—462. In: Ruddiman, W. F. and Wright, J. E., Jr. (Eds) North America and Adjacent Oceans During the Last Deglaciation. Boulder, Colorado, Geological Society of America, The Geology of North America, V. K—3.

New Forests **6**: 23—47, 1992.

Review paper

Genetic variation within European tree species

G. MÜLLER-STARCK[1,2], Ph. BARADAT[3], and F. BERGMANN[1]

[1] *Abteilung Forstgenetik und Forstpflanzenzüchtung, Universität Göttingen, Büsgenweg 2, D-3400 Göttingen, Germany;* [2] *Eidgenössische Forschungsanstalt für Wald, Schnee und Landschaft, Zürcherstrasse 111, CH-8903 Birmensdorf, Switzerland;* [3] *Institut National de la Recherche Agronomique, Centre de Recherches de Bordeaux, Pierroton, F-33610 Cestas, France*

Received 7 December 1990; accepted 23 December 1991

Key words: genetic markers, isoenzymes, terpenes, polyphenols, heterozygosity, geographic differentiation

Application. Genetic variation is a fundamental requirement for the maintenance and long-term stability of forest ecosystems since amount and pattern of genetic variation determine the ability of forest tree species to adapt to the variability of environmental conditions.

Abstract. This paper reviews results of studies on genetic variation in various European coniferous and angiosperm tree species, with emphasis on *Picea abies* and *Pinus sylvestris, Abies alba, Fagus sylvatica*, and *Castanea sativa*. Most of the studies employ enzyme gene markers, but terpenes and other markers are also used. The comparison of data is complicated due to the variety of measures of genetic variation that are reported. Nevertheless, substantial differences in intra- and interpopulational genetic variation can be observed among and also within species. Causes for such heterogeneity are briefly discussed. The necessity for monitoring genetic variation and for standardizing measurement of genetic variation is emphasized.

Introduction

During recent decades, forest ecosystems and forest resources in Central Europe have been threatened particularly by air pollution-induced environmental changes. These and other man-made deteriorating effects on forest ecosystems severely challenge the ability of the constituent forest tree populations to adapt to and to survive in spatially and temporally heterogeneous environmental conditions. This ability to adapt to change is a fundamental requirement for the maintenance of forest ecosystems. Within each tree species, the amount and pattern of genetic variation determine its adaptability and are consequently essential parameters of the long-term stability of forest ecosystems.

The objective of this paper is to review the present state of knowledge on genetic variation within native European forest tree species and to point out causes of heterogeneity in intra- and interpopulational variation. We cite studies which aim at the utilization of biochemical markers, which are expressed environmentally independent. The majority of the studies employ enzyme gene markers detected by means of various electrophoretic methods. Low molecular weight substances such as terpenes and polyphenols are also considered. DNA restriction fragment length polymorphisms have not been applied to date to studies on intrapopulational genetic variation in European tree species.

In reviewing the publications on forest tree species, three major problems arise. Firstly, in many studies an appropriate genetic analysis is missing so that the genetic control and mode of inheritance of markers are not proven unequivocally. Secondly, methods of measuring genetic variation differ among studies, so that direct comparison of results is difficult. Thirdly, in many studies the gene and/or genotype frequencies are not communicated, so that the original data cannot be reanalysed and subjected to a direct comparison. Consequently, this review of genetic variation within European tree species can only describe major trends without detailed characterization and interpretation of differences in genetic structures within and among species.

Brief survey of biochemical and molecular markers in use

(a) *Terpenes and polyphenols*

Terpenes and polyphenols, as secondary metabolites, have a less obvious genetic control than isozymes and they are often used without precise information about their mode of inheritance. The first class of compounds is the more widely used, with the limitation that only coniferous forest species are concerned. They are generally extracted from cortex oleoresin. Improvement of gas-liquid chromatography (GLC) techniques with automatic integration of peak areas considerably enhances the number of hydrocarbons which can be separated and reduces the analytical cost. Polyphenols, which are found in both conifers and broadleaves have been investigated much less. Analytical techniques for monitoring polyphenols are paper chromatography, thin layer chromatography or high performance liquid chromatography (HPLC).

(b) *Immunoproteins*

Immunological techniques can be efficiently used to study chemotaxonomic problems (for review see Prus-Głowacki 1982). Also clinal trends in immunoproteins ("antigenic" proteins) have been found among populations of *Pinus sylvestris* L. (Prus-Głowacki and Rudin 1981; Prus-Głowacki et al. 1985). Since the genetic control and the mode of inheritance of immunoproteins have not yet been elucidated, these substances are not suitable for monitoring genetic variation among individuals and populations, and will not be discussed further in this review.

(c) *Enzyme gene markers*

Isoenzymes quite frequently are found to be controlled by single gene loci, and thus, are very useful for monitoring genetic (allelic) variation. Certain enzyme systems often reveal considerable variation and are detectable in various plant tissues. Usually, alleles at isoenzyme loci are codominant. Occasional lack of isoenzyme banding as a consequence of "null alleles" is equivalent to recessive expression and results in underestimates of heterozygosity.

(d) *DNA markers*

Like enzyme gene markers, restriction fragment length polymorphisms (RFLP's) or other DNA variants allow for monitoring genetic polymorphisms in cases where appropriate genetic analyses have identified allelic variants (for specific problems in genetic analysis see Gillet 1990). The extremely high information potential will favour DNA markers in the future, although many techniques like "fingerprinting" reveal intrapopulational variation which does not necessarily correspond to allelic variation. To our knowledge, there are no results published to date on intra- or interpopulational genetic variation in European tree species using DNA markers. Chloroplast DNA, however, has been utilized to study phylogenetic relationships among species (e.g., Szmidt 1991) and for solving specific problems, like the classification of seedlots which originate from introgression zones (e.g., Szmidt et al. 1988).

Results of experimental studies

The present survey refers to forest tree species which are commonly considered as to be ecologically and/or economically important and are managed according to forestry, and not to agricultural regimes. Studies which mainly refer to characterization of components of mating systems are not included. Investigations involving only one population were taken into consideration only if no other information was available for that particular tree species. Studies using terpene markers, are reviewed separately from those which have used enzyme gene markers.

Results of studies using terpenes or polyphenols

Biology and inheritance

Terpenes

Monoterpenes (C_{10} hydrocarbons) and sesquiterpenes (C_{15} hydrocarbons) have been intensively studied in forest trees. Recent progress has been made in understanding their biosynthesis (Bernard-Dagan et al. 1982; Cori 1983) and the relationships between intensity of their synthesis and architecture of resin ducts (White and Nilsson 1984). Zavarin (1970) used correlations between expression of terpenes to help in understanding biochemical pathways. Squillace (1976) reviewed methods of evaluating the inheritance of these compounds and focused attention on autocorrelations between relative amounts. Birks and Kanowski (1988) consider that these autocorrelations are an important cause of error in determination of inheritance pattern of terpenic compounds and lead to uncertainty about the number of polymorphic loci involved in their biosynthesis. Strong genetic control of terpenes is well established by broad sense heritabilities close to 1 even when grafts are grown at different localities (Baradat and Yazdani 1988). Segregation data based on full-sib families support monogenic inheritance of terpenes in three European species: 3-carene in Norway spruce, *Picea abies* L. Karst. (Esteban et al. 1976); β-phellandrene, β-pinene, myrcene, limonene and 3-carene in Scots pine, *Pinus sylvestris* L. (Yazdani et al. 1982); 3-carene, myrcene, limonene, longifolene and caryophyllene in Maritime pine, *Pinus pinaster* Ait. (Baradat et al. 1972, 1974; Marpeau et al. 1975, 1983).

Pleiotropy of terpene genes is frequent due to common precursors in biosynthetic pathways. This is, for example, the case for 3-carene and terpinolene (Baradat et al. 1972). Linkage relationships have been reported in Maritime pine: genes which control 3-carene, myrcene and limonene belong to the same linkage group (Marpeau et al. 1983).

Polyphenols

Inheritance studies have been done on Norway spruce (Wellendorf and Kaufmann 1977) and on Scots pine (Thielges 1972; Yazdani and Lebreton

1991). The last named authors found a case of pleiotropy, where one locus controlled two flavonoids, prodelphinidin and taxifolin.

Use in the study of genetic variability

Table 1 summarizes results on intraspecific variability in seven European species of conifers using terpenes or polyphenols. The main field of utilization of terpene and phenolic compounds is chemotaxonomy because they display a strong geographic variation. The corresponding data are usually relative concentrations which are mainly processed with classical statistical methods used for quantitative traits. The most widely spread biometrical technics are analysis of variance, discriminant analysis, and cluster analysis.

Terpenes

Many experimental results (see Table 1) suggest that terpene markers correspond to selection pressure in a similar way as isoenzymes can do. For instance, Li and Adams (1988) found in Douglas-fir (*Pseudotsuga menziesii* (Mirb.) Franco) a similar pattern of geographical variation using isozymes than v. Rudloff (1973) found for terpenoid compounds of needles (see also Baradat et al. 1991).

High discriminating power of terpenes at the population level may be a consequence of the fact that different molecular forms correspond to drastic changes of physiological properties of enzymes acting on metabolic pathways. The expected result is a very low effective mutation rate at loci which control terpenes. This property makes these markers useful also for phylogenetic studies (see Table 1).

As basic terpenoid data are quantitative, precise identification of genotypes is possible only when the dominance of a richness allele over a poorness allele, or vice-versa, is not great. Dominance relationships have been determined for β-pinene, 3-carene, myrcene, limonene, longifolene and caryophyllene in Maritime pine (*Pinus pinaster*). These relationships range from high dominance of the poorness allele for caryophyllene to moderate dominance of the richness allele for 3-carene (Bernard-Dagan and Baradat 1977; Baradat and Marpeau 1991b). Most authors describe natural variability for terpenes, without any previous genetic analysis, using percentages of each terpene from individual trees or from mean supplies (one GLC analysis of oleoresin representative of a stand, sampled from series of individuals). Only in a few cases have patterns of variation in terpenes been reported in terms of chemotypes (Gansel and Squillace 1976 for slash pine (*Pinus elliottii* Engelm.), Forrest 1979, 1980 for Scots pine), or gene frequencies (Baradat and Marpeau 1991b).

Table 1. Survey of genetic variation in coniferous tree species on the basis of terpenes (T) and polyphenols (P)

Species and reference	Sampling design	Traits	Levels of variability	Processing of data	Most significant compounds	Genetic conclusions
Abies alba Mill						
Paule et al. 1987	13 prov. (Poland, Czecho-slovakia	T; QT	Within and between prov.	ANOVA	Tricyclene, α-pinene	50% of variance between prov.
Moreau, Pastuska 1991	83 F_0, 159 F_1 trees	T; QT, QL	Within stand	F_{is}	α-Pinene, limonene	$F_{is} = 0.77$ (F_0) $F_{is} = 0.57$ (F_1)
Abies cephalonica Loud.						
Fady 1991	6 prov., 83 halfsib families	T; QT	Within and betw. famil. (within prov.)	ANOVA	Monoterpenes, longifolene	$F_{is} > 0.50$
Cupressus sempervirens						
Schiller 1990	22 prov. (Israel)	T; QT	Within and between prov.	Discr. analysis, cluster analysis	Pinenes, 3-carene, phellandrene, caryophyllene	Israel prov. belong to different origins
Pinus halepensis Mill.						
Schiller, Grunwald 1987	22 prov. (Europe, Israel)	T; QT	Within and between prov.	Discr. analysis, cluster analysis	α-Pinene, 3-carene, myrcene	Structure of geographic variability
Baradat et al. 1989	4 prov. (Italy)	T; QT	Within and between prov.	Discr. analysis, cluster analysis	3-Carene, α-terpinene, caryophyllene	High discriminating power of terpenes between prov.
Pinus nigra Arnold						
Arbez et al. 1974	14 prov. (France, Italy)	T; QT	Between subspecies and geographic races	ANOVA	Limonene	Discrimination between 4 subspecies and between Laricio

Table 1. (Continued)

Species and reference	Sampling design	Traits	Levels of variability	Processing of data	Most significant compounds	Genetic conclusions
Fineschi, Grossoni 1981	2 geographical races	T; QT	Between geographical races	ANOVA	Limonene	Discrimination between Laricio from Corsica and Calabria
Gerber et al. 1991	72 prov. (southern Europe)	T; QT	Between prov. and subspecies	Discr. analysis, cluster analysis	β-Pinene, phellandrene, caryophyllene	Revision of taxonomy
Pinus pinaster Ait.						
Baradat et al. 1979	36 prov. 4 geographical groups	T; QT	Within and between stands and zones	ANOVA and principal component	β-Pinene, 3-carene, caryophyllene	Structure of geographic variability; panmictic equilibrium
Baradat, Marpeau 1991a	105 prov. from natural populations	T; QL	Between stands and geographic races	Discr. analysis, cluster analysis	β-Pinene, 3-carene, caryophyllene	Reconstruction of differences among races
Baradat, Marpeau 1991b	105 prov. from natural populations	QL	Within stands	Test of panmixia	6 Terpenes	Panmixia except for coastal stands
Baradat et al. 1991	7 geographical races	T; QT, QL	Between geogr. races	Discr. analysis, cluster analysis	6 Terpenes, 17 loci, 312 spots total prot.	Similar clustering of races with proteins and terpenes
Idrissi-Hassani, Lebreton 1992	4 geographical races, 80 trees	T; QT QL	Within and between gegraphical races	Frequency description	Kaempferol	Discrimination of races from Morocco and Landes
Pinus sylvestris L.						
Tigerstedt, Hiltunen 1979	146 clones from N-S range	T; QT, QL	Between stands	Regression	3-Carene	Clinal variation
Forrest 1979	17 prov. (Scotland)	T; QL	Between prov.	Frequency of terpenoid patterns	Pinenes, myrcene, phellandrene	Large variability between prov.

Table 1. (continued)

Species and reference	Sampling design	Traits	Levels of variability	Processing of data	Most significant compounds	Genetic conclusions
Forrest 1980	41 prov. (Scotland)	T; QL	Between prov.	Cluster analysis	Pinenes, myrcene	Clustering of prov. into 4 main groups
Yazdani et al. 1985	26 prov. (Sweden)	T; QT	Between prov.	ANOVA and discr. analysis	β-Pinene, 3-carene, limonene	Clinal variation
Yazdani, Nilsson 1986	10 prov. (Sweden)	T; QT	Within and between prov.	ANOVA and discr. analysis	Limonene, sabinene	Clinal variation; 85% of variance between prov.
Lebreton et al. 1990	16 geographical zones (Sweden, France, Spain); 156 indiv.	P; QT, QL	Within and between stands and zones	Regression and frequency analysis	Prodelphinidin, procyanidin, quercetin	High frequency of chemotypes with procyanidin and quercetin in low altitudes populations

QT stands for "quantitative" and QL for "qualitative" traits, ANOVA for "analysis of variance", discr. for "discriminant", prov. for "provenance(s)", F_0 for parents and F_1 for descendants; F_{is} designates the within population inbreeding coefficient (Wright 1969)

Studies on genetic variability involving terpenes generally use analyses of variance, multivariate analyses (principal components, discriminant analysis), and cluster analyses (see Table 1). They generally bear on comparison of provenances and geographic races and results can also partition the total variance between and within populations (e.g., Yazdani and Nilsson 1986 for Scots pine or Paule et al. 1987 for silver fir, *Abies alba* Mill.). Some studies estimate within stand inbreeding coefficients (F_{is}, see Table 1).

Polyphenols

When HPLC analysis techniques are used, interpretation of genetic variability in phenolic compounds is similar to that employed for terpenes, i.e., the detection of genotypic classes underlying quantitative variation in concentrations. Although there are only a small number of studies and a few discriminating compounds, some flavonoids seem to have good taxonomic value for identifying geographic races. This is the case of kaempferol in Maritime pine (Idrissi-Hassani and Lebreton 1992) and for prodelphinidin, procyanidin and quercetin in Scots pine (Lebreton et al. 1991; see Table 1).

Studies on intraspecific genetic variation using enzyme gene markers

Studies utilizing enzyme gene markers quantify genetic variation at specified gene loci according to a variety of measures. There are varying concepts and overlapping definitions especially with respect to the terms differentiation, diversity and heterozygosity (Gregorius 1978, 1987; Müller-Starck and Gregorius 1986). When the term heterozygosity is used, most investigators mean actual (observed) heterozygosity, H_a, or expected heterozygosity, H_e (Nei 1973), which is in reality a measure of genetic diversity. The conditional heterozygosity, H_c (Gregorius et al. 1986), is independent of underlying allele frequencies, but is not yet in general use. Tables 2 and 3 briefly characterize degrees of intrapopulational variation by means of a compilation of the average numbers of alleles per locus, A_L, and the heterozygosities H_a and H_e respectively. The most frequently employed measure for interpopulational variation is G_{ST} (Nei 1973), which is equivalent to F_{ST} (Wright 1943, 1951). Recent studies also refer to the subpopulation differentiation, δ (Gregorius and Roberds 1986) which measures for each population its genetic distance from the remaining populations (lumped together). Additional comments in Tables 2 and 3 briefly summarize main conclusion from the respective studies.

The investigations summarized in Tables 2 and 3 reveal considerable variation among species and among studies within species in observed patterns of genetic diversity within and between populations. In the next section, an attempt is made to describe and classify possible causes for this heterogeneity of results.

Causes for heterogeneity in intra- and interpopulational variation as inferred from enzyme gene markers

Specific features of tree species

European forest tree species can be classified as follows:

(a) Species with large geographic ranges such as *Picea abies*, *Pinus sylvestris* or *Fagus sylvatica* tend to show little genetic differentiation among populations within regions, but greater differentiation among populations derived from different glacial refugia (continuous species).

Table 2. Survey of genetic variation in European conifers based on isoenzymes

Species and reference	No. of pop.	Indiv. per pop.	No. of loci	A_L	H (%)	G_{ST} (%)	Main conclusions about intra- and/or interpopulational genetic variation
Abies alba Mill.							
Bergmann and Kownatzki 1988	45[1]	25+	5	1.7			Great allelic differentiation among pop. (Europe)
Bergmann et al. 1990	45[1]	25+#	10	1.8			Variation in diversity level and allelic differentiation among pop. (Europe)
Kormuťák et al. 1982	4	22#	9	3.3	56.3[a]		Great intrapop. variation, small interpop. differentiation (Slovakia)
Kormuťák 1988	7	29#	5		43.1[a]		Great allelic variation between two regions (Slovakia)
Mejnartowicz 1980	5	> 25+	5	4.7	41.6[e]		Great intrapop., moderate interpop. differentiation (Poland)
Schröder 1989	43	10—50+	4	2.2			Distinct differentiation among regions (Germany)
Larix decidua Mill.							
Lewandowski and Mejnartowicz 1992	11	100+	18	1.9	16.3[e]	2.6	Relatively little differentiation among pop. (Poland)
Picea abies (L) Karst.							
Bergmann 1974	9	200+	4	2.7		4.0	Great differentiation only between pop. in Central Europe and Scandinavia
Bergmann 1975	15	200+	6	2.6			Great differentiation between northern, central and southern European pop.
Bergmann and Gregorius 1979	21	200+	7	2.6	41.1[e]		Differences in intrapop. variat. between marginal and central pop. (Europe)

Table 2. (Continued)

Species and reference	No. of pop.	Indiv. per pop.	No. of loci	A_L	H (%)	G_{ST} (%)	Main conclusions about intra- and/or interpopulational genetic variation
Bergmann 1983	38	200⁺	3				Distinct differentiation between pop. based on private alleles (Central Europe and Scandinavia)
Giannini et al. 1991	9	150⁺	21	1.8	16.5[(e)]	4.2	Minor polymorphism; small interpop. variation (north. Italy)
Konnert and Franke 1990	39	12—45[#]	11	2.2			Interpop. variation between indigenous and artifical pop. (South. Germany)
Lagercrantz and Ryman 1990	70	28—51[#]	22	1.6	11.5[(a)]	5.2	Rel. great intra- and rel. small interpop. variation; diversity differences between pop. (Northeast. and Central Europe)
Lundkvist and Rudin 1977	11	35—80[#]	4	4.3	35.8[(e)]	2.0	Small interpopul. variation among populations (Sweden)
Lundkvist 1979	4	49—64[#]	11	3.8	36.0[(a)]	3.0	Small interpopul. variation among populations (Sweden)
Muona et al. 1990	2	23[#]	14	2.7	20.0[(e)]	12.0	Rel. great intrapop. variation and clear differentiation between pop. (Finland, Slovakia)
Stutz 1990	19	40[#]	6			2.5	Certain differentiation among pop.
Tigerstedt 1973	2	45[#]	4	2.5	43.0[(a)]		Rel. great intrapop. variation in marginal and central pop. (Finland)
Tigerstedt 1979	10	18—107[#]	6			5.0	No differences in intrapop. variation among pop.; higher homozygosity at the species margin
Pinus cembra L.							
Krutovskii and Politov 1991[2)]	1	16[#]	19	1.5	12.8[(a)] 10.9[(e)]		Very small intrapop. variation (Siberia)

Table 2. (Continued)

Species and reference	No. of pop.	Indiv. per pop.	No. of loci	A_L	H (%)	G_{ST} (%)	Main conclusions about intra- and/or interpopulational genetic variation
Szmidt 1982	11	22.5#	8	1.9	26.0(e)	32.0[3]	Very large interpop. variation among European pop.
Pinus halepensis Mill.							
Loukas et al. 1983	3	16—83#	10	2.1	17.1(e)	1.3	Great intrapop. variation and very small interpop. variation
Schiller et al. 1986	19	75+	30	1.2	4.0(e)		Small intrapop. variation; greater differentiation between two races
Pinus leucodermis Ait.							
Morgante and Vendramin 1990	5 2	23# 145+	23 23	1.6 1.4	12.4(e) 11.0(e)	4.0	Little intra- and interpop. variation (Italy (5), Greece (2))
Pinus nigra Arnold							
Bonnet-Masimbert and Bikay-Bikay 1978	40	125+	4	2.4		9.8	Differentiation among the five subspec. in mediterranean area
Fineschi 1984	11	64+	2	2.5			Great differentiation among geographical groups of pop. (Italy)
Nikolić and Tucić 1983	28	14—72+#	4	3.0	27.2(e)	13.5	Great intra- and rel. small interpop. variation; differentiation between subspecies
Pinus pumila (Pall.) Regel							
Krutovskii and Politov 1991[2]	3	56#	17	2.3	24.7(a) 24.8(e)		Rel. large intra- and interpop. variation (Siberia)
Pinus sibirica Du Tour							
Krutovskii et al. 1989	9	37#	14	1.6	16.3(a)	1.6	Little interpop. variation (Eastern Siberia)

Table 2. (Continued)

Species and reference	No. of pop.	Indiv. per pop.	No. of loci	A_L	H (%)	G_{ST} (%)	Main conclusions about intra- and/or interpopulational genetic variation
Pinus sylvestris L.							
Gullberg et al. 1982	3	45#	9	2.9	29.7$^{(a)}$ 30.3$^{(e)}$	1.0	Moderate genetic variation among adjacent pop.
Gullberg et al. 1985	9	60#	11	2.9		2.0	Great intra- and little interpop. variation (Sweden)
Krzakowa 1982[4)]	8	32#	6				Differentiation between north. and south. pop. (Poland)
Mejnartowicz 1979	19	30#	3	3.9	38.6$^{(a)}$	16.0[3)]	Very large intra- and interpop. variation (Poland)
Mejnartowicz and Bergmann 1985	9	17.5#	9	2.6	33.5$^{(e)}$		Great intra- and little interpop. variation (Poland)
Mejnartowicz and Palowski (1989)	6	25#	8	3.3	30.7$^{(a)}$ 37.5$^{(e)}$	7.6	Great intra- and rel. little interpop. variation (Poland)
Müller-Starck 1987, Müller-Starck and Gregorius 1986	9	120*	10	3.1	26.8$^{(a)}$	2.0	Great intra- and little interpop. variation (Bavaria)
Muona and Szmidt 1985	3	133*	14	3.0	25—30$^{(e)}$	0.6	Little interpop. variation (Northern Sweden)
Prus-Glowacki (pers. comm.)	17	30#	8	2.8	35.9$^{(a)}$		Great intra- and interpop. variation (European habitat)
Rudin et al. 1974	3	207#	3	4.3		3.0[3)]	Great intra- and little interpop. variation (Sweden)

\+ Seed lot (macrogametophyte)
* Seed lot (embryo)
\# Adult trees

[1)] Populations in common
[2)] See also Krutovskii et al. 1990
[3)] Loc. cit. Gullberg et al. 1985
[4)] See also Krzakowa and Szweykowski (1979)

A_L stands for "average number of alleles per locus", H for "average heterozygosity per population" (index (a) indicates "actual", index (e) "expected "heterozygosity), G_{ST} for "proportion of total diversity among populations", and pop. for "population(s)" (see text for further explanation).

Table 3. Survey of genetic variation in European angiosperm tree species based on isoenzymes

Species and reference	No. of pop.	Indiv. per pop.	No. of loci	A_L	H (%)	G_{ST} (%)	Main conclusion about intra- and/or interpopulational genetic variation
Castanea sativa Mill.							
Fineschi et al. 1990	13	101*	8	2.1			Differentiation between north. and south. pop. (Italy)
Pigliucci et al. 1990	18	16#	15	1.7	24.0[(a)]	10.0	Little intrapop. variation; clinal variation suggested (Italy)
Villani et al. 1991a	15	25#	13	1.6	21.0[(e)]	8.7	Rel. great interpop. variation; no geographical trends (Italy)
Villani et al. 1991b	13	29#	16	2.3	27.2[(a)]	16.7	Clear differentiation between eastern and west. pop. (Turkey)
Fagus sylvatica L.							
Barrière et al. 1984[2)]	210[1)]	50#	3	2.3			Differentiation among geographical groups of pop. (Europe)
Comps et al. 1987	104[1)]	50#	4	2.2			Differentiation through the Atlantic range
Comps et al. 1990	140[1)]	50#	6	2.3	28.9[(a)]	5.4	Differentiation between continental and Mediterr. regions
Comps et al. 1991	35[1)]	50#	6	2.2		3.6—5.2	Differentiation between low- and highland pop. (Croatia)
Felber and Thiebaut 1984	41[1)]	50#	2	2.5			Differentiation among geographical groups of pop. (Central Europe)
Kim 1980, 1985	2	596*	1	4.0	31.2[(a)]		Great intra- and interpop. variation (Germany, Rumania)
Merzeau et al. 1989	2	12.5#[3)]	4 (2)	2.3			Great intrapop. variation in pollen (South. France)
Müller-Starck 1985, 1989	6	96#	16	2.6	25.1[(a)]		Great intrapop. variation (Germany)

Table 3. (Continued)

Species and reference	No. of pop.	Indiv. per pop.	No. of loci	A_L	H (%)	G_{ST} (%)	Main conclusion about intra- and/or interpopulational genetic variation
Müller-Starck and Ziehe 1991	5	172*	13	2.7	22.2[(a)]	1.6 4.5[4)]	Great intra- and small interpop. variation (Germany)
Thiebaut et al. 1982	13[1)]	50#	2	2.5			Intrapop. variation greater under marginal environments (South. France)
Ficus carica							
Valizadeh 1977	4	90#	2	2.7			Deficiency of heterozygotes, small variation (South. France)
Quercus ilex L.							
Lumaret et al. 1991	37	35#	3	2.8		7.3	Differentiation among North African and South. European pop.
Quercus petraea Liebl.							
Zanetto and Kremer 1991	32	120*	15	3.2	27.5[(e)]	1.7	Large intrapop. and small interpop. variation (West. Europe)
Müller-Starck and Ziehe 1991	5	155+	13	3.1	21.9[(a)]	3.6 8.5[4)]	Large intrapop. and moderate interpop. variation (Germany)
Quercus robur L.							
Müller-Starck and Ziehe 1991	5	166+	13	3.2	21.3[(a)]	1.8 5.5[4)]	Large intrapop. and small interpop. variation (Germany)

* Seeds
\+ 2-year old plants
\# Adult trees

1) Partially populations in common
2) See also Cuguen et al. 1985, 1988
3) Pollen contribution to seed lots from single trees
4) Subpop. differentiation δ (Gregorius and Roberds 1986)

A_L stands for "average number of alleles per locus", H for "average heterozygosity per population" (index (a) indicates "actual", index (e) "expected "heterozygosity), G_{ST} for "proportion of total diversity among populations", and pop. for "population(s)" (see text for further explanation)

(b) Species with large geographic ranges, but subdivided into different subspecies (races) such as *Pinus nigra*, *Pinus halepensis* or *Larix decidua*, reveal little interpopulational variation within subspecies, but great differentiation among subspecies (disjunct species).
(c) Species with small and geographically disjunct ranges like *Abies alba* or *Pinus cembra* tend to show a great interpopulational differentiation and moderate intrapopulational genetic variation (endemic species).
(d) Species with extremely small geographic ranges like *Pinus pumila* reveal relatively great interpopulational differentiation (relic species). Further studies may help to find out why this does not seem to hold for another species of this type, *P. leucodermis*.

Heterogeneity of population types and environments

Various types of populations may be sampled, mostly indigenous ones, but within the same species introduced populations or mixtures of both may also be sampled. Populations may differ in amounts of genetic variation because some may have gone through historical bottlenecks in population sizes, others not. It cannot be excluded that different silvicultural management affect the genetic structures of the respective populations. It is self-evident that environmental conditions can differ substantially even within small areas. The great natural micro- and macro-habitat variation always will complicate standardization of genetic inventories.

Furthermore, intra-specific heterogeneity in genetic variation within populations can be a consequence of sampling at different census stages, biased sampling, or small sample sizes. Due to varying reproduction, effective neighborhood sizes, fertility and/or viability selection (e.g. against inbreds), sampling at the seed stage or subsequent stages may not be representative of the genetic structure in the parental population. Even within the same age class, sampling of only flowering trees will not necessarily represent the entire population. As can be seen from Tables 2 and 3, sample sizes varied considerably among the studies. This is true for the number of investigated populations per region, as well as the number of individuals per population.

Methodical pitfalls

(a) *Choice of enzyme systems*

Heterogeneity in isoenzyme data for the same tree species can result when the numbers and types of enzyme gene loci differ among studies. In the early days of isoenzyme surveys in tree species, enzyme systems such as peroxidases, esterases, phosphatases and aminopeptidases were studied, which later turned out to be very polymorphic. Hence, the respective estimates of the variation measures were relatively high compared to later studies, in which many other — partly invariant — enzyme systems, such as aminotransferases, dehydrogenases and isomerases, were included.

(b) *Different interpretations of genetic control of isoenzymes*

Differences in diversity estimates among studies can also occur if the number of loci controlling an enzyme is interpreted differently by different investigators of the same tree species. For example, the GOT system in Scots pine is assumed to be encoded by two, three or five gene loci depending on the results of different research groups (see Szmidt 1980). Estimates of the genetic diversity are likely to be much greater if isoenzymes, actually under the control of multiple loci, are assumed to be controlled by a single locus. Therefore it is always necessary to identify the correct number of gene loci underlaying enzyme polymorphisms prior to estimating genetic variability in populations. This includes a check on possible ontogenetic and/or environmental instabilities in the expression of enzyme systems on zymograms (e.g., Bergmann et al. 1989).

(c) *Inclusion of monomorphic gene loci*

Some investigators include monomorphic loci in their estimates of genetic variation while others do not. In one case, these estimates will be biased upwards, the other, downwards. The problem is that many investigators do not communicate whether or not monomorphic loci were excluded.

If the extent of intra- and interpopulational genetic variation should be compared in studies on geographical variation, it is not advisable to include monomorphic enzyme loci in samples of enzyme loci: patterns of variation will be the same, but the absolute magnitude of differences will be less when all loci are included. In particular, the "averaging effect" in the calculation of measures of genetic distance and F_{ST} or G_{ST} values would bias the extent of differentiation, if the proportion of monomorphic loci among the loci used in the study is relatively high.

Concluding remarks

In spite of the great variety of environmental conditions and forest stand characteristics in different studies, general tendencies are observed in genetic patterns of geographic variation related to species distribution, i.e., continuous, disjunct, endemic, and relic.

Methodological problems are evident which concern the choice of population and of samples per population, the choice of enzyme systems and loci, the methods used to quantify genetic variation, and in several cases also the verification of the genetic control and inheritance of markers. Generally, it appears that the quantification of interpopulational genetic variation suffers from a lack of appropriate measures: The widely applied F_{ST} or G_{ST} values represent in reality fixation indices which therefore cannot be considered as the most favourable condition for the measurement of genetic differentiation among populations. Furthermore, the partition of a total genetic variation into an intra- and interpopulational additive component appears as a concept which needs to be extended or restated in order to improve the accuracy of measures of genetic variation among populations (for suggestions see Gregorius 1988).

Although there is an increase in forest genetics research in Europe, the majority of published results still refer to coniferous and not to angiosperm tree species. While there is increasing work on *Fagus sylvatica*, there are no studies on *Acer*, *Fraxinus*, *Tilia*, or *Ulmus* species. The choice of which forest tree species to study seems to depend primarily on their economic importance, and not on their ecological significance. Studies on ecologically important species are especially needed in Central Europe, where forest tree populations are severely threatened by air-pollution.

It appears that ecological genetic research in forestry has to be intensified in nearly all European countries. Moreover, present and future information on genetic variation in European tree species needs to be incorporated into strategies for the preservation of genetic resources so that adaptability of forest tree populations to changing environments can be maintained.

Acknowledgments

We are grateful to many colleagues who provided reprints and in some cases also unpublished data. At the same time we apologize that not all contributions could be cited in this review. We greatly appreciate the help of H. Glock, M. Günther, and A. Uteau in the preparation of the manuscript. Finally, we wish to thank unknown referees for very helpful comments and suggestions.

References

Arbez, M., Bernard-Dagan, C. and Fillon, C. 1974. Variabilité intraspécifique des monoterpènes de *Pinus nigra*. Bilan des premiers resultats. Ann. Sci. For. 31(1): 57—70.

Baradat, Ph., Bernard-Dagan, C., Fillon, C. and Marpeau, A. 1972. Les terpènes du Pin maritime: aspects biologiques et génétiques. 2. Hérédité de la teneur en monoterpènes. Ann. Sci. For. 29(3): 507—334.

Baradat, Ph., Bernard-Dagan, C. and Pauly, G. 1974. Les terpènes du Pin maritime: aspects biologiques et génétiques. 3. Hérédité de la teneur en myrcène. Ann. Sci. For. 32(1): 29—54.

Baradat, Ph., Bernard-Dagan, C. and Marpeau, A. 1979. Variation of terpenes within and between populations of Maritime pine, pp. 151—168. In: Rudin, D. (Ed) Proceedings of the IUFRO Conference on Biochemical Genetics of Forest Trees, Umeå, Sweden, 1978.

Baradat, Ph. and Yazdani, R. 1988. Stability of genotypic expression for monoterpene synthesis in clones of Scots pine growing in different sites. Scand. J. Forest Res. 3: 25—36.

Baradat, Ph. and Marpeau, A. 1991 a. Reconstitution of differentiation of Maritime pine on the basis of terpenic data. (submitted to Ann. Sci. For.)

Baradat, Ph. and Marpeau, A. 1991 b. Genetic structure of natural stands of Maritime pine on the basis of terpenic data. (submitted to Ann. Sci. For.).

Baradat, Ph., Lambardi, M. and Michelozzi, M. 1989. Terpene composition for four Italian provenances of Aleppo pine. J. Genet. Breeding 43: 195—200.

Baradat, Ph., Barhman, N. and Petit, R. 1991. Comparison of relationships between geographical races of Maritime pine based on three kinds of gene markers: terpenes, isozymes and total proteins. (submitted to Theor. Appl. Genetics).

Barrière, G., Comps, B., Cuguen, J., N'Tsiba, F. and Thiebaut, B. 1984. The genetical ecological variability of beech (*Fagus sylvatica* L.) in Europe — an alloenzymatic study: genetic isolations of beechwoods, pp. 24—50. In: Proceedings of the First Symposium on Improvement and Silviculture of Beech, Grosshansdorf 1984, IUFRO Project Group P1 10—00.

Bergmann, F. 1974. Genetischer Abstand zwischen Populationen. II. Die Bestimmung des genetischen Abstands zwischen europäischen Fichtenpopulationen (*Picea abies*) auf der Basis von Isoenzym-Genhäufigkeiten. Silvae Genetica 23: 28—32.

Bergmann, F. 1975. Herkunfts-Identifizierung von Forstsaatgut auf der Basis von Isoenzym-Genhäufigkeiten. All. Forst- u. Jagdzeitung 146(10): 191—195.

Bergmann, F. 1983. Ein besonderer Fall geographischer Variation an zwei Enzym-Genloci der Fichte (*Picea abies*), pp. 8—24. In: Verhandlungen 3. Arbeitstagung: FORUM Genetik-Wald-Forstwirtschaft. Universität Göttingen.

Bergmann, F. and Gregorius, H.-R. 1979. Comparison of the genetic diversities of various populations of Norway spruce (*Picea abies*), pp. 99—107. In: Rudin, D. (Ed) Proceedings of the IUFRO Conference on Biochemical Genetics of Forest Trees. Umeå, Sweden, 1978.

Bergmann, F. and Kownatzki, D. 1988. The genetic variation pattern of silver fir (*Abies alba*) in Europe monitored from enzyme gene loci, pp. 21—26. In: Paule, L. and Korpel', Š. (Eds) 5. IUFRO-Tannensymposium. VŠLD, Zvolen, ČFSR.

Bergmann, F., Gregorius, H.-R. and Scholz, F. 1989. Isoenzymes, indicators of environmental impacts on plants or environmentally stable gene markers? pp. 17—25. In: Scholz, F., Gregorius, H.-R. and Rudin, D. (Eds) Genetic Effects of Air Pollutants in Forest Tree Populations. Springer-Verlag, Berlin, Heidelberg, New York, Tokyo.

Bergmann, F., Gregorius, H.-R. and Larsen, J. B. 1990. Levels of genetic variation in

European silver fir (*Abies alba*) — Are they related to the species decline? Genetica 82: 1—10.

Bernard-Dagan, C., and Baradat, Ph. 1977. Utilisation des terpènes comme outil en génétique forestière. Proceeding of EEC-Symposium on Forest Tree Biochemistry, Brux. 25—27, January 1977: 109—132.

Bernard-Dagan, C., Fillon, C., Marpeau, A. and Pauly, G. 1982. Les terpènes du Pin maritime: aspects biologiques et génétiques. 2. Hérédité de la teneur en monoterpènes. Ann. Sci. For. 29: 307—334.

Bernard-Dagan, C., Pauly, G., Marpeau, A., Gleizes, M., Carde, J. P. and Baradat, Ph. 1982. Control and compartmentation of terpene biosynthesis in leaves of *Pinus pinaster*. Physiologie Vegetale 20(4): 775—795.

Birks, J. S. and Kanowski, J. S. 1988. Interpretation of the composition of coniferous resin. Silvae Genetica 37(1): 29—39.

Bonnet-Masimbert, M. and Bikay-Bikay, V. 1978. Variabilité intranspécifique des isozymes de la glutamate-oxaloacetate-transaminase chez *Pinus nigra* Arnold. Interêt pour la taxonomie des sous especès. Silvae Genetica 27: 71—79.

Comps, B., Barrière, G., Merzeau, D. and Letouzey J. 1987. La variabilité alloenzymatique des hêtraies dans les sours-domaines médio- et eu-atlantiques d'Europe. Canad. Journ. For. Res. 17(7): 1043—1049.

Comps, B., Thiebaut, B., Paule, L., Merzeau, D. and Letouzey, J. 1990. Allozyme variability in beechwoods (*Fagus sylvatica* L.) over central Europe: spatial differentiation among and within populations. Heredity 65: 407—417.

Comps, B., Thiebaut, B., Sugar, I., Trinajstic, I. and Plazibat, M. 1991. Genetic variability of the croatian beech stands (*Fagus sylvatica* L.): Spatial differentiation with environment. Ann. des Sci. For. 48:15—28.

Cori, O. M. 1983. Enzymic aspects of the biosynthesis of monoterpenes in plants. Phytochemistry 22: 331—341.

Cuguen. J., Thiebaut, B., N'Tsiba, F. and Barrière, G. 1985. Enzymatic variability of beech stands (*Fagus sylvatica* L.) on three scales in Europe: evolutionary mechanisms, pp. 17—39. In: Jacquart, P., Heim, G. and Antonovics, J. (Eds) Genetic Differentiation and Dispersal in Plants. NATO ASI Series, Springer-Verlag, Berlin, Heidelberg, New York, Tokyo.

Cuguen, J., Merzeau, D. and Thiebaut, B. 1988. Genetic structure of the european beech stands (*Fagus sylvatica* L.): F-statistics and importance of mating system characteristics in their evolution. Heredity 60: 91—100.

Dupouey, J. L., Fougere, V. and Kremer, A. 1990. Variabilité génétique des chênes sessile et pédonculé estimée à l'aide de marqueur morphologique moléculaire. Revue Forestière Francaise 42: 198—204.

Esteban, I., Bergmann, F., Gregorius H.-R. and Huhtinen, O. 1976. Composition and Genetics of monoterpenes from cortical oleoresin of Norway spruce and their significance for clone identification. Silvae Genetica 25(2): 59—66.

Fady, B. 1991. Genetic variability of the Greek fir. (submitted to Ann. Sci. For.)

Felber, F. and Thiebaut, B. 1984. Etude préliminaire sur le polymorphisme enzymatique du hêtre, *Fagus sylvatica* L.: variabilité génétique de deux systèmes de peroxydases en relation avec les conditions écologiques. Acta Oecologica. Oecol. Plant. 5: 133—150.

Fineschi, S. 1984. Determination of the origin of an isolated group of trees of *Pinus nigra* through enzyme gene markers. Silvae Genetica 33(4—5): 169—172.

Fineschi, S. and Grossoni, P. 1981. Contenuto in monoterpeni di oleoresine xilematiche in provenienze diverse di Pino laricio. Italia forestale montana 36(1): 232—259.

Fineschi, S., Malvolti, E., Morgante, M., Vendramin, G. and Paciucci, M. 1990. Erhaltung von Genressourcen bei der Kastanie (*Castanea sativa* Mill.), pp. 155—164. In: Hattemer,

H. H. (Ed) Erhaltung forstlicher Genressourcen. Schriften aus der Forstl. Fak. d. Univ. Göttingen und der Nds. Forstl. Vers. Anst. 98, J.D. Sauerländer's Verlag, Frankfurt a.M.

Forrest, G. I. 1979. Monoterpenic variation in lodgepole pine and Scots pine, pp. 136—150. In: Rudin, D. (Ed) Proceedings of the IUFRO Conference on Biochemical Genetics of Forest Trees. Umeå. Sweden 1978.

Forrest, G. I. 1990. Genotypic variation among native Scots pine populations in Scotland based on monoterpene analysis. Forestry 53: 101—128.

Gansel, C. E. and Squillace, A. E. 1976. Geographic variation in cortical oleoresin of Slash pine. Silvae Genetica 25(5—6): 150—154.

Gerber, S., Arbez, A., Baradat, Ph. and Marpeau, A. 1991. Geographical variation and terpenic composition of *Pinus nigra*. (submitted to Silvae Genetica)

Giannini, R., Vendramin, G. and Morgante, M. 1991. Allozyme variation in Italian populations of *Picea abies* (L.) Karst. Silvae Genetica 40(3—4): 160—166.

Gillet, E. 1990. Probeme der genetischen Analyse von Restriktionsfragment-Längen-Polymorphismen (RFLP), pp. 272—277. In: Marker und Genidentifizierung — Methoden und Ergebnisse. Tagungsberichte Arbeitsgemeinschaft Pflanzenzüchtung 8/9. März 1990, Hannover, Vorträge für Pflanzenzüchtung, H. 18.

Gregorius, H.-R. 1978. The concept of genetic diversity and its formal relationship to heterozygosity and genetic distance. Math. Biosci. 41: 253—271.

Gregorius, H.-R. 1987. The relationship between the concepts of genetic diversity and differentiation. Theor. Appl. Genet. 74: 397—401.

Gregorius, H.-R. 1988. The meaning of genetic variation within and between subpopulations. Theor. Appl. Genet. 76: 947—951.

Gregorius, H.-R. and Roberds, J. H. 1986. Measurement of genetical differentiation among subpopulations. Theor. Appl. Genet. 71: 826—834.

Gregorius, H.-R., Krauhausen, J. and Müller-Starck, G. 1986. Spatial and temporal genetic differentiation among the seed in a stand of *Fagus sylvatica* L. Heredity 57: 255—262.

Gullberg, U., Yazdani, R. and Rudin, D. 1982. Genetic differentiation between adjacent populations of *Pinus sylvestris*. Silva Fennica 16(2): 205—214.

Gullberg, U., Yazdani, R., Rudin, D. and Ryman, N. 1985. Allozyme Variation in Scots pine (*Pinus sylvestris* L.) in Sweden. Silvae Genetica 34(6): 193—201.

Idrissi-Hassani, M. and Lebreton, P. 1992. Les Flavonoïdes du Pin maritime: une approche chimiotaxinomique. Forêt méditerranéenne 13(1): 3–8.

Kim, Z. S. 1980. Veränderungen der genetischen Struktur von Buchenpopulationen durch Viabilitätsselektion im Keimlingsstadium, pp. 1—88. Göttingen Res. Notes in Forest Genetics 3.

Kim, Z. S. 1985. Viability selection at an allozyme locus during development in European beech (*Fagus sylvatica* L.) Silvae Genetica 34(4—5): 181—186.

Konnert, M. and Franke, F. 1990. Nachkommenschaftsprüfung und Generhaltung von Fichtenbeständen im Schwarzwald, pp. 165—177. In: Hattemer, H. H. (Ed) Erhaltung forstlicher Genressourcen. Schriften aus der Forstl. Fak. d. Univ. Göttingen und der Nds. Forstl. Vers. Anst. J. D. Sauerländer's Verlag, Frankfurt a.M.

Kormuťák, A. 1988. Isozyme polymorphism in Slovak populations of silver fir (*Abies alba* Mill.), pp. 27—39. In: Paule, L. and Korpel', Š. (Eds) 5. IUFRO-Tannensymposium. VŠLD, Zvolen, ČFSR.

Kormuťák, A., Benčať, F., Rudin, D. and Seyedyazdani, R. 1982. Isoenzyme variation in the four Slovakian populations of *Abies alba Mill.* Biológia (Bratislava) 37: 433—440.

Krutovskii, K. V., Politov, D. V., Altukhov, Y. P., Milutin, L. I., Kuznetsova, G. V., Iroshnikov, Vorobyev, V. N. and Vorobyeva, N. A. 1989. Genetic variability in Siberian stone pine, *Pinus sibirica* in Tour. IV. Genetic diversity and amount of genetic differentiation between natural populations. Genetika 25(11): 2009—2032.

Krutovskii, K. V., Politov, D. V. and Altukhov, Y. P. 1990. Genetic differentiation by isozyme loci between Eurasian cedar pines. Genetika 26(4): 694—707.

Krutovskii, K. V. and Politov, D. V. 1991. Study of intra- and interspecific genetic differentiation of Eurasian cedar pines (*Pinus cembra*) using isozyme loci and multi-dimensional analysis. In: Molecular Mechanisms of Genetic Processes, Nauka Publishers, Moscow. (in press)

Krzakowa, M. and Szweykowski, J. 1979. Variation of 6-PGD in the populations of Polish Scots pine (*Pinus sylvestris*), pp. 86—98. In: Rudin, D. (Ed) Proceedings of the Conference on Biochemical Genetics of Forest Trees. Umeå, Sweden 1978.

Krzakowa, M. 1982. Genetic differentiation of Scots pine populations. 1. Genotypes. Silva Fennica 16(2): 200—205.

Lagercrantz, U. and Ryman, N. 1990. Genetic structure of Norway spruce (*Picea abies*): Concordance of morphological and allozymic variation. Evolution 44: 38—53.

Lebreton, P., Laracine-Pittet, C., Bayet, C. and Lauranson, J. 1990. Variabilité des polyphénolique et systématique du Pin sylvestre. Ann. Sci. For. 47: 117—130.

Lewandowski, A. and Mejnartowicz, L. 1992. Levels and patterns of allozyme variation in some European larch (*Larix decidua* Mill.) populations. Hereditas (in press).

Li, P. and Adams, W. T. 1989. Range — wide patterns of allozyme variation in Douglas-fir. Can. J. For. Res. 19: 149—161.

Loukas, M., Vergini, Y. and Krimbas, C. B. 1983. Isozyme variations and heterozygosity in *Pinus halepensis* L. Biochemical Genetics 21(5—6): 497—509.

Lumaret, R., Yacine, A., Berrod, A., Romane, F. and T. X. Li. 1991. Mating system and genetic diversity in holm oak (*Quercus ilex* L., *Fagaceae*), pp. 149—153. In: Biochemical markers in the population genetics of forest trees. Fineschi, S., Malvolti, M. E., Cannata, F. and Hattemer, H. H. (Eds) SPB Academic Publishing bv, The Hague, The Netherlands.

Lundkvist, K. 1979. Allozyme frequency distributions in four Swedish populations of Norway spruce (*Picea abies* K.). Hereditas 90: 127—143.

Lundkvist, K. and Rudin, D. 1977. Genetic variation in eleven populations of *Picea abies* as determined by isozyme analysis. Hereditas 85: 67—74.

Marpeau, A. and Baradat, Ph, 1975. Les terpènes du Pin maritime: aspects biologiques et génétiques. 4. Hérédité de la teneur en deux sesquiterpènes: le longifolène le caryophyllène. Ann. Sci. For. 32 (4): 185—204.

Marpeau, A., Baradat, Ph. and Bernard-Dagan, C. 1983. Les terpènes du Pin maritime: aspects biologiques et génétiques. 5. Hérédité de la teneur en limonène. Ann. Sci. For. 40(2): 197—216.

Mejnartowicz, L. 1979. Genetic variation in some isoenzyme loci in Scots pine (*Pinus silvestris* L.) populations. Arboretum Kornickie 24: 91—104.

Mejnartowicz, L. 1980. Polymorphism at the LAP and GOT loci in *Abies alba* Mill. populations. Bulletin de l'Académie Polonaise des Sciences. Série des Sciences Biologiques. C1. V., 27(12): 1063—1070.

Mejnartowicz, L. and Bergmann, F. 1985. Genetic differentiation among Scots pine populations from the lowlands and the mountains in Poland, pp. 253—266. In: Gregorius, H.-R. (Ed) Population Genetics in Forestry. Lecture Notes in Biomathematics, 60. Springer-Verlag, Berlin, Heidelberg, New York, Tokyo.

Mejnartowicz. L. and Palowski, B. 1989. Studies of Scots pine populations in polluted and clean areas, pp. 115—125. In: Scholz, F., Gregorius, H.-R. and Rudin, D. (Eds) Genetic Effects of Air Pollutants in Forest Tree Populations. Springer-Verlag, Berlin, Heidelberg, New York, Tokyo.

Merzeau, D., Di Giusto, F., Comps, B., Thiebaut, B., Letousey, J. and Cuguen, J. 1989.

Genetic control of isozyme systems and heterogeneity of pollen contribution in beech (*Fagus sylvatica* L.). Silvae Genetica 38(5—6): 195—201.

Moreau, M. and Pastuszka P. 1991. Contribution a l'étude de la structure génétique d'un peuplement de sapin pectiné de Basse Vosges. (submitted to Ann. des Sci. For.)

Morgante, M. and Vendramin, G. 1990. Analyse von Genressourcen von *Pinus leucodermis* Ant., einer Art mit kleinem Verbreitungsgebiet, pp. 87—98. In: Hattemer, H. H. (Ed) Erhaltung forstlicher Genressourcen. Schriften aus der Forstl. Fak. d. Univ. Göttingen und der Nds. Forstl. Vers. Anst. J.D. Sauerländer's Verlag, Frankfurt a.M.

Müller-Starck, G. 1985. Genetic differences between "tolerant" and "sensitive" beeches (*Fagus sylvatica* L.) in an environmentally stressed adult forest stand. Silvae Genetica 34(6): 241—247.

Müller-Starck, G. 1987. Genetic differentiation among seed samples provenances of *Pinus sylvestris* L. Silvae Genetica 36(5—6): 232—238.

Müller-Starck, G. 1989. Genetic implications of environmental stress in adult forest stands of *Fagus sylvatica* L., pp. 127—142. In: Scholz, F, Gregorius, H.-R. and Rudin, D. (Eds) Genetic Effects of Air Pollutants in Forest Tree Populations. Springer-Verlag, Berlin, Heidelberg, New York, Tokyo.

Müller-Starck, G. and Gregorius, H.-R. 1986. Monitoring genetic variation in forest tree populations. Proc. 18th IUFRO World Congress, Ljubljana, Yugoslavia, 1986, Div. 2, Vol. II., 589—599.

Müller-Starck, G. and Ziehe, M. 1991. Genetic variation in populations of *Fagus sylvatica, Quercus robur,* and *Q. petraea* in Germany, pp. 125—140. In: Müller-Starck, G. and Ziehe, M. (Eds) Genetic variation in European Populations of Forest Trees. Sauerländer's Verlag, Frankfurt a.M.

Muona, O. and Szmidt, A. E. 1985. A multilocus study of natural populations of *Pinus sylvestris*, pp. 226—240. In: Gregorius, H.-R. (Ed) Population Genetics in Forestry. Lecture Notes in Biomathematics, 60. Springer-Verlag, Berlin. Heidelberg, New York, Tokyo.

Muona, O., Paule, L., Szmidt A. E. and Kärkkäinen, V. 1990. Mating system analysis in a central and northern European population of *Picea abies*. Scand. J. For. Res. 5: 97—102.

Nei, M. 1973. Analysis of gene diversity in subdivided populations. Proc. Nat. Acad. Sci. USA 70(12): 3321—3323.

Nikolić, D. and Tucić, N. 1983. Isoenzyme variation within and among populations of european Black pine (*Pinus nigra* Arnold) Silvae Genetica 32(3—4): 80—89.

Paule L., Yazdani, R. and Gömöry, D. 1987. Monoterpene composition of silver fir foliar oleoresin, pp. 50—66. In: Paule, L. and Korpel', Š. (Eds) 5. IUFRO-Tannensymposium. VŠLD, Zvolen, ČFSR.

Pigliucci, M., F. Benedettelli, S. and Villani, 1990. Spatial patterns of genetic variability in Italian chestnut (*Castanea sativa* Mill.). Can. Journ. Botany. 68: 1962—1968.

Prus-Glowacki, W. and Rudin, D. 1981. Short Note: Studies on antigenic proteins of *Pinus sylvestris* from six Swedish provenances — A pilot study. Silvae Genetica 30(6): 200—203.

Prus-Glowacki, W. 1982. Immunochemical methods in analysis of forest tree proteins. Silva Fennica, Vol. 16(2): 219—226.

Prus-Glowacki, W., Szweykowski, J. J. and Nowak, R. 1985. Serotaxonomical investigation of the European pine species. Silvae Genetica 34(4—5): 162—170.

Rudin, D., Erikson, E., Ekberg, I. and Rasmuson, M. 1974. Studies of allele frequency and inbreeding in Scots pine populations by the aid of the isozyme technique. Silvae Genetica 23(1—3): 10—13.

Rudloff E. von. 1973. Geographical variation in the terpene composition of the leaf oil of Douglas-fir. Pure Appl. Chem. 34: 401—410.

Schiller, G., Conkle, M. T. and Grunwald, C. 1986. Local differentiation among mediterranean populations of Aleppo pine in their isoenzymes. Silvae Genetica 35(1): 11—19.

Schiller, G. and Grunwald, C. 1987. Resin monoterpenes in range-wide provenance trials of *Pinus halepensis* in Israel. Silvae Genetica. 36(3—4): 109—114.

Schiller, G. 1990. Variation in resin composition of the Italian cypress (*Cupressus sempervirens*) grown in Israel. Silvae Genetica 39(3—4): 89—95.

Schröder, S. 1989. Die Weißtanne in Süddeutschland: Genetische Variation, Kline, Korrelationen. Allg. Forst- und Jagdztg., 160: 100—104.

Squillace A. E. 1976. Analysis of monoterpenes of conifers by gas liquid chromatography, pp. 129—157. In: Miksche, J. P. (Ed) Modern methods in Forest Genetics. Springer-Verlag, Heidelberg, Berlin, New York, Tokyo.

Stutz, H. P. 1990. Eigenschaften und Ursachen der genetischen Differenzierung der Fichte im Wallis (Schweiz), pp. 99—114. In: Hattemer, H. H. (Ed) Erhaltung forstlicher Genressourcen. Schriften aus der Forstl. Fak. d. Univ. Göttingen und der Nds. forstl. Vers. Anst. J. D. Sauerländer's Verlag, Frankfurt a.M.

Szmidt, A. 1982. Genetic variation in isolated populations of stone pine (*Pinus cembra*). Silva Fennica 16(2): 196—200.

Szmidt, A. E., El-Kassaby, Y.-A., Sigurgeirsson, A., Alden, T., Lindgren, D. and Hällgren, J.-E. 1988. Classifying seedlots of *Picea sitchensis* and *P. glauca* in zones of introgression using restriction analysis of chloroplast DNA. Theor. Appl. Genetics 76: 841—845.

Szmidt, A. E. 1990. Phylogenetic and applied studies on chloroplast genome in forest conifers. Biochemical markers in the population genetics of forest trees, pp. 185—196. In: Fineschi, S., Malvolti, M. E., Cannata, F. and Hattemer, H. H. (Eds) Biochemical markers in the population genetics of forest trees. SPB Academic Publishing bv, The Hague, The Netherlands.

Thiebaut, B., Lumaret, R. and Vernet, Ph. 1982. The bud enzymes of beech (*Fagus sylvatica* L.). Genetic distinction and analysis of polymorphism in several french populations. Silvae Genetica 31(2—3): 51—60.

Thielges, B. A. 1972. Intraspecific variation in foliage polyphenole of *Pinus* (subsection *sylvestris*). Silvae Genetica 21(3—4): 114—119.

Tigerstedt, P. M. A. 1973. Studies on isozyme variation in marginal and central populations of *Picea abies*. Hereditas 75: 47—60.

Tigerstedt, P. M. A. 1979. Genetic adaptation of plants in the subartic environment. Holarctic Ecology 2: 264—268.

Tigerstedt P. M. A., Hiltunen, R., Chung, M. S. and Moren, E. 1979. Inheritance and genetic variation of monoterpenes in *Pinus sylvestris*, pp. 29—39. In: Rudin, E. (Ed) Proceedings of the IUFRO Conference on Biochemical Genetics of Forest Trees, Umeå, Sweden 1978.

Valizadeh, M. 1977. Esterase and acid phosphatase polimorphism in the fig tree (*Ficus carica* L.) Biochemical Genetics 15: 1037—1048.

Villani, F., Benedettelli, S., Paciucci, M., Cherubini, M. and Pigliucci, M. 1991a. Genetic variation and differentiation between natural populations of chestnut (*Castanea sativa* Mill.) from Italy, pp. 91—103. In: Fineschi, S., Malvolti, M. E., Cannata, F. and Hattemer, H. H. (Eds) Biochemical markers in the population genetics of forest trees. SPB Academic Publishing bv, The Hague, The Netherlands.

Villani, F., Pugliucci, M., Benedettelli, S. and Cherubini, M. 1991b. Genetic vs. geographical and climatic differentiation of Turkish chestnut. (*Castanea sativa* Mill.) Heredity 66: 131—136.

Wellendorf H. and Kaufmann, U. 1977. This layer chromatography of fluorescent phenolic compounds in needles: A review of current activities in *Picea*, pp. 203—226. In: Proceedings of EEC Symposium on Forest Tree Biochemistry, Bruxelles, Belgium.

White, E. C. and Nilsson, J. E. 1984. Genetic variation in resin canal frequency and relationships to terpene production in foliage of *Pinus contorta*. Silvae Genetica 33(2—3): 79—84.

Wright, S. 1943. Isolation of distance. Genetics 28: 114—138.

Wright, S. 1951. The genetic structure of populations. Ann. Eugenics 15: 323—354.

Wright, S. 1969. Evolution and the Genetics of Populations. Vol. 2. The Theory of Gene Frequencies. Univ. of Chicago Press, Chicago.

Yazdani R. and Lebreton, P. 1991. Inheritance pattern of the flavonic compounds in Scots pine. Silvae Genetica 40(2): 57—59.

Yazdani, R., Nilsson, J. E. and Ericsson, T. 1985. Geographical variation in the relative proportion and monoterpenes in cortical oleorosis of *Pinus sylvestris* in Sweden. Silvae Genetica 34(6): 201—208.

Yazdani, R. and Nilsson, J. E. 1986. Cortical monoterpene variation in natural populations of *Pinus sylvestris* in Sweden, Scand. J. For. Res. 1: 85—93.

Yazdani, R., Rudin, R., Aldén, T., Lindgren, L., Harbom, B. and Ljung, K. 1982. Inheritance pattern of five monoterpenes in Scots pine. Hereditas 97: 261—272.

Zanetto, A. and Kremer A. 1991. Allozyme polymorphism of sessile oak populations in western Europe. (submitted to Ann. Sci. For.)

Zavarin E. 1970. Qualitative and quantitative co-occurence of terpenoids as a tool of elucidation to their biosynthesis. Phytochemistry 9: 1049—1963.

New Forests **6**: 49–66, 1992.

Review paper

Patterns of genetic diversity in Australian tree species

G. F. MORAN
Division of Forestry and Forest Products, CSIRO, PO Box 4008, Canberra ACT Australia 2600

Received 8 October 1990; accepted 29 November 1991

Key words: isozymes, conservation, geographic range, domestication, breeding systems

Application. Patterns of allozyme variation in widespread Australian trees often reveal patterns of genetic variation that are similar to those based on other traits including growth. Therefore, in the domestication phase of such species, a cost efficient strategy is to initially carry out an isozyme survey of a limited number of populations covering the geographic range. If the study shows broad geographic patterns within a species, then more efficient sampling strategies could be devised for intensive seed collections and large scale provenance trials.

Abstract. Australia has a large endemic tree flora with many of the genera largely confined to the southern hemisphere. The two dominant genera are *Eucalyptus* and *Acacia*. Isozyme studies of patterns of genetic diversity in populations of these species are reviewed. Generally, Australian tree species have high levels of allozyme variation with most of this variation within rather than between populations. The species with the most genetic differentiation between populations are those with regional distributions but with small disjunct populations. Many of the species show no discernible relationship between current population sizes and genetic diversity. A number of species with widespread distributions exhibit similar clusters of populations both on isozymes and other traits. Such clusters often correspond to large geographic regions. This pattern suggests that preliminary low intensity isozyme surveys could help to define more efficient sampling strategies for intensive seed collections and subsequent field trials of many tree species.

Introduction

Population genetic studies of Australian tree species have aimed at providing the information needed in domestication and breeding programs and for conservation of genetic resources. Much of the research has been on the amount and distribution of genetic diversity in forest eucalypt species and variation in mating system parameters. A major requirement for natural populations is the development of efficient strategies for *in situ* conservation of genetic resources (Moran and Hopper 1987). In contrast, the major goals for eucalypt breeding programs have been the maximum capture of desirable genetic diversity during domestication and the production of seed of high genetic quality in seed orchards. In the last few years research has also focused on acacias, especially species for overseas plantations in the tropics.

Since there are a very large number of tree species in Australia, genetic research on their conservation has been directed at developing general strategies for *in situ* conservation of genetic resources. To achieve this for eucalypts, the extent and patterns of genetic diversity in natural populations have been studied for not only major commercial forest species but also for a number of rare species. The genetic structure of populations of Australian species was largely unknown until the application of isozymes in the 1970's (Brown et al. 1975; Phillips and Brown 1977). In addition to population and conservation genetics research, isozymes have been used to investigate biosystematics of eucalypts (Moran et al. 1990a; Prober et al. 1990a) and rainforest species of the Lauraceae (Moran et al. 1990a).

Studies of isozyme variation within Australian tree species are reviewed in this paper. Genetic diversity within populations and genetic differentiation among populations within species are assessed. Patterns of genetic diversity are sought particularly in relation to the geographic distribution of populations. The isozyme data are examined for the effect of population size and gene migration on patterns of genetic diversity.

Australian tree flora

In Australia there are about 2500 native tree species of which fewer than 40 are gymnosperms. About half the total number are rainforest species. There are no population genetic data on any of the rainforest or gymnosperm tree species, and they will not be considered further in this review. Six gymnosperm and less than 200 angiosperm tree species in Australia are of importance in commercial forestry. Outside rainforests, many of the vegetation communities in Australia are dominated by tree species belonging to only a few genera, namely *Eucalyptus*, *Acacia* and *Casuarina* (Bridgewater 1987). Most of the commercial forest species are eucalypts (Table 1) and occur primarily in the southern temperate region of Australia.

The present Australian flora is characterised by high endemism. The Australian flora is of Gondwanan origin and is considered today to contain two components. One is an evolutionarily conservative relict element and the other a derived autochthonous (Australian) element. The autochthonous element is considered to have undergone rapid evolutionary changes towards scleromorphy under geographic isolation and intense competition in nutrient deficient soils (Barlow 1981). These changes have resulted in (1) distinctive genera endemic to the Australian

Table 1. Approximate numbers of Australian tree species in each of the three major genera, including numbers of species of commercial importance and numbers investigated in isozyme studies

Genera	Total	Commercial use	Isozyme studies
Eucalyptus	570	50	25
Acacia	250	20	10
Casuarina	22	3	2

region and (2) species richness particularly in the southwest and southeast of the continent. A significant fraction of these species have small geographic ranges. The tree species that have been studied isozymically belong primarily to this Australian element in the flora.

The temperate floras of eastern and western Australia have been geographically isolated for many millions of years (Nelson 1981). Despite this separation most generic and subgeneric elements of the major tree groups are common to both regions. The profound biogeographical changes to the flora in the Quaternary (Galloway and Kemp 1980) were probably more marked on tree distributions in eastern Australia than in southwestern Australia, because in the latter land surfaces are more ancient and stable.

Factors influencing patterns of genetic diversity

The amount and distribution of genetic diversity within tree species could be influenced by factors such as tree size, longevity, fecundity, breeding systems and geographic distribution (Hamrick and Godt 1990). In this review only the influence of breeding systems and geographic distribution on patterns of genetic diversity are considered.

Breeding systems

Most forest tree species investigated have been found to have mating systems featuring high levels of outcrossed progeny (Muona 1990), and the Australian species are no exception. In Australia most angiosperm trees are animal-pollinated and have hermaphroditic flowers. Exceptions are casuarina species which are wind-pollinated and primarily dioecious. The twelve eucalypt species, that have been examined, all have a mixed-mating system with predominant outcrossing but a significant fraction of inbreeding (Table 2) (Moran and Bell 1983; Brown et al. 1985; Sampson et al. 1989). In contrast, progeny of four acacia species are nearly entirely the result of outcrossing (Moran et al. 1989b; Muona et al. 1991) and evidence suggests they have a more complete self-incompatibility system than eucalypts (Kenrick and Knox 1989). Banksias and *Grevillea robusta* are largely bird and mammal-pollinated, and appear to have high outcrossing rates. The apparent difference in breeding systems between eucalypts and acacias might result in differences in population genetic structure between species of the two groups.

Geographic distribution of populations

For purposes of comparison, the species considered in this review were divided into three main types (see Moran and Hopper 1987): (1) widespread species with a range of 600 km in at least one direction; (2) regional species with a range between 150 and 600 km; and (3) localised species with a small number of populations usually of limited size and endemic to a restricted geographic area of less than 100 km. Over these geographic ranges species can exhibit a variety of population structures from continuous distributions, to many disjunct populations, to a single small population.

Experimental procedures and genetic analyses

Sampling

In this review isozyme studies included were only those for species for which the populations sampled covered the geographic range. In fact, for a

Table 2. Estimates of outcrossing rates (t) in natural populations of Australian tree species

Genera/species	Number		t		Source
	Species	Populations	Mean	Range	
Eucalyptus sp.	12	29	0.74	0.59—0.83	Moran and Bell 1983 Brown et al. 1985 Sampson 1988
Acacia sp	4	6	0.94	0.89—0.96	Muona et al. 1991 Moran et al.1989b
Banksia sp.	2	3	0.88	0.80—0.96	Carthew et al. 1988
Grevillea robusta		2	0.92		Harwood et al. 1990

number of species of regional and localised distributions (such as *E. caesia, E. parvifolia* and *E. paliformis*, see Table 3) all known populations were sampled. Generally, 15 to 20 loci were assayed per species. For most of the species, seed was collected from 10 trees per population and a total of 50 progeny per population assayed for isozyme loci. Details of starch gel electrophoretic procedures are presented by Moran and Bell (1983) for eucalypts, for acacias by Moran et al. (1989b) and for casuarinas by Moran et al. (1989c). Cellulose acetate electrophoretic procedures for acacias and eucalypts are described in Coates (1988).

Genetic parameters

Three parameters were used to describe levels of genetic variation within populations: percentage of polymorphic loci, the mean number of alleles per locus, and genetic diversity. The percentage of polymorphic loci (*P*, 0.99 criterion) is the proportion of loci polymorphic in each population averaged over all populations sampled. The number of alleles per locus (A) was calculated for each population and a mean value found by averaging over all populations. The mean genetic diversity (H_e) was calculated for each population by averaging across all loci and then calculating a mean across all populations. At the species level the mean number of alleles per locus (A_s) was calculated by summing the alleles over all loci in all populations and dividing by the number of loci.

The estimated number of migrants exchanged between populations per generation (*Nm*) were calculated by the private alleles method (Slatkin 1985) or by using Wright's F_{ST} (Slatkin 1987). The levels and distribution of genetic diversity between and within populations and geographic regions were calculated using the gene diversity statistics of Nei (1973).

Genetic diversity within populations

Mean estimates of allozyme variation at the population level for Australian tree species are shown in Table 3. Overall mean estimates, both across all species, and for eucalypts alone, are very similar to those for gymnosperms (Loveless and Hamrick 1984; Hamrick and Godt 1990). Conifers have been considered to have high genetic variability compared to other plants (Muona 1990). It appears that Australian angiosperm trees have similarly high levels. Generally, widespread forest eucalypts have substantial amounts of allozyme variation. However, within each of these widespread species there was no obvious patterns to population levels of allozyme variation in relation to environmental factors such as latitude,

Table 3. Estimates of allozyme variation at the population level for 19 tree species

Species[1]	n	A	P	H_e	Source
Eucalypts					
			Widespread		
E. cloeziana	17	2.21	79.8	0.205	Turnbull (1980)
E. delegatensis	24	2.52	78.9	0.239	Moran (1991)
E. grandis	12	2.09	56.5	0.167	Moran and Bell (1983)
E. saligna	7	2.10	64.3	0.239	Moran and Bell (1983)
			Regional		
E. caesia	13	1.31	29.0	0.068	Moran and Hopper (1983)
E. crucis	10	1.90	60.9	0.187	Sampson et al. (1988)
E. diversicolor	13	1.61	48.7	0.152	Coates and Sokolowski(1989)
E. lane-poolei	7	2.26	80.0	0.246	Sampson (1988)
E. pulverulenta	4	1.41	27.0	0.068	Peters et al. (1990)
E. rhodantha	6	1.98	56.9	0.168	Sampson (1988)
			Localised		
E. johnsoniana	3	1.29	27.0	0.084	Moran and Hopper (1987)
E. lateritica	2	2.16	75.0	0.278	Moran and Hopper (1987)
E. paliformis	6	1.44	36.7	0.137	Prober et al. (1990)
E. parvifolia	8	2.29	70.6	0.213	Prober et al. (1990)
E. pendens	7	1.80	57.0	0.156	Moran and Hopper (1987)
E. suberea	3	1.85	60.0	0.170	Moran and Hopper (1987)
Mean	8.3	1.89	56.8	0.174	
Acacias					
			Widespread		
A. mangium	11	1.14	12.7	0.017	Moran et al. (1989a)
A. melanoxylon	27	2.29	69.5	0.208	J. Playford (unpubl.)
			Localised		
A. anomala	6	2.00	61.1	0.209	Coates (1988)
Casuarinas					
			Widespread		
C. cunninghamiana	20	1.99	74.5	0.211	Moran et al. (1989b)
Overall mean	9.8	1.88	56.3	0.170	

[1] Widespread = with a range of 600 km in at least one direction, regional = range between 150 and 600 km, localised = geographic area of less than 100 km.

n = Number of populations sampled, A = mean number of alleles per locus, P = mean percentage of polymorphic loci (0.99 criterion), H_e = unbiased expected panmictic heterozygosity.

altitude, soil type etc. All measures of genetic variation were very low for *E. caesia, E. pulverulenta* and *E. johnsoniana* (Table 3).

Acacia mangium, which occurs in north Queensland, Papua New Guinea and Indonesia (Moran et al. 1989a) has remarkably low levels of allozyme variation, yet is rapidly becoming one of the most widely planted hardwood species in the tropics. In sharp contrast, *Acacia melanoxylon*, the major commercial acacia in Australia, has high mean values for the three parameters (Table 3). In addition, the estimates of H_e for individual populations increase with increasing latitude (J. Playford unpublished). On the other hand, the populations of *Casuarina cunninghamiana* from eastern Australia exhibit decreasing heterozygosity with increasing latitude (Moran et al. 1989c). It is of interest that results from provenance trials for both species indicate that the faster growing populations come from areas with the most heterozygous populations (Tasmanian For. Comm., pers. comm.; El-Lakany 1990).

Size of populations

In Australian tree species, the number of plants per population can vary enormously particularly for eucalypts and acacias. Although populations in some species may be small because of the advent of European agricultural practices, a substantial number of tree species naturally have very small populations. Theory suggests that populations with few individuals will experience a reduction in genetic variation over several generations. This reduction will be enhanced if the populations are disjunct rather than continuously distributed since they will be less subject to gene flow from other populations (Lande and Barrowclough 1987). In *E. pendens* the current population sizes vary from 27 to about 3000 yet there is no evidence of reduction in levels of genetic variation with smaller population sizes (Table 4). Similar results have been found in a number of other localised eucalypts (Moran and Hopper 1987; Prober et al. 1990b). The eucalypts showing significant within-species correlations between population size and levels of genetic diversity are *E. caesia* and *E. crucis* (Moran and Hopper 1983; Sampson et al. 1988). Both species appear to have occupied small, stable but very isolated niches for long evolutionary periods. Thus the data suggest that the effect of small population size on genetic diversity patterns in Australian trees may be greatest when there has been little chance of migration between populations. Such is not the case in *E. pendens*.

For species, like *E. parvifolia* and perhaps *E. rhodantha*, some of the populations are small because of tree clearance for agriculture. With generation times of about 100 years for these species, reductions in

Table 4. Population sizes and estimates of allozyme variation for *E. pendens* (from Moran and Hopper 1987)

Population	N	A	P	H_o	H_e
North Badgingarra	27	1.69	63	0.183	0.208
Central Badgingarra	33	1.63	44	0.204	0.191
Isolated Hill	35	1.75	56	0.156	0.148
Alexander Morrison	56	1.88	75	0.113	0.131
South Badgingarra	83	1.73	50	0.145	0.126
Coonawarra Downs	1000	1.44	44	0.105	0.123
Williams Hill	3000	1.75	69	0.163	0.163

N = Population size, H_o = observed heterozygosity, see Table 3 for descriptions of A, P, and H_e.

population sizes have not yet affected patterns of genetic variation. Of course, it is the effective population size (N_e) rather than N that can affect gene frequencies. Estimates of N_e in *E. parvifolia* (Prober et al. 1990b) and *E. pulverulenta* (Peters et al. 1990) are often much smaller than N.

In some species levels of genetic diversity could be due to bottlenecks in earlier times rather than current population sizes. For instance, the lack of allozyme variation in populations of *A. mangium* is thought to be due to bottlenecks occurring during Pleistocene glaciations (Moran et al. 1989a). The species occurs on coastal tropical lowlands in northern Australia, Indonesia and Papua New Guinea. During interglacial periods in the Quaternary these environments were largely underwater and the species was restricted to small refuge populations.

Gene migration and genetic diversity

The low genetic differentiation between populations of conifers (Hamrick and Godt 1990; Muona 1990) is commonly attributed, in large part, to effective gene flow brought about by wind dispersal of pollen. Most eucalypts and acacias are primarily insect-pollinated but it is not known how much gene migration there is between populations and what role it plays in their genetic differentiation. Estimates of gene flow in some regional and localised eucalypts were made using the private allele method of Slatkin (1985) (Table 5). Nm, the estimated mean number of migrants exchanged between local populations, is low in *E. caesia* and *E. crucis*, which might be expected since these species have very isolated populations. In contrast, estimates were comparatively high for species, for which distances between populations had been apparently increased by recent

Table 5. Estimates of gene flow (*Nm*) between populations for six eucalypt species

Species	Average interpop. distance (km)	Average pop. size (plants)	Average number private alleles	*Nm*	Source
E. paliformis	2	15000	0	6.16[1]	Prober et al. (1990)
E. parvifolia	20	440	0.018	4.71	Prober et al. (1990)
E. rhodantha	41	50	0.002	359.72	Sampson (1988)
E. caesia	89	160	0.073	0.95	Sampson (1988)
E. crucis	90	70	0.089	0.54	Sampson (1988)
E. lane-poolei	117	80	0.035	2.28	Sampson (1988)

[1] All estimates by private allele method (Slatkin 1985), except for *E. paliformis* calculated by F_{ST} method (Slatkin 1987).

Nm = mean number of migrants between populations per generation.

clearing for agriculture (*E. parvifolia* and *E. lane-poolei*). Thus these estimates of gene migration along with the genetic diversity measures may be indicative of more continuous distributions in these species in the recent past.

Distribution of genetic diversity

The proportion of genetic diversity between populations averaged over 21 Australian species is 26.4% (Table 6). The average G_{ST} across 17 eucalypts is 18.1%. The latter is more than two times greater than the average reported for wind-pollinated conifers (G_{ST} = 6.8%, Hamrick and Godt 1990) and northern hemisphere wind-pollinated angiosperms (G_{ST} = 7.5%, Muona 1990). It appears that animal-pollinated trees from Australia are more highly differentiated between populations than wind-pollinated species. It is not clear whether this greater genetic differentiation is due primarily to their mode of reproduction or to other characteristics of Australian trees. Nevertheless, Australian trees like most other tree species studied have most of their allozyme diversity within populations. For eucalypts, the G_{ST} values range from 3.9% for *E. paliformis* to 61.4% for *E. caesia*. The species *E. nitens*, which has a large geographic range but several disjunct regions within it, has a much higher G_{ST} value than other widespread species. Of the two widespread acacias, *A. melanoxylon* has more than ten times the diversity of *A. mangium* but about the same proportion of genetic diversity among populations (Table 6).

When the eucalypts are grouped by geographic range the widespread species have the highest values for H_T and H_S whilst regional species have the largest G_{ST} values (Table 6). If the regional species are further divided into those with continuous distributions and those with disjunct distributions, the disjunct group has a much higher G_{ST} value (38.6%) than the continuous group (11.2%). Thus eucalypts with regional but disjunct distributions have the highest proportion of diversity between populations. However, *E. johnsoniana* with a localised distribution also exhibits considerable genetic differentiation probably because of a combination of very small populations and the occurrence of very few multilocus genotypes within populations.

Geographic regions in widespread species

Although many commercial forest species are widespread and have large populations, the spatial distribution of populations is usually a mosaic determined by environmental requirements and competition from other species. Although the majority of genetic diversity appears to reside within populations, species vary considerably in the proportion of diversity due to differences among populations (G_{ST}) (Table 6). Does variation in G_{ST} reflect the extent of geographic disjunctions or regions within the distributions of species?

In *E. delegatensis*, a major commercial forest species of south-eastern Australia, samples from population throughout the geographic range were assayed isozymically (Fig. 1; Moran 1991). A cluster analysis based on the population allelic frequencies separated the populations into two distinct groups which corresponded to the geographic areas or regions of Tasmania and the mainland. This genetic separation of geographic regions supports similar findings based on morphological characters (Boland and Dunn 1985) and growth rate (Moran et al. 1990b). Apparently, genetic differences in the genomes between the two regions have been so extensive that most characters, including isozymes, show distinct geographic separation. Similarly, in a rangewide isozyme study of 22 populations of *E. nitens* (Cook 1989), a widespread species with disjunct regions, a cluster analysis showed strong genetic separation of the major geographic regions. Analysis of morphological characters (Cook 1989) and growth data in *E. nitens* (Pederick 1979) showed similar clustering of populations into geographic regions.

Moran et al. (1989c), in an extensive population study of isozyme

Table 6. Estimates of total genetic diversity and distribution of diversity within and between populations in Australian tree species

Species	A_s	H_T	H_S	G_{ST} (%)
Eucalypts				
		Widespread		
E. cloeziana	2.71	0.230	0.205	11.0
E. delegatensis	4.10	0.272	0.238	12.5
E. grandis	2.93	0.190	0.167	12.0
E. nitens	3.35	0.202	0.140	30.2
E. saligna	2.86	0.260	0.239	8.0
Mean	3.19	0.231	0.198	14.7
		Regional		
E. caesia	1.94	0.176	0.068	61.4
E. crucis	3.30	0.291	0.202	24.4
E. diversicolor	2.09	0.168	0.152	9.7
E. lane-poolei	3.20	0.292	0.245	13.7
E. pulverulenta	1.88	0.100	0.070	30.0
E. rhodantha	2.60	0.221	0.184	10.1
Mean	2.99	0.208	0.154	24.9
		Localised		
E. johnsoniana	1.88	0.139	0.084	39.6
E. lateritica	2.56	0.318	0.278	12.6
E. paliformis	1.52	0.141	0.136	3.9
E. parvifolia	3.41	0.228	0.212	7.0
E. pendens	2.88	0.170	0.156	8.2
E. suberea	2.44	0.197	0.170	13.7
Mean	2.45	0.199	0.173	14.2
Acacias				
		Widespread		
A. mangium	1.73	0.025	0.017	31.1
A. melanoxylon	5.20	0.334	0.208	37.7
		Localised		
A. anomala[1]	3.07	0.237	0.224	5.6
Casuarinas				
		Widespread		
C. cunninghamiana	4.32	0.287	0.211	26.4

[1] Sexual populations only, Coates (1988).

A_s = mean number of alleles per locus, H_T = total genetic diversity, H_S = mean genetic diversity within populations, G_{ST} = mean proportion of total diversity due to differences between populations.

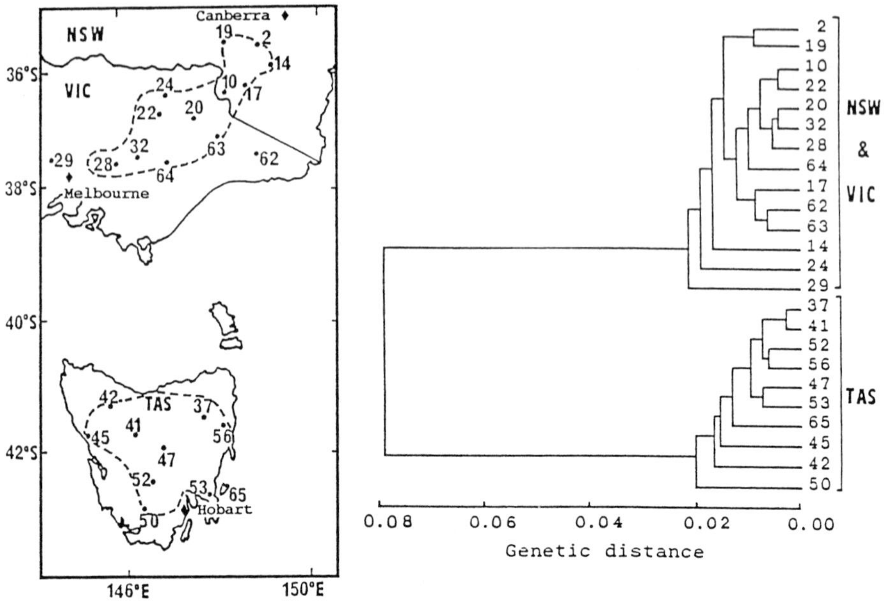

Fig. 1. Geographic distribution of sampled populations of *Eucalyptus delegatensis* in south east Australia (NSW and VIC) and Tasmania and genetic clustering of populations based on isozyme gene frequencies. Dotted line shows main distribution of species. NSW = New South Wales, VIC = Victoria, and Tas = Tasmania. Cluster based on Nei's (1978) unbiased genetic distance and the UPGMA algorithm.

variation in *Casuarina cunninghamiana,* found that a north-western population (population 20 in Fig. 2) was very distinct from the rest of the populations. This result agreed with previous morphological evidence of a north-western geographic form (Boland et al. 1984). Further, populations from the northeast region, which clustered together based on isozyme variation, have also shown distinctly superior growth rates in provenance trials (El-Lakany 1990).

Maslin and Pedley (1982) mapped the distribution of *Acacia holosericea* across the whole of the top of Australia. Thirty-three populations were assessed for isozyme variation and a cluster analysis performed on data from 17 loci (Moran and Thomson unpublished) (Fig. 3). The cluster analysis showed that there were three very distinct groups corresponding to three geographic areas (A, B and C), and that within each of the groups isozyme variation within and between populations was very small.

Following this isozyme study, Thomson (pers. comm.) assessed early domestication trials of some of the same populations in Niger and other African countries and found that populations from the different Australian

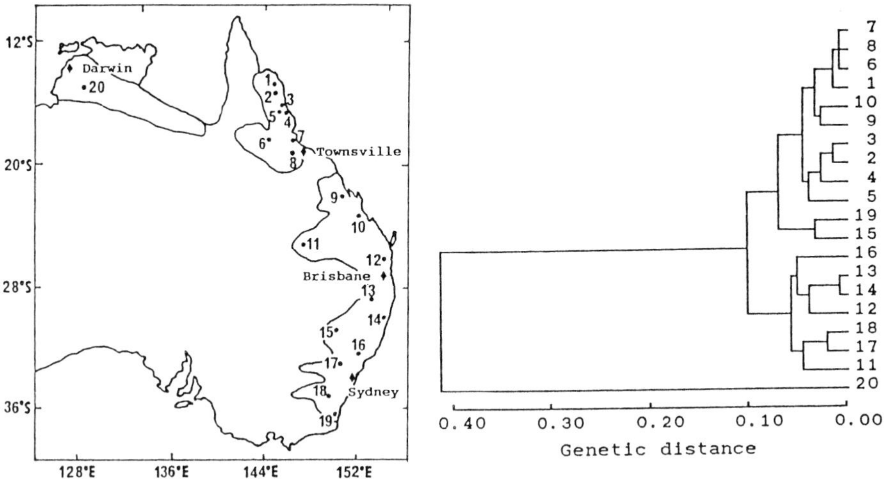

Fig. 2. Geographic distribution of sampled populations of *Casuarina cunninghamiana* in Australia and genetic clustering of populations based on isozyme gene frequencies. Distribution of species is shown by the line bounding population locations. Cluster based on Nei's (1978) unbiased genetic distance and the UPGMA algorithm.

geographic areas were distinct morphologically in leaves and fruit. More importantly the populations from the drier inland area (C) had better growth than those from area A. Populations from within an area were very uniform in their growth. An initial low level isozyme assessment of the species could have refined considerably the sampling strategies for field seed collections and testing for growth performance.

In a recent population study of isozyme variation in *Acacia melanoxylon* by J. Playford (pers. comm.) at CSIRO has found populations grouping into three clusters, which correspond to geographic regions within the widespread distribution of the species. Early results of provenance trials in Tasmania indicate large growth differences between populations of *A. melanoxylon* from these different geographic regions.

A common thread of the four studies discussed above is that a significant fraction of the genetic differentiation between populations can be apportioned to meaningful geographic regions. The data in Fig. 4 demonstrate that more than 50% of the G_{ST} in these four species can be attributed to differences between regions, the rest to differences among populations within regions. Such geographic patterns of genetic differentiation do not occur for all widespread species, with *E. grandis* and *E. saligna* being examples (Burgess and Bell 1983). However the studies do indicate that isozymes can have an important role in defining more

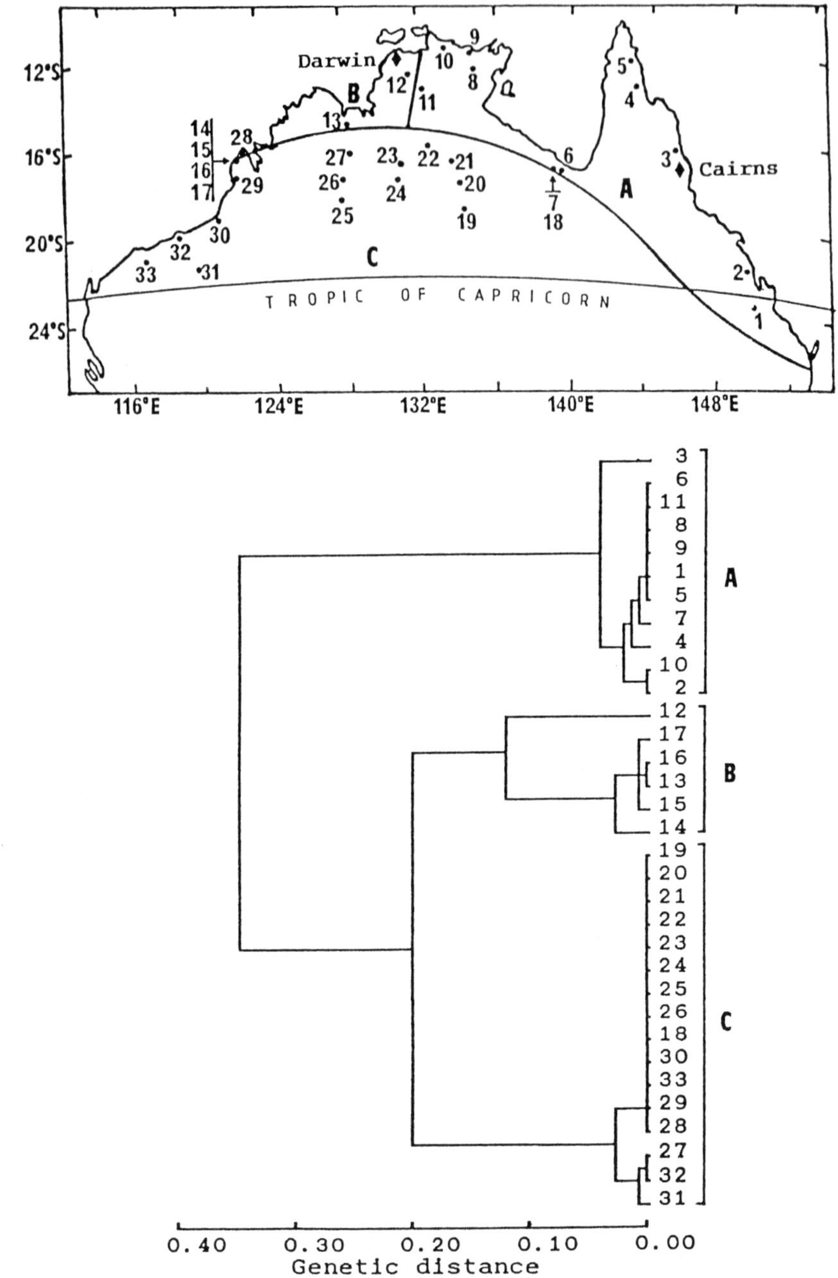

Fig. 3. Geographic distribution of sampled populations of *Acacia holosericea* across northern Australia and genetic clustering based on isozyme gene frequencies. Dotted line shows southern limit of species distribution. Cluster based on Nei's (1978) genetic distance and the UPGMA algorithm. See text for further details.

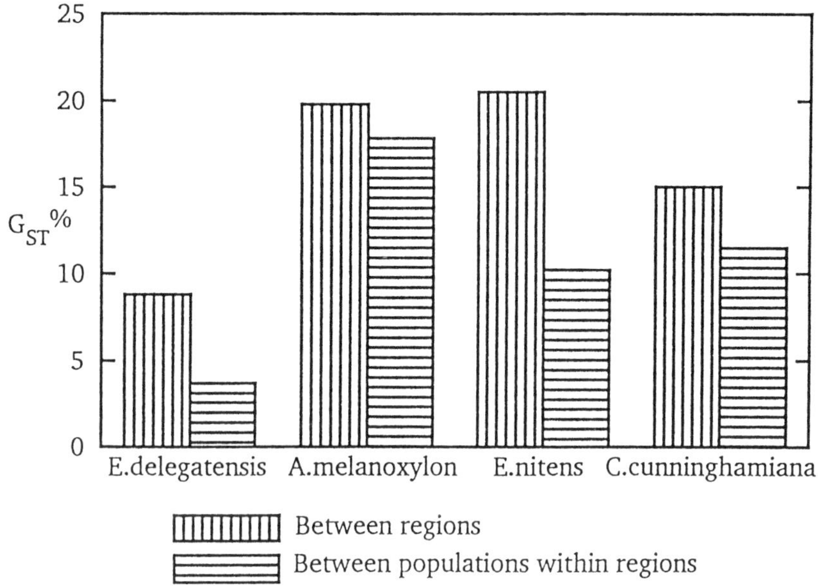

Fig. 4. Apportionment of the genetic diversity between populations into (1) a between regions component and (2) a between populations within regions component for four tree species.

efficient strategies for initial genetic evaluation of species during domestication.

A related area in which biochemical markers will continue to play an important role is in the evaluation of the genetic resources of tree species that are already sampled and in plantations. The key issues are how much of the genetic resources have been captured and from what part of the natural range were the genetic resources sampled. An example is *Grevillea robusta*, an Australian species with a regional distribution, that has been domesticated in India, Sri Lanka and Africa for over a hundred years. Land races now occur in these countries. Harwood et al. (1990) found that populations from the northwest part of the natural range clustered on the basis of isozyme data at 20 loci. The same alleles causing the genetic separation of this geographic region are absent in the land races surveyed so far. This result suggests that material from the northwest region was not included in the introductions of this species.

Conclusions

Tree species in Australia generally have high levels of allozyme variation. Species with widespread ranges have the highest levels of allozyme variation. Geographic range appears to partly explain the current patterns of genetic diversity in Australian species. Thus species with disjunct distributions exhibit greater genetic differentiation between populations. For widespread species, patterns of genetic diversity are most apparent across geographic areas which have been disjunct for long evolutionary periods. For regional species genetic differentiation appears to be greatest when populations are not only small but have been isolated for a long evolutionary time.

Many commercial widespread species exhibit clusters of populations for isozyme and other characters which correspond to large geographic regions. Isozymes thus have a role in recognizing broad patterns of genetic diversity in such species. Such analyses will provide more efficient strategies for the initial phase of domestication of tree species. An initial small isozyme survey across the range of a widespread, but previously uninvestigated, species may often prove to be a cost effective approach. Because of the relatively large isozyme differentiation observed among populations of widespread Australian species, it is often possible to determine what genetic diversity was captured during domestication and from where it came. This is less possible in other outcrossing trees such as in conifers, because of the relative lack of population divergence observed in these species.

Acknowledgments

I wish to thank J. C. Bell for invaluable assistance in the preparation of all parts of this manuscript. Constructive comments on drafts were provided by A. H. D. Brown, M. I. H. Brooker, W. T. Adams and anonymous reviewers.

References

Barlow, B. A. 1981. The Australian flora: Its origin and evolution, pp. 25—75. In: George, A. S. (Ed) Flora of Australia, Vol. 1. Australian Government Publishing Service, Canberra.

Boland, D. J. and Dunn, A. 1985. Geographic variation in alpine ash (*Eucalyptus delegatensis* R. T. Baker). Aust. For. Res. 15: 155—171.

Boland, D. J., Brooker, M. I. H., Chippendale, G. M., Hall, N., Hyland, B. P. M., Johnston,

R. D., Kleinig, D. A. and Turner, J. D. 1984. Forest Trees of Australia. Thomas Nelson and CSIRO, Melbourne.

Bridgewater, P. B. 1987. The present Australian environment — terrestrial and freshwater, pp. 69—100. In: Dyne, G. R. and Walton, D. W. (Eds) Fauna of Australia. General Articles. Vol 1A. Australian Government Publishing Service, Canberra.

Brown, A. H. D., Barrett, S. C. H. and Moran, G. F. 1985. Mating system estimation in forest trees: models, methods and meanings, pp. 32—49. In: Gregorius, H. R. (Ed) Population Genetics in Forestry. Springer-Verlag, Berlin.

Brown, A. H. D., Matheson, A. C. and Eldridge, K. G. 1975. Estimation of the mating system of *Eucalyptus obliqua* L'Herit. by using allozyme polymorphisms. Aust. J. Bot. 23: 931—949.

Burgess, I. P. and Bell, J. C. 1983. Comparative morphology and allozyme frequencies of *Eucalyptus grandis* Hill ex Maiden and *E. saligna* Sm. 1983. Aust. For Res. 13: 133—149.

Carthew, S. M., Ayre, D. J. and Whelan, R. J. 1988. High levels of outcrossing in populations of *Banksia spinulosa* R.Br and *Banksia paludosa* Smith. Aust. J. Bot. 36: 217—223.

Coates, D. J. 1988. Genetic diversity and population genetic structure in the rare Chittering grass wattle *Acacia anomala*. Aust. J. Bot. 36: 273—286.

Coates, D. J. and Sokolowski, R. E. 1989. Geographic patterns of genetic diversity in karri (*Eucalyptus diversicolor*). Aust. J. Bot. 37: 145—156.

Cook, I. 1989. Morphological and isozyme variation in *Eucalyptus nitens*. M.Sc. Thesis, Faculty of Agriculture and Forestry, University of Melbourne.

El-Lakany, M. H. 1990. Provenance trials of *Casuarina glauca* and *C. cunninghamiana* in Egypt, pp. 12—22. In: El-Lakany, M. H. and Turnbull, J. W. (Eds) Second International Casuarina Workshop, IUFRO, Cairo.

Galloway, R. W. and Kemp, E. M. 1980. Late cainozoic environments in Australia, pp. 51—80. In: Keast, A. (Ed). Ecological Biogeography of Australia. Junk, The Hague.

Hamrick, J. L. and Godt, M. J. W. 1990. Allozyme diversity in plant species, pp. 44—64. In: Brown, A. H. D., Clegg, M. T., Kahler, A. L. and Weir, B. S. (Eds) Plant Population Genetics, Breeding and Genetic Resources. Sinauer, Sunderland.

Harwood, C. E., Bell, J. C. and Moran, G. F. 1991. Isozyme studies and genetic variation in *Grevillea robusta* A. Cunn. pp. Proceedings, ICRAF/CSIRO International Workshop on *Grevillea robusta*. Nairobi, Kenya, August 1990.

Kenrick, J. and Knox, R. B. 1989. Quantitative analysis of self-incompatibility in trees of seven species of *Acacia*. J. Hered. 80: 240—245.

Lande, R. and Barrowclough, G. F. 1987. Effective population size, genetic variation and their use in population management, pp. 69—124. In: Soule, M. (Ed) Viable Populations for Conservation. CUP, Cambridge.

Loveless, M. D. and Hamrick, J. L. 1984. Ecological determinants of genetic structure in plant populations. Ann. Rev. Ecol. Syst. 15: 65—95.

Maslin, B. R. and Pedley, L. 1982. The distribution of *Acacia* (Leguminosae: Mimosoideae) in Australia. Part 1. Species distribution maps. WA Herbarium Research Notes 6: 1—128.

Moran, G. F. 1992. Spatial variation in the genetic structure and mating system of *Eucalyptus delegatensis*. Silvae Genetica (in press).

Moran, G. F. and Bell, J. C. 1983. *Eucalyptus*, pp. 423—441. In: Tanksley, S. D. and Orton, T. J. (Eds) Isozymes in Plant Genetics and Breeding, Part B. Elsevier, Amsterdam.

Moran, G. F. and Hopper, S. D. 1983. Genetic diversity and insular population structure of the rare granite rock species *Eucalyptus caesia* Benth. Aust. J. Bot. 31: 161—172.

Moran, G. F. and Hopper, S. D. 1987. Conservation of the genetic resources of rare and widespread eucalypts in remnant vegetation, pp. 151—162. In: Saunders, D. A., Arnold, G. W., Burbidge, A. A. and Hopkins, A. J. M. (Eds) Nature Conservation: The Role of Remnants of Native Vegetation. Surrey Beatty and Sons: NSW.

Moran, G. F., Bell, J. C. and Turnbull, J. W. 1989c. A cline in genetic diversity in river she-oak *Casuarina cunninghamiana*. Aust. J. Bot. 37: 169—180.

Moran, G. F., Bell, J. C. and Prober, S. 1990a. The utility of isozymes in the systematics of some Australian tree groups. Aust. Syst. Bot. 3: 47—57.

Moran, G. F., Forrester, R. I. and Rout, A. R. 1990b. Early growth of *Eucalyptus delegatensis* provenances in four field trials in south-eastern Australia. NZ. J. For. Science 20: 148—161.

Moran, G. F., Muona, O. and Bell, J. C. 1989a. *Acacia mangium*: a tropical forest tree of the coastal lowlands with low genetic diversity. Evolution 43: 231—235.

Moran, G. F., Muona, O. and Bell, J. C. 1989b. Breeding systems and genetic diversity in *Acacia auriculiformis* and *A. crassicarpa*. Biotropica 21: 250—256.

Muona, O. 1990. Population genetics in forest tree improvement. pp. 282—298. In: Brown, A. H. D., Clegg, M. T., Kahler A. L., and Weir, B. S. (Eds) Plant Population Genetics, Breeding and Genetic Resources. Sinauer, Sunderland, Massachusetts.

Muona, O., Moran, G. F. and Bell, J. C. 1991. Hierarchical patterns of correlated mating in *Acacia melanoxylon*. Genetics 127: 619—626.

Nei, M. 1973. Analysis of gene diversity in subdivided populations. Proc. Nat. Acad. Sci. 70: 3321—3.

Nei, M. 1978. Estimation of average heterozygosity and genetic distance from a small number of individuals. Genetics 89: 583—590.

Nelson, E. C. 1981. Phytogeography of southern Australia, pp. 733—760 In: Keast, A. (Ed) Ecological Biogeography of Australia. W. Junk, The Hague.

Pederick, L. A. 1979. Natural variation in shining gum (*Eucalyptus nitens*) Aust. For. Res. 9: 41—63.

Peters, G. B., Lonie J. S. and Moran, G. F. 1990. The breeding system, genetic diversity and pollen sterility in *Eucalyptus pulverulenta*, a rare species with small disjunct populations. Aust J. Bot. 38: 559—570.

Phillips, M. A. and Brown, A. H. D. 1977. Mating system and hybridity in *Eucalyptus pauciflora*. Aust. J. Biol. Sci. 30: 337—344.

Prober, S., Bell, J. C. and Moran, G. F. 1990a. A phylogenetic approach to understanding rarity in three 'green ash' eucalypts (Myrtaceae). Plant Syst. Evol. 172: 99—118.

Prober, S. M., Tompkins, C., Moran, G. F. and Bell, J. C. 1990b. The conservation genetics of *Eucalyptus paliformis* L. Johnson et Blaxell and *E. parvifolia* Cambage, two rare species from south-eastern Australia. Aust. J. Bot. 38: 79—95.

Sampson, J. F. 1988. The Population Genetic Structure of *Eucalyptus rhodantha* Blakely and Steedman and its allies *Eucalyptus crucis* Maiden and *Eucalyptus lane-poolei* Maiden. Ph.D. Thesis, University of Western Australia.

Sampson, J. F., Hopper, S. D. and James, S. H. 1988. Genetic diversity and the conservation of *Eucalyptus crucis*. Maiden. Aust. J. Bot. 36: 447—460.

Sampson, J. F., Hopper, S. D. and James, S. H. 1989. The mating system and populations genetic structure in a bird-pollinated mallee, *Eucalyptus rhodantha*. Hered. 63: 383—393.

Slatkin, M. 1985. Rare alleles as indicators of gene flow. Evolution 39: 53—65.

Slatkin, M. 1987. Gene flow and the geographic structure of natural populations. Science 236: 787—792.

Turnbull, J. W. 1980. Geographic variation in *Eucalyptus cloeziana*. Ph. D. Thesis, ANU, Canberra.

New Forests **6**: 67–94, 1992.

Review paper

Isozyme variation in tropical trees: patterns of genetic organization

M. D. LOVELESS

Department of Biology, The College of Wooster, Wooster, Ohio 44691, USA

Key words: tropics, isozymes, tropical trees, genetic variation, population structure

Received 9 October 1990; accepted 10 April 1992

Application: Isozymes provide a method for assaying levels of genetic variation in natural populations of tropical forest trees and measuring population processes of importance to ecologists, conservationists, and forest managers. Patterns of variation throughout a species' range and levels of population differentiation can indicate the spatial scale over which germplasm resources should be collected. Populations with large differences in electrophoretic genotypes have probably been subjected to different historical or selective processes, and might therefore be good sources of genetic diversity for other traits useful in management or breeding.

Isozymes also provide genetic markers for estimating mating systems, pollen neighborhoods, progeny variability, population structure, and seed dispersal. These aspects of the natural history of tropical tree species are critical for developing forest conservation strategies, as well as for efforts directed at the management or breeding of economically-important tropical trees.

Abstract. Tropical forests have long been of interest to biologists because of their high species diversity and their complicated patterns of community organization. The traditional notion of the rain forest as an ancient, unvarying archive of species has been challenged by a host of recent ecological studies which have demonstrated that tropical trees are diverse in their reproductive biology and dynamic in their population structure.

Data from 97 isozyme studies on the genetics of tropical woody species demonstrate that cultivated taxa maintain higher percentages of polymorphic loci and higher mean heterozygosities than native tropical species. Levels of within-population variation in tropical taxa are as high or higher than in plants in general, previously reviewed by Hamrick and Godt (1989). Levels of genetic variation differ significantly among species with different geographic ranges, life forms, and taxonomic affinities. Levels of population differentiation, measured by G_{ST} values, are significantly different only between species with different seed dispersal modes. Outcrossing rates in 16 tropical tree species showed a preponderance of highly outcrossed breeding systems. Genetic evidence suggests that gene flow among local populations is high, but geographically separated populations show moderate levels of genetic differentiation.

Ecological and historical processes provide the mechanisms influencing the genetic architecture of plant species. Where appropriate, data on isozyme variation in tropical trees are related to breeding systems, seed dispersal mechanisms, demography, and patterns of environmental heterogeneity. Profitable avenues for future research are suggested.

Introduction

Tropical climates, which occur between 23°30′N and S latitude (Jacob 1988), encompass a wide variety of plant communities (Hartshorn 1983, 1988), many of which are diverse in their woody floras. Within this geographic region, temperature and the amount and seasonality of rainfall define habitat types (UNESCO 1978). The FAO has estimated that there are about 19 million km^2 of potentially forested area in the global tropics, of which 58% were estimated to still be in closed forest in the mid-1970s (Sommers 1976; UNESCO 1978). Of this potentially forested region, 42% is categorized as dry forest lifezone, 33% is tropical moist forest, and 25% is wet or rain forest (Lugo 1988).

The species diversity of these tropical habitats is very high. Raven (1976, in Mooney 1988) estimated that 65% of the 250,000 or more plant species of the earth are found in tropical regions. Of this floristic assemblage, a large fraction are woody species. In the well-collected tropical moist forest of Barro Colorado Island, Panamá, 39.7% (481 of 1212 species) of the native phanerogams are woody, arborescent species (Croat 1978). Another 21.9% are woody vines and lianas. Southeast Asian Dipterocarp forests may contain 120—200 species of trees per hectare (Whitmore 1984), and recent surveys in upper Amazonia recorded from 89 to 283 woody species $\geqslant 10$ cm dbh per hectare (Gentry 1988). Tropical communities thus represent a global woody flora of significant scope.

The species richness of tropical forests has prompted much speculation about the evolutionary processes by which this diversity has arisen (Leigh 1982). A number of hypotheses have been advanced to explain the origin and maintenance of high tropical species diversity (Janzen 1970; Connell 1971, 1978; Hubbell 1980; Becker et al. 1985), and some of these hypotheses have implications for the genetic organization of tropical tree populations, as well. Tree species diversity is of particular importance, since the tree flora represents the structural habitat within which other species exist. The persistence of evolutionarily viable populations of tropical trees is crucial to the preservation of the tropical forest habitat. From this perspective, the genetic organization and breeding structure of tropical trees are central to an understanding of tropical diversity at all levels, as well as to the long-term conservation of that diversity.

Genetic studies of economically useful tropical plantation species such as teak (*Tectona grandis*), Caribbean pine (*Pinus caribbea*), and rubber (*Hevea brasiliensis*) have been carried out using provenance trials, measuring quantitative traits under various growth conditions (Hedegart 1976; Alika 1980). Selection on quantitative traits has also been used to develop cultivars of tropical fruits (avocado, citrus, papaya, dates, etc.). These studies document the existence of substantial genetic variation in at least some tropical woody species, but managed populations of economically valuable species do not typically provide information on the population structure or evolutionary processes underlying variation in natural populations.

Most economically valuable tropical timber species are currently harvested from intact forests rather than from plantations. Large-scale plantation silviculture is probably not an ecologically viable option for many tropical hardwoods, although it may be appropriate in some settings for pulp, firewood, or other non-sawn uses (Schmidt 1987; Whitmore 1987). In addition, the long generation times and specialized reproductive ecology of many of these species make traditional breeding programs difficult (Bawa 1976).

In the past several years, isozymes have provided another technique for documenting genetic diversity, population structure, and aspects of ecological genetics in natural populations of tropical woody species (Ashton 1977; Gan et al., 1977, 1981; Hamrick and Loveless 1986a; Heywood and Fleming 1986; O'Malley and Bawa 1987). Temperate forest geneticists have been quick to understand the value of these techniques for studies of population genetics and tree breeding programs (Brown and Moran, 1981; Adams 1983). However, because of the limitations of traditional forest genetic methods for tropical species, isozymes have a special potential to increase our understanding of tropical tree genetics. They provide a relatively standardized way of measuring and comparing variability among species and between populations of a single species (Hamrick et al. 1979; Loveless and Hamrick 1984; Hamrick and Godt 1989). They permit assessments of genetic relatedness among taxa, thus providing a tool for examining evolutionary processes and speciation patterns in tropical species (Gottlieb 1981). They also provide genetic markers for measuring breeding structure, mating systems, pollen movement, seed dispersal, and spatial genetic structure within populations (Shaw and Allard 1981; Ritland and El-Kassaby 1985). The relatively low cost and technical ease of these isozyme methods makes them especially appropriate for population studies requiring large numbers of samples. The major obstacle to using isozymes for studies of tropical trees has been the difficulty of transporting samples to the laboratory. However, vacuum-dried leaf or seed tissue gives good results (Hamrick and Loveless 1986a), and freshly germinated seeds are also widely used, especially for mating system studies. Electrophoresis laboratories are increasingly becoming available at field stations and university facilities throughout the tropics, making isozyme studies logistically much easier than in the past.

In this paper, I will examine what is known about genetic variability, genetic structure, and mating systems in tropical woody species as a result of recent isozyme studies. I will consider various ecological factors which characterize tropical species and which are likely to contribute to this genetic structure, and I will discuss future directions for the use of biochemical markers in research on tropical tree genetics.

Characteristics of tropical forests

In contrast to temperate forests, tropical lowland forests are usually high-diversity assemblages where many different species attain canopy status, none of them showing numerical dominance. Low-diversity forest assemblages dominated by only one or a few species do occur throughout the lowland tropics (Hart et al. 1989; Connell and Lowman 1989), often in disturbed sites, but most tropical moist forests are complex communities of many tree species. Within a given forest, some species may be relatively common while others are rare (Hubbell 1979; Hubbell and Foster 1983, 1986b; Bawa and Ashton 1991). The spatial distribution of individuals within a population will also have implications for genetic structure. Reproductively mature individuals may be widely separated from conspecifics (Baker 1970; Janzen 1970; Connell 1971; Bawa 1976; Bawa and Ashton 1991). Such dispersion could result in selfing, or it could promote long-distance pollen flow. Species with high population densities or clumped distributions may be more likely to mate with nearby, possibly related individuals (Hamrick and Loveless 1986b; Bawa and Ashton 1991). Dispersion patterns will thus interact in complicated ways with breeding biology, creating a wide variety of genetic predictions for population structure in tropical trees (Hamrick and Loveless 1986b; Howe 1989).

Tropical trees also have diverse and, in many cases, poorly understood pollination and seed dispersal mechanisms, many of which represent facultative mutualisms with various insect or vertebrate pollinators and dispersers (Baker 1959; Stiles 1975; Heithaus et al. 1975; Frankie 1975; Bawa 1976; Frankie et al. 1983; Baker et al. 1983; Bawa et al. 1985b; Bawa and Ashton 1991). As a result, studies of pollination and reproductive biology, as well as genetic breeding systems, are critical to management, genetic improvement, and conservation of tropical forest species (Bawa 1976). The behavior of pollen vectors and seed dispersers will have important consequences for the spatial and genetic structure of the tree species they visit, and strategies for forest management and conservation must consider the resource and habitat requirements necessary to maintain viable populations of these critical members of the rain forest ecosystem (Gilbert 1980; Howe 1989; Bawa and Ashton 1991).

Lowland tropical forests have also been traditionally characterized as environmentally stable (Corner 1954; Federov 1966). This climatic stability, maintained over long eras of geological time, was hypothesized to provide a refuge for tropical biotypes, reducing the probability of species extinctions and permitting the accumulation of species (reviewed in Colinvaux 1987). However, recent evidence has challenged the notion of an ancient, unvarying tropical forest. Although tropical forests did not suffer the same climatic extremes as the temperate forests of North America and Eurasia, data suggest that there were substantial climatic changes during the Pleistocene (Livingston 1975; Colinvaux 1987) which have important implications for the genetic organization of modern tropical plant species.

The low population density and hyperdispersion of many tropical trees, especially as reproductively mature individuals, led early investigators to suggest that selfing would be the predominant breeding system in these species (Corner 1954; Baker 1959; Federov 1966). In a synthesis of many of the traditional assumptions about the ecology and evolution of the tropical forest, Federov (1966) devised a genetic argument to account for speciation among tropical woody plants. He argued that low population densities, hyperdispersion of adults, habitat homogeneity, lack of climatic seasonality, and erratic and asynchronous flowering within tree populations all contributed to extensive reproductive isolation, even among individual trees within a population. The gradual accumulation of mutations within these genetically isolated individuals, he hypothesized, created conditions in which genetic drift, rather than selection, was the principle mechanism of tropical speciation. In genetic terms, Federov's view of tropical forests thus predicted that individual trees should be highly homozygous as a result of constant inbreeding, but that even nearby subpopulations of the same species should differ genetically (albeit at very low levels) as a result of accumulated mutations and genetic drift.

Ashton (1969) drew on his own extensive experience in southeast Asian Dipterocarp forests to counter the genetic arguments implicit in Federov's paper. Ashton argued that selection and adaptation, rather than drift, were responsible for tropical species diversity. He agreed that autogamy was likely to be the predominant breeding system among tropical trees, but he also noted the importance of dioecy and other sexual systems in the rain forest flora. He argued that outcrossing, even over short distances, still permitted constant, low levels of gene flow among members of a contiguous population. His assertion that tropical trees show considerable morphological constancy, even among widely separated populations, implies that they should have only low levels of genetic variation. Like Federov, he hypothesized that there would be considerable genetic differentiation among populations. But unlike Federov, Ashton attributed this differentiation to a combination of limited seed dispersal, clumped adult distributions, a variety of mating systems with both selfing and local outcrossing and selective adaptation of populations to different rain forest habitats. Ashton's rain forest would thus contain moderate amounts of genetic variability, organized into locally selected populations, probably with strong family structure.

Over the past 25 years, research has given us new insights into the ecology of tropical species. Many of the generalizations about tropical forests on which Federov and Ashton based their arguments have been revised or reinterpreted. It seems clear that, contrary to previous suggestions, tropical trees possess many mechanisms to enhance or enforce outcrossing, including dioecy, protandry, heterostyly, and genetic self-incompatibility (Bawa 1974; Bawa and Opler 1975; Zapata and Arroyo 1978; Baker et al. 1983; Bawa et al. 1985a, 1985b). Isozyme analyses have also begun to provide data to test predictions about genetic variation and population organization in these species. Although our knowledge of the ecology, genetics, and history of tropical forests remains fragmentary, electrophoretic techniques hold considerable potential for increasing our understanding of these aspects of tropical tree biology.

Isozyme data on tropical woody species

Isozyme data on tropical woody species came from a data base accumulated by J. L. Hamrick and M. J. W. Godt for their recent reviews (Hamrick and Godt 1989; Hamrick et al., this symposium). Additional studies were identified by searching the appropriate literature and from unpublished results. Data were available for 97 taxa, three of which were different studies on *Pinus oocarpa*. For most of the summary data, all three studies were retained in the data base as independent entries.

Each taxon was classified by variables which have been shown to be useful in explaining patterns of variation (Table 1) (Hamrick et al. 1979; Loveless and Hamrick 1984; Hamrick and Godt 1989). Measures of genetic variation were taken from the literature or recalculated as necessary (Hamrick and Godt 1989). The genetic variation in a species was measured by P, the percentage of isozyme loci which were polymorphic at any level, and H, the genetic diversity, or expected heterozygosity under Hardy-Weinberg equilibrium. For the 16 studies of cultivated species in this analysis, these measures were usually made on named varieties which

Table 1. Categories used to analyze patterns of genetic variability in tropical woody plants

Trait	Categories	Comments
Domestication	Cultivar	Grown principally for human exploitation; not sampled from natural habitats
	Native species	Sampled from wild, indigenous individuals in natural habitat
Life form	Shrub	Multiple stems or characteristic low habit
	Tree species	Single stem, > 3m at maturity
Geographic Range	Endemic	Known from only one or a few populations, in only one locality
	Narrow	Found over several to many populations in a relatively restricted area, parts of one or a few countries or political units
	Regional	Found throughout a geographic region or a large portion of a continent
	Widespread	Found throughout a continent or on more than one continental mass
Successional Stage	Early	Quickly invading newly disturbed habitats; usually shade-intolerant
	Late	Shade tolerant, persisting in closed forest
Seed Dispersal	Abiotic	Dispersed by wind, water, or gravity
	Biotic	Dispersed by animals, either attached to body or ingested and voided intact
Taxonomy	*Pinus*	Tropical pines
	Other species	In this study, all species other than *Pinus* were angiosperms

had been propagated asexually, not from collections in natural populations. Values of P and H are thus measurements averaged over all accessions tested.

For the 81 studies of native species, values of P and H were usually calculated separately from each natural population. Where more than one population was surveyed, composite calculations of these parameters could also be made for each species. However, because studies of some species only examined a single population, all the interpretive analyses of native species in this review were performed using population-level statistics. Where more than one population was surveyed, these values are the means of measures in all the available populations. In seven of 81 studies of non-cultivated species, species-level calculations were available, but population calculations were not. For these studies, the available species-level measures were used instead of mean population values. As a result, averages for different groups in the following analysis may be slightly inflated. Since Hamrick and Godt (1989) have shown that there is a very strong correlation between measures of genetic diversity at the species level and measures within individual populations, this should not seriously affect the results.

Measures of population differentiation were calculated, where possible, using G_{ST} statistics of Nei (1973). G_{ST} measures the proportion of the total gene diversity at polymorphic loci which is due to differences among populations. G_{ST} values were available for 37 species of native tropical woody taxa.

Mean values for *P*, *H*, and G_{ST} for the various groups (categories) of tropical woody species are given in Table 2. Differences between mean values for the groups were tested using a two-tailed *t*-test or, where appropriate, single-factor analysis of variance. A significance level of 0.05 was applied for both tests. Results of the statistical analyses are also shown in Table 2.

Table 3 summarizes estimates of mating systems for tropical woody species measured using isozymes. Although only 18 species have been analyzed, almost all of them from the neotropics, these data provide initial genetic information on the reproductive biology of tropical trees.

Results and discussion

Amount of within-population genetic variation

Only a relatively few tropical woody species have been studied using isozymes. In Table 2, species with similar ecological characteristics are grouped, and mean values are given for genetic parameters within these groups. Groups of species which differ in traits that may influence genetic organization are then compared to suggest factors that might be important in influencing genetic diversity and population structure.

Cultivated species showed significantly higher levels of variation than native species, both in average number of polymorphic loci (P) and in mean heterozygosity (H)(Table 2). Since cultivar studies usually investigate accessions already known to be genetically distinct, at least with respect to quantitative traits, this high level of variability among cultivars is not unexpected. In addition, P and H for cultivars are measured at the species level (across all accessions), rather than at the population level, as is the case for native tropical species. Hamrick and Godt (1989) reported that, on average, species level measures of P and H were 47% and 37% higher, respectively, than corresponding population measures of *P* and *H*. If

Table 2. Within-population genetic variation in tropical trees

	Average % polymorphic loci (P)			Average gene diversity (H)			Average population differentiation (G_{ST})		
	P ± s.e.	*n*[a]	Prob[b]	H ± s.e.	*n*	Prob	G_{ST} ± s.e.	*n*	Prob
Cultivars[c]	60.5 ± 7.5	16		0.216 ± 0.042	15		no data		
Native woody taxa	39.0 ± 2.7	81	0.0029	0.129 ± 0.011	74	0.0049	0.109 ± 0.019	37	
Endemic[d]	18.7 ± 4.6	15		0.045 ± 0.012	14				
Narrow	34.3 ± 4.4	29		0.109 ± 0.015	28				
Regional	53.9 ± 3.9	31		0.181 ± 0.015	28				
Widespread	36.0 ± 6.2	6	0.0001	0.204 ± 0.068	4	0.0001			
Shrubs[d]	23.1 ± 3.6	31		0.081 ± 0.017	29		0.133 ± 0.038	13	
Tree species	49.0 ± 3.2	50	0.0001	0.160 ± 0.012	45	0.0001	0.096 ± 0.021	24	n.s.
Shrubs excluding endemics[d]	28.7 ± 5.0	18		0.116 ± 0.028	16		0.075 ± 0.019	7	
Trees excluding endemics	49.3 ± 3.2	48	0.0013	0.161 ± 0.012	44	n.s.	0.096 ± 0.021	24	n.s.
Endemic shrubs[d]	15.3 ± 4.1	13		0.038 ± 0.011	13		0.201 ± 0.072	6	
Non-endemic shrubs	28.7 ± 5.0	18	n.s.	0.116 ± 0.028	16	0.0219	0.075 ± 0.019	7	n.s.

Table 2. (Continued)

	Average % polymorphic loci (P)			Average gene diversity (H)			Average population differentiation (G_{ST})		
	P ± s.e.	*n*[a]	Prob[b]	H ± s.e.	*n*	Prob	G_{ST} ± s.e.	*n*	Prob
Early successional species[d]	32.0 ± 4.5	26		0.109 ± 0.021	24		0.111 ± 0.047	10	
Late successional species	42.4 ± 3.4	55	n.s.	0.139 ± 0.012	50	n.s.	0.109 ± 0.019	27	n.s.
Abiotic seed dispersal[d]	37.4 ± 3.3	54		0.120 ± 0.014	47		0.138 ± 0.026	25	
Biotic seed dispersal	42.4 ± 4.8	27	n.s.	0.146 ± 0.016	27	n.s.	0.050 ± 0.008	12	0.0186
Pinus[d, e]	43.6 ± 5.7	11		0.118 ± 0.016	11		0.094 ± 0.034	6	
Other tree species	50.5 ± 3.8	37	n.s.	0.174 ± 0.014	32	0.0345	0.097 ± 0.025	18	n.s.

[a] *n* indicates the number of species in this category.

[b] Test of mean differences between categories based on either a two-tailed *t*-test (two categories) or F-test (more than two categories); n.s. means non-significant difference ($P > 0.05$).

[c] Values of P and H for cultivars are species-level statistics.

[d] Only native species are considered in this comparison.

[e] Only one of the three studies on *Pinus oocarpa* (Matheson *et al.* 1989) was included in this analysis (see text for details).

values for tropical native species in this study are adjusted upward by these amounts, to hypothetical *P* and *H* values of 57.5% and 0.174, the differences between cultivated and native species are much less pronounced, and probably would not be significant. Values for genetic variation in tropical native species (Table 2) are consistent with values of *P* (34.2%) and *H* (0.113) calculated for plant species over all (Hamrick and Godt 1989). They are somewhat higher than values for tropical species (including tropical herbaceous taxa) of $P = 32.7\%$ and $H = 0.109$ reported in Hamrick and Godt (1989). Thus, in contrast to predictions (Corner 1954; Baker 1959; Federov 1966), tropical woody plants are at least as genetically variable as plants in general. The tropical woody taxa in this analysis are slightly less variable than long-lived woody perennials surveyed by Hamrick and Godt (1989). This difference may be due, in part, to the smaller proportion of gymnosperms in the tropical woody data set. Gymnosperms represented 14% of the species surveyed here, while they comprised 50% of the taxa summarized by Hamrick and Godt (1989). Gymnosperms (at least temperate species) generally have high levels of isozyme variation (Hamrick et al. 1981).

As in previous surveys (Hamrick et al. 1979; Hamrick et al. 1981; Hamrick and Godt 1989), geographic range is a major correlate of within-population genetic variation in these tropical tree species. Endemic tropical species showed significantly less variation than species with broader distributions. Successively more widely distributed species showed progressively more genetic variation, although only a few truly widespread (globally distributed) species were included in the survey. Tropical endemic species ($P = 18.7\%$, $H = 0.045$) are less variable than endemics in general ($P = 26.3\%$; $H = 0.063$) reported by Hamrick and Godt (1989). More widely distributed tropical taxa show levels of variation similar to plants overall.

Hamrick and Murawski (1991) recently compared levels of genetic diversity between 16 rare and 16 common woody species from the same locality in Panamá. Values of *P* and *H* in the group of rare species were quite high ($P = 42.1\%$; $H = 0.142$), but significantly less than in the common species ($P = 60.9\%$; $H = 0.211$). They suggest that this is the result of lower effective population sizes in the rare taxa. Nonetheless, the high heterozygosity of even these rare tree species suggests that they are effective at moving pollen among individuals (O'Malley and Bawa 1987; Bawa and Ashton 1991).

Tropical shrubby species or species with a shorter stature had significantly lower values of *P* and *H* than did canopy species. The reasons for these differences are not clear. Perhaps shrubby species show patchier distributions, with higher local densities. Thus, shrubby species might have more family structure and more local inbreeding. Patchiness in phenology and flower density might also limit pollen movement among nearby patches, thus reducing gene exchange between adjacent areas, as suggested by Augspurger (1980) for *Hybanthus prunifolius*. Previous surveys (Hamrick 1979; Hamrick et al. 1979; Hamrick et al. 1981) found a significant relationship between plant longevity and levels of genetic variability. If it is correct to assume that small trees and shrubs are shorter-lived than canopy trees, the genetic data summarized here are consistent with these previous surveys. However, there is currently no evidence to suggest that shrubby lifeforms are less long-lived than canopy species. Tropical understory taxa may reproduce vegetatively or resprout after damage, and thus may have longevities similar to those of larger trees.

The shrub data in this dataset included three separate studies of species groups of shrubby endemics, two from Hawaii (Lowrey and Crawford 1985, and Helenurm and Ganders 1985) and one from Panamá (Sytsma and Schaal 1985). To explore the possibility that these groups of endemic shrubby taxa were biasing the overall comparison between shrubs and trees, the data were reanalyzed after excluding all endemic taxa from both lifeform groups (Table 2). Mean values of H were then no longer significantly different for shrubs and trees, but P was still significantly lower in shrubby species.

In contrast to previous studies for plant species as a whole, successional status was not related to genetic variability in tropical species. This may be due, in part, to the difficulty of assigning tropical taxa to well-defined successional categories (Hartshorn 1978; Hubbell and Foster 1986a, 1986b; Clark and Clark 1987; Lieberman et al. 1989; but see Whitmore 1989). Hamrick et al. (1981) suggested that in temperate climates, late-successional habitats are more biotically complex and environmentally heterogeneous than early-successional sites, thus selecting for more genetic variation in species of late-successional seres. These generalizations are not necessarily true of tropical forest gap environments (Hubbell and Foster 1986b; Denslow 1987; Lieberman et al. 1989; Brokaw and Scheiner 1989), which may be as biotically complicated and challenging as closed old-forest habitats. Species invading treefall gaps often persist to become components of the mature forest (Hartshorn 1978; Hubbell and Foster 1986a, 1986b), and thus must respond to the challenges of both successional and climax environments. Whether there are really characteristic genetic differences in tropical species of different successional habitats must await further studies, especially of pioneer taxa like *Cecropia*.

As was the case in earlier surveys, mechanism of seed dispersal did not correlate well with amounts of within-population genetic variation. Type of seed dispersal did, however, contribute importantly to genetic differentiation among populations (Table 2), as discussed in the next section.

Finally, data were grouped taxonomically to determine if tropical pines showed the same high levels of variation which have been described for temperate gymnosperms. Hamrick et al. (1981) surveyed data from 28 studies of 20 species of temperate gymnosperms, and found an average *P* of 67.7% of isozyme loci polymorphic, with a mean H of 0.207. In the current study, however, Central American pine species showed no higher levels of *P*, and significantly lower levels of H, than tropical angiosperm woody species. This may be due, at least in part, to the more local ranges and lower population sizes of tropical pines relative to their temperate congeners. Tropical pines are confined to high altitude habitats in the mountains of Central America. These localities are thought to have experienced repeated and substantial altitudinal shifts in vegetation zonation as a result of Pleistocene climatic changes (Bush and Colinvaux 1990; Piperno et al. 1990). The resulting repeated isolation and reconsolidation of these pine populations is likely to have had a major impact on their current population structure.

Only one arborescent, non-cultivated monocot has been examined electrophoretically. *Astrocaryum mexicanum* (Palmae), in México, has a *P* of 31.8% and an H of 0.135 (L. Eguiarte, pers. comm), making it slightly less variable than most tropical dicot trees.

These data make it clear that, in contrast to traditional views of the tropics as a museum of ancient, static species, tropical woody plants maintain high levels of within-population genetic variation which should permit adaptation and differentiation by natural selection.

Population differentiation

Average G_{ST} values for various groups of tropical taxa are given in Table 2. The mean G_{ST} for all 37 taxa was 0.109, indicating that, on average, 11% of the total genetic variation within a tropical species is due to differences among populations. Previous studies (Loveless and Hamrick 1984; Hamrick and Godt 1989) have shown that values of G_{ST} are significantly different in plants of different taxonomic groups, life forms, regional distribution, breeding systems, pollination syndromes, seed dispersal mechanisms, and successional habitats. In general, highly subdivided populations of plants have values of G_{ST} of 0.20 or more, while temperate pines, with large and often continuous populations, tend to have G_{ST} values of 0.05 or less.

Of those life history categories examined in this review, only seed dispersal mechanisms showed significant differences in G_{ST} among groups. Abiotically dispersed species showed more than twice as much population differentiation as biotically-dispersed species. In this analysis, abiotic dispersal included both wind and gravity dispersal, and gave an average

G_{ST} value of 0.138. Previous reviews of the plant literature separated these two abiotic dispersal mechanisms. Hamrick and Godt (1989) found wind-dispersed plants to have lower levels of population differentiation (G_{ST} = 0.143), while gravity-dispersal was associated with relatively strong population divergence (G_{ST} = 0.277). In tropical canopy trees, wind dispersal has the potential to move some diaspores over long distances (Augspurger 1986; Augspurger and Franson 1987), thus possibly reducing population differentiation. But even most gravity-dispersed tropical species did not show high levels of population differentiation, perhaps because many initially gravity-dispersed fruits and seeds may be moved about secondarily by water or by animals.

The value of G_{ST} for biotically-dispersed species is much lower for this data set than for analogous categories of dispersal in previous reviews. Overall, plants with animal-mediated seed dispersal, including those which are ingested or which are animal-attached, showed G_{ST} values of 0.223 to 0.257 in Hamrick and Godt's analysis (1989), compared to a very low value of 0.050 in the tropical taxa considered here. The results from this study suggest that animal dispersal may be moving seeds over reasonably large distances in tropical forests, thus reducing population genetic differentiation. Such implied long-distance dispersal is consistent with ecological data on seed dispersal in at least some tropical species (Howe 1986; Fleming and Heithaus 1981; Price and Jenkins 1986). Hamrick and Loveless (1986b) and Howe (1989) analyzed the potential effects of different dispersal regimes, both abiotic and biotic, on plant population genetic structure. If animal vectors do consistently disperse seeds of tropical trees long distances from their parents, the genetic architecture of tropical forests should reflect this fact.

One factor which may confound the data summarized here is the wide variety of spatial scales over which G_{ST} values were measured in different studies. Some studies measured differentiation among subpopulations separated by a kilometer or less; others, between populations in different parts of a species' range. As a result, the overall data set is heterogeneous, and is probably not subject to the simple interpretations implied in the categories of Table 1. While this is likely to be true of any review synthesizing genetic literature, it may be exacerbated in these tropical studies by our frequent lack of knowledge of species distributions, and by the relative inaccessibility of many collecting localities.

Within a specific locality and a particular forest type, however, measures of population differentiation can provide useful comparative information about the genetic organization of populations of different species (Hamrick et al. 1991). Loveless and Hamrick (1987) and Hamrick and Loveless (1989) measured genetic variation in from 3 to 6 local populations of

16 tropical shrub and tree species from Barro Colorado Island (BCI), Panamá. From 1 to 4 populations were sampled within the 50-ha Forest Dynamics Plot (FDP) (Hubbell and Foster 1986a), and additional samples were made at more distant localities on the island. This hierarchical sampling design permitted calculations of G_{ST} within the FDP and between populations on BCI as a whole. The average G_{ST} among population on the FDP was 0.048, and among populations on the island as a whole, was 0.055. Comparable low estimates of G_{ST} (0.021 and 0.057) were obtained by Heywood and Fleming (1986) for two species of *Piper*, an understory shrub, sampled on a similarly local scale in Costa Rica. Luis Eguiarte (pers. comm.) also found little genetic differentiation between 5 local populations of *Astrocaryum mexicanum*, an understory palm species, at Los Tuxtlas, México. These taxa vary widely in breeding systems and seed dispersal mechanisms, but their very low levels of genetic divergence suggest that gene flow may be quite high among populations at the local level. Hamrick and Loveless (1989) found a significant correlation (r_s) between G_{ST} values for 14 species on BCI and ranked estimates of their genetic mobility, based on pollination and seed dispersal mechanisms. More mobile species had lower G_{ST} values ($r_s = 0.679$, $P < 0.01$). In this community analysis, shrubs on BCI showed a higher mean G_{ST} than trees, again suggesting that understory species are more limited in their pollen or seed movement than canopy taxa. Hamrick and Loveless (1989) also calculated *Nm* (the number of migrants per generation) from G_{ST} values (Wright, 1951) and from measures of private alleles (alleles found in only one population; Slatkin 1985; Barton and Slatkin 1986). Although values varied widely for different species, the average *Nm* for the two methods was 5.12 and 5.38, respectively, signifying a large amount of between-population migration. This provides additional evidence for significant gene flow among local populations in these species.

In a more detailed analysis, Loveless and Hamrick (in prep.) also sampled populations of nine of these 16 species from locations 0.5 km to 400 km away from BCI. The mean G_{ST} for the entire sampling array was 0.075. But for four species which included samples from the most distant population collections, the mean G_{ST} was 0.127. G_{ST} values for these widely-sampled species were then subdivided to identify the fraction of the total gene diversity contributed by populations within the Forest Dynamics plot, by collections from elsewhere on BCI, or from additional, more distant populations. This hierarchical analysis indicated that from 8.5% to 74% of the total G_{ST} value for these nine species was contributed by genetic variation sampled in the more distant populations. Other studies of geographically separate populations also indicate, not surprisingly, that distant populations are genetically different. One study which does not show this trend is that of *Bertholletia excelsa*, or Brazil nut (Buckley et al. 1988), where two populations 150 kilometers apart showed a G_{ST} value of just 0.036. The authors suggest that these populations could have been planted by Amerindians, and thus have had a similar genetic origin. Overall, however, the data for tropical woody species suggest that, while gene flow on the local level seems to prevent differentiation among nearby populations, geographically distant populations are, in fact, genetically divergent. This result has important implications for planning germplasm collections of tropical woody species, since it suggests that regionally or widely distributed tropical species will be genetically distinct in different locations, and should be collected throughout their ranges to assure adequate sampling of the genome (Allard 1970).

Mating systems

Estimates of the proportions of outcrossed progeny (t) in tropical trees based on isozyme data are given in Table 3. Of 16 tree species for which isozyme studies have been carried out, 14 are largely outcrossing (estimates of $t > 0.80$). These data are consistent with measures of G_{ST} and local gene flow discussed earlier. They also agree with evidence on breeding systems derived from morphological studies and controlled crosses, which suggest that a large proportion of tropical trees are adapted for outcrossing. Bawa et al. (1985) found that 80% of the hermaphroditic rain forest tree species on which they did controlled crosses were self-incompatible. Genetic self-incompatibility has also been documented in Southeast Asia lowland forest species (Kaur et al. 1977) and in tropical dry forests (Bawa 1974; Zapata and Arroyo 1978). Values of t close to 1.0 for species of hermaphroditic, mass-flowering trees in Table 3 are consistent with the possibility of genetic self-incompatibility, as well. If this is so, then estimates of t slightly less than 1.0 may be the result of "effective selfing", or consanguineous mating due to local population structure, rather than true self-fertilization (Ritland and Jain 1981; Ritland 1984; O'Malley and Bawa 1987; Hamrick and Murawski 1990). This is almost certainly the case for *Brosimum alicastrum*, which is dioecious but shows an outcrossing rate of only 0.876 (Table 3).

Outcrossing rates in the four species of *Pinus* reported here are somewhat lower than is typical for temperate pines (Matheson et al. 1989). The authors offer several explanations for this result. First, these species could in fact have greater amounts of true selfing, although their breeding system and floral morphology appear to be similar to temperate pine species. Secondly, there could be less inbreeding depression in these taxa, thus generating higher proportions of selfed progeny in seed collec-

Table 3. Isozyme estimates of mating systems in tropical trees

Species		Outcrossing rate[a]	Source
Trees			
Bertholletia excelsa		0.85 (0.03)	O'Malley et al. 1988
Pithecellobium pedicilare		0.951 (0.02)	O'Malley and Bawa 1987
Tachigali versicolor		0.937	Loveless et al. in prep.
Cavanillesia platanifolia	1987	0.569 (0.02)	Murawski and Hamrick 1990
	1988	0.347 (0.03)	
Quararibea asterolepis		1.01 (0.01)	Murawski and Hamrick 1990
Beilschmedia pendula		0.918	Hamrick and Murawski 1990
Brosimum alicastrum		0.876	Hamrick and Murawski 1990
Platypodium elegans		0.921	Hamrick and Murawski 1990
Trichilia tuberculata		1.08	Hamrick and Murawski 1990
Acacia crassicarpa		0.93—0.99 (0.03) [2]	Moran et al. 1989

Table 3. *(Continued)*

Species	Outcrossing rate[a]	Source
Acacia auriculiformis	0.92—0.93 (0.02—0.04) [2]	Moran et al. 1989
Astrocaryum mexicanum	"close to 1.0"	Eguiarte, pers. comm.
Pinus caribea var *bahamensis*	0.85—0.93 (0.02—0.04) [2]	Matheson et al. 1989
Pinus caribea var *hondurensis*	0.899—0.92 (0.03—0.06) [2]	Matheson et al. 1989
Pinus oocarpa	0.81—0.96 (0.02—0.04) [5]	Matheson et al. 1989
Pinus maximinoi	0.65 (0.06)	Matheson et al. 1989
Shrubs		
Piper amalgo	0.578 (0.04)[b] [5]	Heywood and Fleming 1986
Psychotria faxlucens	"high but significantly lower than 1.0"	Pérez, pers. comm.

[a] Mutilocus outcrossing estimates. Standard errors are shown in parentheses. Where more than one population was surveyed, number of populations is shown in square brackets.
[b] t is an average of single locus estimates from 5 populations.

tions. Finally, the populations may be genetically substructured, such that biparental inbreeding generates effective selfing in the progeny. Since the characteristics of the field populations are not known, it is difficult to distinguish between these alternatives. Relative population isolation and prolonged partial inbreeding could result in loss of deleterious alleles, and thus less inbreeding depression among extant individuals (Lande and Schemske 1985). The patchy population structure and historical isolation of these montane pines could have thus produced a gradual modification of levels of inbreeding depression.

In virtually all of the mating system studies in Table 3, pollen allele frequencies were significantly heterogeneous among individual maternal trees, indicating substantial spatial variation in the mating system even within a local population. O'Malley and Bawa (1987) interpret the presence of albino seedlings in several progeny arrays of *Pithecellobium pedicilare* as evidence of consanguineous matings within a structured population. Hamrick and Murawski (1990) also interpret heterogeneity in the pollen pool as evidence of population structure. They found a significant negative correlation between relative adult density and F_{ST} values averaged over individual maternal trees for 6 moist tropical forest species from Panamá. Adults of species at low densities showed more heterogeneity in pollen allele frequencies among individual trees than adults of species flowering in higher densities, each of which would be more likely to receive pollen from several, probably local, individuals.

A study on the mating system of *Cavanillesia platanifolia* (Murawski et al. 1990) provides excellent evidence for the probable role of local spatial distribution among flowering adults in structuring the breeding system of tropical trees. *Canvanillesia* is the only tree species in Table 2 which seems to have a truly mixed mating system. It represents a particularly interesting case, since the two measures of outcrossing correspond to years in which local flowering densities and spatial patterns of flowering individuals differed. The authors attribute the low outcrossing level in 1988 to lower densities of flowering adults in that year. Changes in the breeding system of individual trees in different years were consistent with the interpretation that spatial patterns of flowering strongly influence individual outcrossing rates and progeny genotypes.

Isozyme studies have also begun to generate direct evidence of long-distance pollen movement in tropical trees which confirms predictions based on pollinator behavior. In *Tachigali versicolor*, a group of 5 trees flowering on BCI in 1984 was separated by 500 m from other flowering individuals. The clumped adults were homozygous for a single IDH allele, but 4% of their progeny carried an alternate allele. The frequency of the alternate allele within the entire island population as a whole was estimated to be 0.16, and pollen migration rates were estimated using the relationship $m = (q_0 - q_1)/(q - q_0) = -0.04/-0.16 = 0.25$ (Loveless et al., in prep.). Thus, 25% of the pollinations in the progeny of these isolated trees is estimated to result from pollen movement over a distance of at least 500 m.

In another, more detailed study of spatial breeding structure in a tropical canopy tree, Hamrick and Murawski (1990) analyzed paternity for progeny of flowering adults of *Platypodium elegans* over three different years. Average pollen movement was 374 m, 420 m, and 369 m over the three years of the study, but pollen movement of >750 m was common. Within individual trees, however, near neighbors often sired up to 10% of the offspring, and 50% or more of an individual's progeny were sired by 10 or fewer fathers. In any given year, a single individual sired between 8% and 15% of all the progeny in the study area, but different subsets of trees flowered in different years, thus preventing a single tree from dominating the genotypes of offspring produced in different seasons. These data strongly support the idea that populations of tropical trees are genetically structured by spatial distribution, flowering phenology, and the behavior of pollinators and seed dispersers. Such studies promise to provide us with considerable insight into the processes which affect genetic organization in tropical species.

Directions for future research

Isozyme techniques have the potential to greatly increase our understanding of genetic organization in tropical species. Only a fraction of the diverse tropical woody flora has been investigated genetically. Many authors have argued that an understanding of genetic organization is a crucial prerequisite to reasoned decisions about conservation or management strategies in tropical forests (Allard 1970; Bawa 1976; Gilbert 1980; Whitmore 1980; Brown and Moran 1981; Hamrick 1983; Namkoong 1986; Bawa and Ashton 1991; Hamrick et al. 1991).

We need to begin by documenting geographic patterns of genetic organization in tropical species. Such data are available for at least some economically important temperate trees (see other papers in this volume). But no similar comprehensive surveys exist for any widely-distributed tropical tree species. Without a better understanding of genetic variation on a regional scale, we have no context within which to interpret local patterns and processes, and we have no basis on which to evaluate the genetic implications of forest conversion or forest conservation.

In addition to these much-needed geographic studies, we need to explore the genetic implications of ecological processes within local populations. Isozymes provide a tool for measuring patterns of pollen movement by different pollinators and for better understanding the sizes and shapes of breeding neighborhoods. Experimental studies should be devised to measure the effects of adult spatial pattern and population density on breeding structure, neighborhood size, and variation in progeny genotypes among trees. We need to quantify year-to-year variation in reproductive ecology in order to understand how these processes influence the lifetime reproductive output of individuals. An extensive theoretical and experimental framework for these studies exists in the temperate plant population literature, and these methodologies need to be extended to tropical plant populations.

If our interest is in understanding and documenting the long-term genetic organization of natural populations, we also need to examine the genetic effects of dispersal, predation, and habitat heterogeneity on population structure. Isozyme markers can be used in tandem with ecological studies to examine how patterns of mortality and survivorship modify initial dispersion patterns to generate or to reduce genetic organization among surviving adults.

Understanding genetic structure in plant populations is of considerable scientific interest. But beyond this, information on the evolutionary genetics of tropical trees should help us to assess the effects of habitat fragmentation, alterations in population density, and changes in pollinator and seed disperser assemblages on the genetics of tropical trees. Such studies may also help us identify species which, because of their natural history, are at particular risk (Bawa and Ashton 1991; Hamrick et al. 1991). Information on population biology and genetic structure is critical to any effort to restore degraded or damaged habitats by species reintroductions, or to manage tropical forests for sustainable uses (Fontaine 1987; Namkoong 1987; Schmidt 1987). Tropical forest management options could be improved by integrating a thorough understanding of the biology and genetics of economically valuable species with traditional management techniques now practiced throughout the global tropics (Gómez-Pompa and Kaus 1990; Anderson 1990). Information on patterns of genetic differentiation could suggest appropriate collection sites for seeds or seedlings in a planting or management scheme (Viana 1990). A knowledge of adult densities needed to promote effective outcrossing and seed set, spatial patterns which promote natural seed dispersal, or regeneration niches in which seedlings have the highest survivorship, could improve the ability of commercial or indigenous forest managers to manipulate forests for continuous yield of valuable products. By increasing the long-term, sustainable value of tropical forests, we have the potential to avoid at least some short-term, irretrievable forest conversion and the loss of diversity which that inevitably entails.

Conclusions

The data reviewed here suggest that populations of tropical trees and shrubs are genetically variable. Virtually all the available isozyme data on population structure in these species suggest that gene flow, whether by pollen or by seed dispersal, is substantial. Local population structure is present, albeit at a low level, and selective pressures are likely to vary in different microhabitats. But levels of gene migration appear to be sufficient to prevent extensive differentiation over short distances, unless there is strong selection. Except for taxa which are local endemics, there is little evidence to support the idea that populations of tropical woody species are genetically isolated, homozygous, and subject to extensive genetic drift. Instead, these species are genetically variable, and are thus likely to adapt to local edaphic, climatic, or biotic conditions. Tropical tree species studied to date are largely outcrossing, and populations are genetically structured, probably by spatial, ecological, and phenological factors. Progeny arrays from individual adults are genetically variable, both within a single season and among different years, and individuals are also likely to differ in their reproductive contributions to the population in different seasons.

These genetic data are in substantial agreement with ecological information being obtained about breeding systems, pollination biology, seed dispersal mechanisms, demography, and evolutionary history of tropical plant species. Nonetheless, only a few tropical woody species have been examined to date, and genetic patterns are likely to differ in different populations. The enormous variety of life history patterns in tropical plants provides a broad spectrum for comparative studies.

Isozyme data are of considerable importance to the conservation and management of genetically-viable populations of tropical species (Hamrick and Murawski 1991). Effective population sizes for tropical trees (and tropical herbs as well; see Young 1988) may be quite large. Hamrick and Murawski (1990) suggest that the effective breeding unit for a common tropical tree species could be on the order of 25 — 50 ha. For rare, widely-dispersed taxa, a much larger area would be necessary in order to include trees which commonly share the same gene pool and to preserve the integrity of the pollinator and seed disperser populations on which these taxa depend (Gilbert 1980; Howe 1989; Bawa and Ashton 1991). Genetically sound conservation and forest management require a robust understanding of the processes by which tropical species organize genetic variation in local populations and the patterns of this variation among populations throughout the ranges of these species.

Acknowledgments

J. L. Hamrick and M. J. W. Godt graciously provided me with summary data sheets on genetic diversity in tropical species, for which I thank them. J. L. Hamrick, D. M. Murawski, K. S. Bawa, D. Piñero, L. Eguiarte, and E. Alvarez-Buylla supplied manuscripts and unpublished data. This review profited from conversations with J. L. Hamrick, P.-M. Forget, and M. A. Wilson and from the comments of J. L. Hamrick, W. T. Adams, and several anonymous reviewers.

References

Adams, W. T. 1983. Application of isozymes in tree breeding. pp. 381—400. In: Tanksley, S. D. and Orton, T. J. (Eds), Isozymes in Plant Genetics and Breeding, Part A. Elsevier Science Publishers, Amsterdam.

Alika, J. E. 1980. Genetic variation among nigerian *Hevea* provenances. Silvae Genetica 29: 201—205.

Allard, R. W. 1970. Population structure and sampling methods. pp. 97—107. In: Frankel, O. H. and Bennett, E. (Eds), Genetic Resources in Plants — Their Exploration and Conservation. IBP Handbook No. 11, F. A. Davis Co., Philadelphia.

Anderson, A. B. 1990. Deforestation in Amazonia: dynamics, causes, and alternatives. pp. 1—23. In: Anderson, A. B. (Ed), Alternatives to Deforestation: Steps Toward Sustainable Use of the Amazon Rain Forest. Columbia University Press, N.Y.

Ashton, P. S. 1969. Speciation among tropical forest trees: some deductions in the light of recent evidence. Biol. J. Linn. Soc. London 1: 155—196.

Ashton, P. S. 1977. A contribution of rain forest research to evolutionary biology. Ann. Missouri Bot. Gard. 64: 694—705.

Augspurger, C. K. 1980. Mass-flowering of a tropical shrub (*Hybanthus prunifolius*): Influence on pollinator attraction and movement. Evolution 34: 475—488.

Augspurger, C. K. 1986. Morphology and dispersal potential of wind-dispersed diaspores of neotropical trees. Am. J. Bot. 73: 353—363.

Augspurger, C. K. and Franson, S. E. 1987. Wind dispersal of artificial fruits varying in mass, area, and morphology. Ecology 68: 27—42.

Baker, H. G. 1959. Reproductive methods as factors in speciation in flowering plants. Cold Spring Harbor Symp. Quant. Biol. 24: 177—199.

Baker, H. G. 1970. Evolution in the tropics. Biotropica 2: 101—111.

Baker, H. G., Bawa, K. S., Frankie, G. W. and Opler, P. A. 1983. Reproductive biology of plants in tropical forests. pp. 183—215. In: Golley, F. B. (Ed), Tropical Rain Forest Ecosystems: Structure and Function. Ecosystems of the World 14A. Elsevier Scientific Publishers, Amsterdam.

Barton, N. H. and Slatkin, M. 1986. A quasi-equilibrium theory of the distribution of rare alleles in a subdivided population. Heredity 56: 409—415.

Bawa, K. S. 1974. Breeding systems of tree species of a lowland tropical community. Evolution 28: 85—92.

Bawa, K. S. 1976. Breeding of tropical hardwoods: an evaluation of underlying bases, current status and future prospects. pp. 43—59. In: Burley, J. and Styles, B. T. (Eds), Tropical Trees: Variation, Breeding and Conservation. Academic Press, London.

Bawa, K. S. and Ashton, P. S. 1991. Conservation of rare trees in tropical rain forests: a genetic perspective. pp. 62—71. In: Falk, D. and Holsinger, K. (Eds), Biology of Rare Plants: Implications for Conservation and Management. Oxford University Press, London.

Bawa, K. S., and Opler, P. A. 1975. Dioecism in tropical forest trees. Evolution 29: 167—179.

Bawa, K. S., Perry, D. R., and Beach, J. H. 1985a. Reproductive biology of tropical lowland rain forest trees. I. Sexual systems and incompatibility mechanisms. Amer. J. Bot. 72: 331—345.

Bawa, K. S., Bullock, S. H., Perry, D. R., Coville, R. E., and Grayum, M. H. 1985b. Reproductive biology of tropical lowland rain forest trees. II. Pollination systems. Amer. J. Bot. 72: 346—356.

Becker, P., Lee, L. W., Rothman, E. D., and Hamilton, W. D. 1985. Seed predation and the coexistence of tree species: Hubbell's models revisited. Oikos 44: 382—390.

Bijlsma, R., Allard, R. W. and Kahler, A. L. 1986. Nonrandom mating in an open-pollinated maize population. Genetics 112: 669—680.

Brokaw, N. V. L. and Scheiner, S. M. 1989. Species composition in gaps and structure of a tropical forest. Ecology 70: 538—541.

Brown, A. H. D. and G. F. Moran. 1981. Isozymes and the genetic resources of forest trees. pp. 1—10. In: Conkle, M. T. (Ed), Isozymes of North American Forest Trees and Forest Insects. USDA Forest Serv., Pac. Southwest Forest and Range Exp. Sta., Berkeley, CA, Gen. Tech. Rep. PSW-48.

Buckley, D. P., O'Malley, D. M., Apsit, V., Prance, G. T., and Bawa, K. S. 1988. Genetics of Brazil nut (*Bertholletia excelsa* Humb. & Bonpl.; Lecythidaceae). I. Genetic variation in natural populations. Theor. Appl. Genet. 76: 923—928.

Bush, M. B. and Colinvaux, P. A. 1990. A pollen record of a complete glacial cycle from lowland Panamá. Jour. Vegetation Sci. 1: 105—118.

Clark, D. B. and Clark, D. A. 1987. Population ecology and microhabitat distribution of *Dipteryx panamensis*, a neotropical rain forest emergent tree. Biotropica 19: 236—244.

Colinvaux, P. 1987. Amazon diversity in light of the paleoecological record. Quat. Sci. Rev. 6: 93—114.

Connell, J. H. 1971. On the role of natural enemies in preventing competitive exclusion in some marine animals and in rain forest trees. pp. 298—312. In: den Boer, P. J. and Gradwell, G. R. (Eds), Dynamics of Populations. Cen. Ag. Publ. and Doc., Wageningen.

Connell, J. H. 1978. Diversity in tropical rain forests and coral reefs. Science 199: 1302—1310.

Connell, J. H. and Lowman, M. D. 1989. Low-diversity tropical rain forests: some possible mechanisms for their existence. Am. Nat. 134: 88—119.

Corner, E. J. H. 1954. The evolution of tropical forest. pp. 34—46. In: Huxley, J. S., Hardy, A. C. and Ford, E. B. (Eds), Evolution as a Process. Allen and Unwin, London.

Croat, T. C. 1978. Flora of Barro Colorado Island. Stanford University Press, Stanford, CA.

Denslow, J. S. 1987. Tropical rainforest gaps and tree species diversity. Ann. Rev. Ecol. Syst. 18: 431—451.

Ellstrand, N. C. and Foster, K. W. 1983. Impact of population structure on the apparent outcrossing rate of grain sorghum (*Sorghum bicolor*). Theor. Appl. Genet. 66:323—327.

Federov, A. A. 1966. The structure of tropical rainforest and speciation in the humid tropics. J. Ecol. 60: 147—170.

Fleming, T. H. and Heithaus, E. R. 1981. Frugivorous bats, seed shadows, and the structure of tropical forests. Biotropica 13 (Supl.): 45—53.

Fontaine, R. G. 1986. Management of humid tropical forests. Unasylva 38: 16—21.

Frankie, G. W. 1975. Tropical forest phenology and pollinator plant coevolution. pp. 192—209. In: Gilbert, L. E. and Raven, P. H. (Eds), Coevolution of Animals and Plants. University of Texas Press, Austin.

Frankie, G. W., Haber, W. A., Opler, P. A., and Bawa, K. S. 1983. Characteristics and organization of the large bee pollination system in the Costa Rican dry forest. pp. 411—447. In: Jones, C. E. and Little, R. J. (Eds), Handbook of Experimental Pollination Ecology. Van Nostrand and Reinhold Co., N.Y.

Gan, Y.-Y., Robertson, F. W., Ashton, P. S., Soepadmo, E., and Lee, D. W. 1977. Genetic variation in wild populations of rain forest trees. Nature 269: 323—325.

Gan, Y.-Y., Robertson, F. W., and Soepadmo, E. 1981. Isozyme variation in some rain forest trees. Biotropica 13: 20—28.

Gentry, A. H. 1988. Tree species richness of upper Amazonian forests. Proc. Natl. Acad. Sci. USA 85: 156—159.

Gilbert, L. E. 1980. Food web organization and the conservation of neotropical diversity. pp. 11—33. In: Soulé, M. E. and Wilcox, B. A. (Eds), Conservation Biology: An Evolutionary-Ecological Perspective. Sinauer Associates, Inc., Sunderland, MA.

Gómez-Pompa, A. and Kaus, A. 1990. Traditional management of tropical forests in Mexico. pp. 45—64. In: Anderson, A. B. (Ed), Alternatives to Deforestation: Steps Toward Sustainable Use of the Amazon Rain Forest. Columbia University Press, N.Y.

Gottlieb, L. D. 1981. Electrophoretic evidence and plant populations. Prog. Phytochem. 7: 1—46.

Hamrick, J. L. 1979. Genetic variation and longevity. pp. 84—113. In: Solbrig, O. T., Jain, S., Johnson, G. B. and Raven, P. H. (Eds), Topics in Plant Population Biology. Columbia University Press, N.Y.

Hamrick, J. L. 1983. The distribution of genetic variation within and among natural plant populations. pp. 335—348. In: Schonewald-Cox, C. M., Chambers, S. M., MacBryde, B., and Thomas, L. (Eds), Genetics and Conservation. Benjamin Cummings, Menlo Park, CA.

Hamrick, J. L. and Godt, M. J. W. 1989. Allozyme diversity in plant species. pp. 44—64. In: Brown, A. H. D., Clegg, M. T., Kahler, A. L., and Weir, B. S. (Eds), Plant Population Genetics, Breeding, and Genetic Resources. Sinauer Associates, Inc., Sunderland, MA.

Hamrick, J. L. and Loveless, M. D. 1986a. Allozyme variation in tropical trees: techniques and preliminary results. Biotropica 18: 201—207.

Hamrick, J. L. and Loveless, M. D. 1986b. The influence of seed dispersal mechanisms on the genetic structure of plant populations. pp. 211—223. In: Estrada, A. and Fleming, T. H. (Eds), Frugivores and Seed Dispersal. Dr W. Junk Publishers, Dordrecht.

Hamrick, J. L. and Loveless, M. D. 1989. Genetic structure of tropical populations: associations with reproductive biology. pp. 129—146. In: Bock, J. H. and Linhart, Y. B. (Eds), The Evolutionary Ecology of Plants. Westview Press, Boulder, CO.

Hamrick, J. L. and Murawski, D. A. 1990. The breeding structure of tropical tree populations. Plant Species Biology 5: 157—165.

Hamrick, J. L. and Murawski, D. A. 1991. Levels of allozyme diversity in populations of uncommon neotropical tree species. Jour. Tropical Ecology 7: 395—399.

Hamrick, J. L., Linhart, Y. B., and Mitton, J. B. 1979. Relationships between life history characteristics and electrophoretically-detectable genetic variation in plants. Ann. Rev. Ecol. Syst. 10: 173—200.

Hamrick, J. L., Mitton, J. B., and Linhart, Y. B. 1981. Levels of genetic variation in trees: influence of life history characteristics. pp. 35—41. In: Conkle, M. T. (Ed), Isozymes of North American Forest Trees and Forest Insects. USDA Forest Serv., Pac. Southwest Forest and Range Exp. Sta., Berkeley, CA, Gen. Tech. Rep. PSW-48.

Hamrick, J. L., Godt, M. J. W., Murawski, D. A., and Loveless, M. D. 1990. Correlations between species traits and allozyme diversity: implications for conservation biology. pp. 75—86. In: Falk, D. and Holsinger, K. (Eds), Biology of Rare Plants: Implications for Conservation and Management. Oxford University Press, London.

Hart, T. B., Hart, J. A., and Murphy, P. G. 1989. Monodominant and species-rich forests of the humid tropics: causes for their co-occurrence. Am. Nat. 133: 613—633.

Hartshorn, G. S. 1978. Tree falls and tropical forest dynamics. pp. 617—638. In: Tomlinson, P. B. and Zimmerman, M. H. (Eds), Tropical Trees as Living Systems. Cambridge Univ. Press, Cambridge.

Hartshorn, G. S. 1983. Plants. pp. 118—157. In: Janzen, D. H. (Ed), Costa Rican Natural History. University of Chicago Press, Chicago.

Hartshorn, G. S. 1988. Tropical and subtropical vegetation of Meso-America. pp. 365—390. In: Barbour, M. G. and Billings, W. D. (Eds), North American Terrestrial Vegetation. Cambridge University Press, Cambridge.

Hedegart, T. 1976. Breeding systems, variation and genetic improvement of teak (*Tectona grandis* L. F.). pp. 109—123. In: Burley, J. and Styles, B. T. (Eds), Tropical Trees: Variation, Breeding and Conservation.

Heithaus, E. R., Fleming, T. H. and Opler, P. A. 1975. Foraging patterns and resource utilization of seven species of bats in a seasonal tropical forest. Ecology 56: 841—853.

Helenurm, K. and Ganders, F. R. 1985. Adaptive radiation and genetic differentiation in Hawaiian *Bidens*. Evolution 39: 753—765.

Heywood, J. S. and Fleming, T. H. 1986. Patterns of allozyme variation in three Costa Rican species of *Piper*. Biotropica 18: 208—213.

Howe, H. F. 1986. Seed dispersal by fruit-eating birds and mammals. pp. 123—190. In: Murray, D. R. (Ed) Seed Dispersal. Academic Press, Sydney.

Howe, H. F. 1989. Scatter- and clump-dispersal and seedling demography: hypothesis and implications. Oecologia 79: 417—426.

Hubbell, S. P. 1979. Tree dispersion, abundance, and diversity in a tropical dry forest. Science 203: 1299—1309.

Hubbell, S. P. 1980. Seed predation and the coexistence of tree species in tropical forests. Oikos 35: 214—229.

Hubbell, S. P. and Foster, R. B. 1983. Diversity of canopy trees in a neotropical forest and implications for conservation. pp. 25—41. In: Sutton, S. L., T. C. Whitmore, and S. Chadwick (Eds), Tropical Rain Forest: Ecology and Management. Blackwell Scientific Publications, London.

Hubbell, S. P. and Foster, R. B. 1986a. Commoness and rarity in a neotropical forest: implications for tropical tree conservation. pp. 205—231. In: Soulé, M. E. (Ed), Conservation Biology: The Science of Scarcity and Diversity. Sinauer Associates, Inc., Sunderland, MA.

Hubbell, S. P. and Foster, R. B. 1986b. Canopy gaps and the dynamics of a neotropical forest. pp. 77—96. In: Crawley, M.J. (Ed), Plant Ecology. Blackwell Scientific Publications, Oxford.

Jacob, M. 1988. The Tropical Rain Forest: A First Encounter. Springer-Verlag, Berlin.

Janzen, D. H. 1970. Herbivores and the number of tree species in tropical forests. Am. Nat. 104: 501—528.

Kaur, A., Ha, C. O., Jong, K., Sands, V. E., Chan, H. T., Soepadmo, E., and Ashton, P. A. 1977. Apomixis may be widespread among trees of the climax rain forest. Nature 271: 440—442.

Lande, R. and Schemske, D. W. 1985. The evolution of self-fertilization and inbreeding depression in plants. I. Genetic models. Evolution 39: 24—40.

Leigh, E. G., Jr. 1982. Introduction: Why are there so many kinds of tropical trees? pp. 63—66. In: Leigh, E. G., Jr. (Ed), The Ecology of a Tropical Forest: Seasonal Rhythms and Long-Term Changes. Smithsonian Institution Press, Washington, D.C.

Lieberman, M., Lieberman, D., and Peralta, R. 1989. Forests are not just swiss cheese: canopy stereogeometry of non-gaps in tropical forests. Ecology 70: 550—552.

Livingston, D. A. 1975. Late Quaternary climatic change in Africa. Ann. Rev. Ecol. Syst. 6: 249—280.

Loveless, M. D. and Hamrick, J. L. 1984. Ecological determinants of genetic structure in plant populations. Ann. Rev. Ecol. Syst. 15: 65—95.

Loveless, M. D. and Hamrick, J. L. 1987. Distribución de la variación en especies de árboles tropicales. Rev. Biol. Trop., 35 (Suppl. 1): 165—175.

Lowrey, T. K. and Crawford, D. J. 1985. Allozyme divergence and evolution in *Tetramolopium* (Compositae: Astereae) on the Hawaiian Islands. Syst. Bot. 10: 64—72.

Lugo, A. E. 1988. Estimating reductions in the diversity of tropical forest species. pp. 58—70. In: Wilson, E. O. (Ed) Biodiversity. National Academy Press, Washington, D.C.

Matheson, A. C., Bell, J. C., and Barnes, R. D. 1989. Breeding systems and genetic structure in some Central American pine populations. Silvae Genetica 38: 107—113.

Mooney, H. A. 1988. Lessons from Mediterranean-climate regions. pp. 157—165. In: Wilson, E. O. (Ed) Biodiversity. National Academy Press, Washington, D.C.

Muraski, D. M., Hamrick, J. L., Hubbell, S. P. and Foster, R. B. 1990. Mating systems of two bombacaceous trees of a neotropical moist forest. Oecologia 82: 501—506.

Namkoong, G. 1986. Genetics and the forests of the future. Unasylva 38: 2—18.

Nei, M. 1973. Analysis of gene diversity in subdivided populations. Proc. Natl. Acad. Sci. USA 70: 3321—3323.

O'Malley, D. M. and Bawa, K. S. 1987. Mating system of a tropical rain forest tree species. Am. J. Bot. 74: 2243—2249.

Piperno, D. R., Bush, M. B. and Colinvaux, P. A. 1990. Paleoenvironments and human occupation in late-glacial Panamá. Quat. Res. 33:108—116.

Price, M. V. and Jenkins, S. H. 1986. Rodents as seed consumers and dispersers. pp. 191—235. In: Murray, D. R. (Ed), Seed Dispersal. Academic Press, N.Y.

Raven, P. H. 1976. Ethics and attitudes. pp. 155—181. In: Simmons, J. B., Beyer, R. I., Brandham, P. E., Lucas, G. and Parry, V. T. H. (Eds), Conservation of Threatened Plants. Plenum, N.Y.

Ritland, K. 1984. The effective proportion of self-fertilization with consanguineous matings in inbred populations. Genetics 106: 139—152.

Ritland, K. and El-Kassaby, Y. A. 1985. The nature of inbreeding in a seed orchard of Douglas fir as shown by an efficient multilocus model. Theor. Appl. Genet. 71: 375—384.

Ritland, K. and Jain, S. 1981. A model for the estimation of outcrossing rate and gene frequencies using *n* independent loci. Heredity 47: 35—52.

Schmidt, R. 1987. Tropical rain forest management. Unasylva 39: 2—17.

Shaw, D. V. and R. W. Allard. 1981. Analysis of mating system parameters and population structure in Douglas-fir using single-locus and multilocus methods. pp. 18—22. In: Conkle, M. T. (Ed), Isozymes of North American Forest Trees and Forest Insects. USDA Forest Serv., Pac. Southwest Forest and Range Exp. Sta., Berkeley, CA, Gen. Tech. Rep. PSW-48.

Slatkin, M. 1985. Rare alleles as indicators of gene flow. Evolution 39: 53—65.

Sommers, A. 1976. Attempt at an assessment of the world's tropical moist forests. Unasylva 28: 5—27.

Stiles, F. G. 1975. Ecology, flowering phenology and hummingbird pollination of some Costa Rican *Heliconia* species. Ecology 56: 285—301.

Sytsma, K. J. and Schaal, B. A. 1985. Genetic variation, differentiation, and evolution in a species complex of tropical shrubs based on isozyme data. Evolution 39: 582—593.

UNESCO. 1978. Tropical Forest Ecosystems: A State of Knowledge Report. 683 pp.

Viana, V. M. 1990. Seed and seedling availability as a basis for management of natural forest regeneration. pp. 99—115. In: Anderson, A. B. (Ed), Alternatives to Deforestation: Steps Toward Sustainable Use of the Amazon Rain Forest. Columbia University Press, N.Y.

Whitmore, J. L. 1987. Plantation forestry in the tropics of Latin America: a research agenda. Unasylva 39: 36—41.

Whitmore, T. C. 1980. The conservation of tropical rain forest. pp. 303—318. In: Soulé, M. E. and Wilcox, B. A. (Eds), Conservation Biology: An Evolutionary-Ecological Perspective. Sinauer Associates, Inc., Sunderland, MA.

Whitmore, T. C. 1984. Tropical Rain Forests of the Far East. Clarendon Press, Oxford.

Whitmore, T. C. 1989. Canopy gaps and the two major groups of forest trees. Ecology 70: 536—538.

Wright, S. 1951. The genetical structure of populations. Ann. Eugenics 15: 323—354.

Young, H. J. 1988. Neighborhood size in a beetle pollinated tropical aroid: effects of low density and asynchronous flowering. Oecologia 76: 461—466.

Zapata, T. R. and Arroyo, M. T. K. 1978. Plant reproductive ecology of a secondary deciduous tropical forest in Venezuela. Biotropica 10: 221—230.

New Forests **6**: 95–124, 1992.

Review paper

Factors influencing levels of genetic diversity in woody plant species

J. L. HAMRICK, MARY JO W. GODT and
SUSAN L. SHERMAN-BROYLES
Departments of Botany and Genetics, University of Georgia, Athens, Georgia 30602, USA

Received 9 October 1990; accepted 13 September 1991

Key words: allozyme, electrophoresis, life history characteristics, polymorphic loci, population genetics, trees

Application. Knowledge of the life history and ecological characteristics of woody plant species allow predictions of the level and distribution of genetic diversity within and among populations. Generalizations developed from such analyses can be used to develop sampling strategies for the preservation of genetic diversity.

Abstract. The plant allozyme literature was reviewed to: (1) compare genetic diversity in long-lived woody species with species representing other life forms, and (2) to investigate whether the levels and distribution of genetic diversity in woody species are related to life history and ecological characteristics. Data from 322 woody taxa were used to measure genetic diversity within species, and within and among populations of species. Woody species maintain more variation within species and within populations than species with other life forms but have less variation among populations. Woody species with large geographic ranges, outcrossing breeding systems, and wind or animal-ingested seed dispersal have more genetic diversity within species and populations but less variation among populations than woody species with other combinations of traits. Although life history and ecological traits explain a significant proportion (34%) of the variation among species for the genetic parameters measured, a large proportion of the interspecific variation is unexplained. The specific evolutionary history of each species must play an important role in determining the level and distribution of genetic diversity.

Introduction

Much of the early research in forest genetics involved studies of geographic patterns of quantitative genetic variation. Such studies demonstrated that morphological and physiological traits of most woody species have high levels of genetic variation that is associated with populations of different geographic origin (Libby et al. 1969). Although fewer studies are available, high levels of quantitative genetic variation have also been found within populations of woody plants (e.g., Kinloch and Stonecypher 1969;

Stonecypher et al. 1973; Wells and Snyder, 1976; Hamrick 1976). From a population genetics standpoint, however, studies of quantitative genetic traits have several drawbacks. First, levels of genetic diversity can not be determined directly from natural populations because of environmental influences on quantitative traits. Second, the polygenic inheritance of most morphometric traits precludes the calculation of genetic diversity parameters (e.g., percent polymorphic loci, heterozygosity). Third, many quantitative traits are expressed only after several years of growth.

The development of electrophoretic techniques provided a more direct method to estimate genetic diversity within and among plant populations. Earlier reviews of the plant allozyme literature found significant correlations between the ecological and life history characteristics of species and the level and distribution of allozyme diversity (Hamrick et al. 1979; Nevo et al. 1984; Loveless and Hamrick 1984; Hamrick and Godt 1989). These reviews concluded that woody plants maintain high levels of variation within species and within populations and have relatively less genetic diversity among populations than non-woody species. The number of isozyme studies of trees has increased significantly since these earlier reviews. It is now possible to thoroughly investigate whether differences in the ecological and life history characteristics of woody plants are associated with allozyme diversity.

In this paper we survey published electrophoretic data on woody plant species to examine variation at three levels: within species, within populations and among populations within species. We first compare the genetic diversity of long-lived woody species (trees and a few long-lived shrubs and treelets) with that of species with different life forms. We then compare woody species with different attributes to determine whether significant relationships exist between genetic diversity and ecological and life history characteristics. The format of this paper deliberately follows that of Hamrick and Godt (1989) to allow the interested reader to easily compare results.

Methods

Data were obtained from studies published from 1968 to 1990 that reported estimates of allozyme variation for gymnosperms and angiosperms. Only papers with genetic interpretations of electrophoretic banding patterns were included. Deficiencies in the data base precluded calculating all the genetic parameters for every study reviewed. For example, reports based on a single population could not be used to estimate variation within species. Studies restricted to polymorphic loci in multiple populations were limited to calculations of diversity among populations. Those studies that surveyed monomorphic and polymorphic loci in multiple populations provided data to calculate all of the genetic parameters. Taxa at the subspecies level or above were considered. Species that were the focus of more than one study were represented more than once in the data.

Species characteristics

Each species surveyed was classified as an annual or a perennial. If perennial, the species was classified as long- or short-lived and then further subclassified as herbaceous or woody. Each long-lived woody species was classified for seven ecological and life history traits: 1. taxonomic status (gymnosperm or angiosperm); 2. regional distribution (boreal-temperate, temperate, temperate-tropical or tropical); 3. geographic range (endemic, narrow, regional or widespread); 4. breeding system [selfed, mixed mating (either animal- or wind-pollinated), or outcrossed (either animal- or wind-pollinated)]; 5. seed dispersal mechanism (gravity, gravity and animal-attached, explosive, wind, animal-ingested or animal-attached); 6. mode of reproduction (sexual or sexual and asexual); and 7. successional status (early, mid or late). Classification of each species was based on descriptions in the original papers or was determined by consulting floras.

Genetic parameters

Four genetic parameters were calculated at the within species and the within population levels: percent polymorphic loci, mean number of alleles per locus, effective number of alleles per locus and genetic diversity (Weir 1989).

The percentage of polymorphic loci within species (P_s) was calculated by dividing the number of loci polymorphic in at least one population by the number of loci analyzed. The mean number of alleles per locus within species (A_s) was determined by summing the alleles observed over all loci and populations and dividing by the total number of loci. Genetic diversity within species (H_{es}) was calculated for each locus (monomorphic and polymorphic) by

$$H_{es} = 1 - \Sigma \bar{p}_i^2$$

where $\bar{p}_i$ is the mean frequency of the ith allele. Mean genetic diversity was obtained by averaging the H_{es} values over all loci. The effective number of alleles (A_{es}) within each species was calculated by

$$A_{es} = 1/(1 - H_{es})$$

The percent polymorphic loci within populations (P_p) was calculated as the proportion of loci polymorphic in each population averaged over all populations. The number of alleles per locus within populations (A_p) was determined for each population and a mean value was obtained by averaging over all populations. Genetic diversity within populations (H_{ep}) was calculated for each locus (monomorphic and polymorphic) and population by

$$H_{ep} = 1 - \Sigma\, p_i^2$$

where p_i is the frequency of the ith allele in each population. The mean H_{ep} value was obtained for each locus by averaging over all loci and populations. The effective number of alleles within populations was calculated by

$$A_{ep} = 1/(1 - H_{ep}).$$

Variation among populations was estimated with Nei's genetic diversity statistics (Nei 1973). Total genetic diversity (H_T) and mean diversity within populations (H_s) were estimated from the polymorphic loci. The proportion of genetic diversity residing among populations (G_{ST}) was determined by

$$G_{ST} = (H_T - H_S)/H_T$$

for each locus. Mean G_{ST} values over all polymorphic loci were calculated for each species.

Statistical analyses

Statistical analyses generally followed procedures used by Hamrick et al. (1979), Loveless and Hamrick (1984) and Hamrick and Godt (1989). Means and standard errors of the means were calculated for each category and genetic parameter. Levels of statistical significance among the categories of each trait were determined by performing separate one-way analyses of variance using GLM procedures of SAS (SAS, 1987) with trait categories treated as class variables. A least squares means procedure (SAS, 1987) was employed in a pairwise fashion to determine where differences among categories were significant. Multiple regression models were constructed to determine characteristics that best described variation in H_{es}, H_{ep} and G_{ST}.

Multiple regression models were constructed by beginning with the univariate model having the largest R^2 value and sequentially entering variables. Variables were entered such that each subsequent variable accounted for a decreasing proportion of the total variance. The proportion of variation accounted for by each factor was calculated from the sequential sums of squares, or the additional amount of variation accounted for by adding another factor to the model. This analysis provided insights into the relative influence of the seven traits on the genetic parameters.

Since many of the seven life history and ecological traits were potentially correlated (e.g. gymnosperms are outcrossing, long-lived species with wind-dispersed seeds), principal components analyses (SAS 1987) were used to determine whether particular combinations of traits explained significant amounts of variation in genetic diversity.

Results

Data were analyzed from 968 entries. An entry was data extracted from a single genetic study of a species. (Some species were represented in several papers and some papers analyzed several species). A total of 662 species representing 220 genera were reviewed. Approximately one-third of the studies provided complete data sets. Sample sizes for analyses of differences among life forms varied from 580 to 705 for the genetic parameters analyzed. The data for woody species came from 322 entries which included 213 species representing 54 genera. For the woody plants, effective sample sizes for the various genetic parameters ranged from 190 to 200.

Variation among life forms

On average, 51% of the isozyme loci are polymorphic in plant species. Mean genetic diversity within species is 0.15 (Table 1). Within plant populations, about 35% of the isozyme loci are polymorphic and mean genetic diversity is 0.11. Approximately 77% of the total diversity at polymorphic loci occurs within populations of woody species.

Long-lived woody species have a higher proportion of polymorphic loci, more alleles per locus, higher effective number of alleles per locus and more genetic diversity than other life forms (Table 1A). Comparatively, the mean genetic diversity of woody species is 15% higher than annual species, 42% greater than herbaceous perennial species, and 55% higher than short-lived woody species. The higher genetic diversity of long-lived woody species is primarily due to a significantly higher propor-

Table 1. Levels of allozyme variation within species, within populations and among populations of species belonging to different life forms

A. Diversity within species

Categories	N^1	Mean no. populations	Mean no. loci	P_s	A_s	A_{es}	H_{es}
				***	***	***	***
All species	655	12.3	17.3	51.3	1.97	1.20	0.150
		(1.0)	(0.3)	(1.1)	(0.04)	(0.01)	(0.004)
Annual	226	18.1	16.2	49.2b[2]	2.02b	1.22ab	0.154b
		(2.5)	(0.5)	(1.9)	(0.07)	(0.02)	(0.008)
Short-lived perennial							
Herbaceous	185	9.1	16.9	43.4c	1.75c	1.16c	0.125c
		(1.0)	(0.5)	(1.9)	(0.05)	(0.01)	(0.007)
Woody	23	23.4	24.2	43.6bc	1.66bc	1.13bc	0.111bc
		(7.9)	(2.0)	(4.3)	(0.10)	(0.02)	(0.016)
Long-lived perennial							
Herbaceous	30	7.8	15.1	39.3bc	1.64c	1.15bc	0.124c
		(1.1)	(0.8)	(3.3)	(0.07)	(0.01)	(0.011)
Woody	191	9.2	18.1	65.0a	2.22a	1.24a	0.177a
		(1.3)	(0.6)	(1.9)	(0.06)	(0.01)	(0.006)

Standard errors are in parentheses.

Table 1. (Continued)

B. Diversity within populations

Categories	N[1]	Mean no. populations	Mean no. loci	P_p	A_p	A_{ep}	H_{ep}
				***	***	***	***
All species	669	12.3	17.3	34.6	1.52	1.15	0.113
		(1.0)	(0.3)	(1.0)	(0.02)	(0.01)	(0.004)
Annual	226	18.1	16.2	29.4b[2]	1.45b	1.14b	0.101b
		(2.5)	(0.5)	(1.6)	(0.03)	(0.01)	(0.007)
Short-lived perennial							
Herbaceous	204	9.1	16.9	28.3b	1.39b	1.12b	0.098b
		(1.0)	(0.5)	(1.5)	(0.03)	(0.01)	(0.006)
Woody	17	23.4	24.2	31.7b	1.54ab	1.11b	0.096b
		(7.9)	(2.0)	(4.9)	(0.10)	(0.02)	(0.015)
Long-lived perennial							
Herbaceous	24	7.8	15.1	21.5b	1.32b	1.09b	0.082b
		(1.1)	(0.8)	(2.2)	(0.03)	(0.01)	(0.009)
Woody	196	9.2	18.1	49.3a	1.76a	1.20a	0.148a
		(1.3)	(0.6)	(1.8)	(0.04)	(0.01)	(0.006)

Table 1. (Continued)

C. Diversity among populations

Categories	N[1]	Mean no. populations	Mean no. loci	H_T	H_s	G_{ST}
				NS	***	***
All species	584	12.3	17.3	0.302	0.224	0.228
		(1.0)	(0.3)	(0.005)	(0.005)	(0.010)
Annual	186	18.1	16.2	0.322a[2]	0.196b	0.355a
		(2.5)	(0.5)	(0.010)	(0.010)	(0.021)
Short-lived perennial						
Herbaceous	164	9.1	16.9	0.301ab	0.219b	0.253b
		(1.0)	(0.5)	(0.010)	(0.010)	(0.018)
Woody	14	23.4	24.2	0.287ab	0.243ab	0.155bc
		(7.9)	(2.0)	(0.025)	(0.025)	(0.038)
Long-lived perennial						
Herbaceous	25	7.8	15.1	0.307ab	0.228ab	0.278ab
		(1.1)	(0.8)	(0.020)	(0.021)	(0.033)
Woody	195	9.2	18.1	0.283b	0.253a	0.084c
		(1.3)	(0.6)	(0.009)	(0.008)	(0.008)

[1] N, number of entries; P_s & P_p, percent polymorphic loci; A_s & A_p, number of alleles per locus; A_{es} & A_{ep}, effective number of alleles per locus; H_{es} & H_{ep}, genetic diversity; H_T, total genetic diversity; H_s, genetic diversity within populations; G_{ST}, proportion of the total diversity among populations. See text for more complete definitions.

[2] Means followed by the same letter in a column are not significantly different at the 5% probability level.

Levels of significance: *, $p < 0.05$; **, $p < 0.01$; ***, $p < 0.001$; NS, not significant.

tion of polymorphic loci. In contrast, the relatively high genetic diversity of annual species is primarily due to the high number of alleles (3.09) at polymorphic loci. This results, in part, from the inclusion in this category of highly variable crop species belonging to genera such as *Zea*, *Hordeum*, and *Triticum*.

Long-lived woody species have more genetic diversity within their populations than other life forms (Table 1B). The genetic diversity for trees is 38% higher than the value for annuals and 51% to 80% higher than that of the other life forms. Trees also exhibit little among population genetic diversity ($G_{ST} = 0.085$) relative to other life forms (Table 1C). Among population diversity for the long-lived woody species is 24% of that of annuals, and only 30% to 54% of the values for the other categories. In summary, woody species have more genetic diversity and partition genetic diversity within and among populations differently than species with different life forms.

Variation within woody species

Differences among the categories of geographic range and breeding systems were significant ($p < 0.05$) for the four genetic parameters (Table 2). Taxonomic status and regional distribution categories were significantly different for the proportion of polymorphic loci and the number of alleles per locus. Significant differences occurred among the categories of seed dispersal and mode of reproduction for the effective number of alleles and genetic diversity, while none of the genetic parameters differed significantly among successional stage categories.

Compared to woody angiosperms, gymnosperms had a significantly higher proportion of polymorphic loci but differences in genetic diversity were not significant. Since angiosperm and gymnosperm species maintain nearly the same number of alleles per polymorphic locus (2.86 vs. 2.93) the relative similarity in genetic diversity within species must be due to a higher proportion of low frequency alleles in the gymnosperms.

Geographic range had a significant effect on levels of genetic diversity within species (Table 2). Endemic species had approximately 30% of the genetic diversity of widespread species and 46—47% of the genetic diversity of narrowly and regionally distributed species. Narrowly and regionally distributed species had 64—66% of the genetic diversity of widespread species.

Woody species with boreal-temperate distributions tended to have higher proportions of polymorphic loci than species from lower latitudes. In terms of the effective number of alleles per locus and genetic diversity there were no differences among regions. This suggests that a high propor-

Table 2. Levels of allozyme variation within species of long-lived woody species with different attributes

Categories	N^1	Mean no. populations	Mean no. loci	P_s	A_s	A_{es}	H_{es}
Taxonomic status				**	*	NS	NS
Gymnosperms	89	8.9	17.3	71.1a[2]	2.38a	1.22a	0.169a
		(1.0)	(0.9)	(2.6)	(0.09)	(0.015)	(0.008)
Angiosperms	102	9.5	18.9	59.5b	2.10b	1.26a	0.183a
		(2.4)	(0.8)	(2.6)	(0.07)	(0.02)	(0.011)
Geographic range				***	*	***	***
Endemic	20	8.5	21.8	42.5b	1.82b	1.09c	0.078c
		(2.1)	(2.0)	(7.6)	(0.15)	(0.02)	(0.016)
Narrow	45	5.3	21.1	61.5a	2.08ab	1.22b	0.165b
		(0.7)	(1.0)	(3.4)	(0.09)	(0.02)	(0.012)
Regional	115	11.6	16.8	55.7a	1.87a	1.23ab	0.169ab
		(2.2)	(0.8)	(2.3)	(0.05)	(0.01)	(0.008)
Widespread	11	5.4	14.0	67.8a	2.11a	1.39a	0.257a
		(1.3)	(1.7)	(7.2)	(0.18)	(0.08)	(0.039)
Regional distribution				**	**	NS	NS
Boreal-temperate	26	8.0	15.8	82.5a	2.58a	1.28ab	0.206a
		(1.2)	(1.6)	(3.2)	(0.14)	(0.03)	(0.014)
Temperate	122	12.2	17.7	63.5b	2.27a	1.22b	0.166a
		(2.1)	(0.8)	(2.3)	(0.07)	(0.01)	(0.008)

Standard errors are in parentheses.

Table 2. (Continued)

Categories	N^1	Mean no. populations	Mean no. loci	P_s	A_s	A_{es}	H_{es}
Temperate-tropical	5	1.5	16.6	62.2ab	1.89ab	1.21ab	0.169a
		(0.4)	(1.6)	(11.5)	(0.40)	(0.06)	(0.038)
Tropical	38	2.9	21.3	57.9b	1.87b	1.28a	0.191a
		(0.4)	(1.4)	(5.0)	(0.10)	(0.04)	(0.022)
Breeding system				***	**	**	***
Selfing	1	1.0	13.0	11.0b	1.15ab	1.03ab	0.025cb
		—	—	—	—	—	—
Mixed-animal	11	11.8	18.7	29.9b	1.51b	1.12b	0.075c
		(3.2)	(2.7)	(9.6)	(0.22)	(0.06)	(0.030)
Mixed-wind	1	4.0	17.0	50.0ab	—	—	—
		—	—	—	—	—	—
Outcrossing-animal	51	4.6	19.5	63.2a	2.18a	1.30a	0.211a
		(0.8)	(1.0)	(3.7)	(0.11)	(0.03)	(0.016)
Outcrossing-wind	128	11.3	17.5	69.1a	2.31a	1.23a	0.173b
		(2.0)	(0.8)	(2.0)	(0.07)	(0.01)	(0.007)
Seed dispersal				NS	NS	**	**
Gravity	23	7.8	20.8	61.9a	2.48a	1.26b	0.144bc
		(1.5)	(1.1)	(5.9)	(0.22)	(0.03)	(0.027)
Gravity-attached	10	7.3	16.8	63.2a	2.09a	1.14c	0.115d
		(2.0)	(1.6)	(2.1)	(0.17)	(0.03)	(0.021)
Attached	26	5.1	18.5	62.7a	2.16a	1.28ab	0.204ab
		(2.3)	(1.2)	(3.9)	(0.08)	(0.30)	(0.016)
Explosive	2	2.3	17.0	40.5a	1.61a	1.18abc	0.133abcd
		(1.0)	(5.0)	(18.9)	(0.56)	(0.11)	(0.065)

Table 2. (continued)

Categories	N^1	Mean no. populations	Mean no. loci	P_s	A_s	A_{es}	H_{es}
Ingested	27	21.0	14.0	67.8a	2.07a	1.34a	0.231a
		(11.7)	(1.8)	(4.7)	(0.11)	(0.04)	(0.022)
Wind	103	8.4	18.3	66.2a	2.24a	1.21bc	0.160cd
		(0.9)	(0.9)	(2.8)	(0.09)	(0.02)	(0.009)
Mode of reproduction				NS	NS	***	**
Sexual	176	7.6	18.8	64.4a	2.23a	1.23b	0.170b
		(0.7)	(0.6)	(2.0)	(0.06)	(0.01)	(0.007)
Sexual and asexual	15	45.2	8.7	72.7a	2.12a	1.39a	0.251a
		(25.1)	(2.1)	(7.2)	(0.19)	(0.07)	(0.036)
Successional status				NS	NS	NS	NS
Early	8	2.7	18.1	44.6b	1.67b	1.17a	0.137a
		(1.1)	(2.8)	(7.6)	(0.13)	(0.04)	(0.031)
Mid	58	7.4	19.5	65.6a	2.18a	1.23a	0.171a
		(0.8)	(1.3)	(4.0)	(0.09)	(0.03)	(0.014)
Late	125	10.3	17.6	66.0a	2.27a	1.25a	0.182a
		(1.8)	(0.7)	(2.16)	(0.08)	(0.01)	(0.008)

[1] N, number of entries; P_s, percent polymorphic loci; A_s, number of alleles per locus; A_{es}, effective number of alleles per locus; H_{es}, genetic diversity. See text for more complete definitions.

[2] Means followed by the same letter in a column are not significantly different at the 5% probability level.

Levels of significance: *, $p < 0.05$; **, $p < 0.01$; ***, $p < 0.001$; NS, not significant.

tion of the polymorphic loci of boreal-temperate species have alleles that are either in high or low frequencies.

Since there was only one entry each in the selfing and the mixed mating, wind-pollinated categories, comparisons could only be made among the mixed mating, animal-pollinated species and the two outcrossing categories. The mean genetic diversity of mixed mating, animal-pollinated species was 36% of that for the outcrossed, animal-pollinated species and 43% of that for outcrossed, wind-pollinated species. The mean genetic diversity for the latter category was 82% as large as that for the outcrossed, animal-pollinated species. The difference in genetic diversity between mixed mating species and outcrossed species is primarily due to differences in the proportion of polymorphic loci.

Species whose seeds are ingested or attached to animals have the highest genetic diversity values. Species with gravity, gravity-attached (seeds that drop to the ground and may subsequently become attached to animals) and wind-dispersed seeds have less genetic diversity (Table 2). The genetic diversity of species with these seed dispersal mechanisms range from 30—69% of the value for species with ingested seeds. The number of entries in the explosive category were too few for valid comparisons. Since there were no significant differences in percent polymorphic loci among categories, differences in genetic diversity were probably due to differences among species in the proportion of loci with intermediate allele frequencies.

Tree species with both sexual and asexual modes of reproduction maintained higher levels of genetic diversity than species limited to sexual reproduction. There were no significant differences in percent polymorphic loci between these two categories which suggests that species reproducing both sexually and asexually have more intermediate allele frequencies. Although the differences were not significant, species of mid and late successional stages had higher genetic diversities than species of the early stages of succession.

The seven ecological and life history traits explained 34% of the variation in genetic diversity within species. Geographic range accounted for 44.5% of the variation explained by the model, breeding system accounted for 21% and seed dispersal and successional status added 16% and 10%, respectively.

Variation within populations of woody species

Differences among the categories of geographic range, regional distribution, breeding system and mode of reproduction were significant ($p < 0.05$) for all four genetic parameters (Table 3). Categories for taxonomic

Table 3. Levels of allozyme variation within populations of long-lived woody species with different attributes

Categories	N^1	Mean no. populations	Mean no. loci	P_p	A_p	A_{ep}	H_{ep}
Taxonomic status				*	*	NS	NS
Gymnosperms	102	8.9	17.3	53.4a[2]	1.83a	1.20a	0.151a
		(1.0)	(0.9)	(2.4)	(0.58)	(0.01)	(0.008)
Angiosperms	96	9.5	18.9	45.1b	1.68b	1.20a	0.143a
		(2.4)	(0.8)	(2.5)	(0.05)	(0.02)	(0.010)
Geographic range				***	***	***	***
Endemic	26	8.5	21.8	26.3c	1.48b	1.08c	0.056d
		(2.1)	(2.0)	(3.9)	(0.09)	(0.01)	(0.010)
Narrow	61	5.3	21.1	44.3b	1.61b	1.19b	0.143c
		(0.7)	(1.0)	(2.8)	(0.05)	(0.02)	(0.010)
Regional	103	11.6	16.8	69.2a	2.31a	1.26b	0.194b
		(2.2)	(0.8)	(2.2)	(0.08)	(0.02)	(0.009)
Widespread	9	5.4	14.0	74.3a	2.56a	1.33a	0.228a
		(1.3)	(1.7)	(7.7)	(0.31)	(0.04)	(0.026)
Regional distribution				***	***	**	**
Boreal-temperate	30	8.0	15.8	68.4a	2.07a	1.28a	0.204a
		(1.2)	(1.6)	(3.6)	(0.08)	(0.28)	(0.015)
Temperate	106	2.9	21.3	49.2b	1.81b	1.20b	0.145b
		(0.4)	(1.4)	(2.2)	(0.06)	(0.02)	(0.008)
Temperate-tropical	9	1.5	16.6	43.6bc	1.62bc	1.17b	0.134b
		(0.4)	(1.6)	(6.7)	(0.14)	(0.04)	(0.030)

Standard errors are in parentheses.

Table 3. (Continued)

Categories	N^1	Mean no. populations	Mean no. loci	P_p	A_p	A_{ep}	H_{ep}
Tropical	53	2.9	21.3	39.8c	1.51c	1.16b	0.125b
		(0.4)	(1.4)	(3.5)	(0.05)	(0.02)	(0.012)
Breeding system				***	**	*	***
Selfing	0	—	—	—	—	—	—
Mixed-animal	11	11.8	18.7	17.2b	1.21b	1.06b	0.035b
		(3.2)	(2.7)	(6.3)	(0.09)	(0.02)	(0.013)
Mixed-wind	1	4.0	17.0	50.0ab	1.31ab	1.12ab	0.110ab
		—	—	—	—	—	—
Outcrossing-animal	66	4.6	19.5	47.6a	1.72a	1.22a	0.163a
		(0.8)	(1.0)	(2.7)	(0.05)	(0.02)	(0.011)
Outcrossing-wind	120	11.3	17.5	53.0a	1.84a	1.21a	0.154a
		(2.0)	(0.8)	(2.2)	(0.05)	(0.01)	(0.008)
Seed dispersal				*	NS	NS	**
Gravity	40	7.8	20.8	47.6ab	1.69ab	1.20ab	0.141bc
		(1.5)	(1.1)	(3.2)	(0.06)	(0.02)	(0.012)
Gravity-attached	11	7.3	16.8	33.8bc	1.57ab	1.16b	0.104bc
		(2.0)	(1.6)	(4.4)	(0.10)	(0.04)	(0.017)
Attached	7	5.1	18.5	48.1abc	1.94a	1.17ab	0.144abc
		(2.3)	(1.2)	(8.9)	(0.10)	(0.35)	(0.027)
Explosive	6	2.3	17.0	26.3c	1.26b	1.09b	0.072c
		(1.0)	(5.0)	(9.7)	(0.13)	(0.05)	(0.036)

Table 3. (continued)

Categories	N^1	Mean no. populations	Mean no. loci	P_p	A_p	A_{ep}	H_{ep}
Ingested	18	21.0	14.0	60.3a	1.90a	1.29a	0.208a
		(11.7)	(1.8)	(4.7)	(0.12)	(0.05)	(0.022)
Wind	117	8.4	18.3	50.9a	1.79a	1.20b	0.149b
		(0.9)	(0.9)	(2.5)	(0.05)	(0.01)	(0.008)
Mode of reproduction				*	*	**	**
Sexual	191	7.6	18.8	48.7b	1.75b	1.20b	0.144b
		(0.7)	(0.6)	(1.8)	(0.04)	(0.01)	(0.006)
Sexual and asexual	7	45.2	8.7	67.7a	2.15a	1.37a	0.251a
		(25.1)	(2.1)	(8.0)	(0.21)	(0.07)	(0.037)
Successional Status				NS	NS	NS	NS
Early	3	2.7	18.1	56.9a	1.75a	—	—
		(1.1)	(2.8)	(11.6)	(0.19)		
Mid	61	7.4	19.5	50.9a	1.81a	1.19a	0.152a
		(0.8)	(1.3)	(3.0)	(0.06)	(0.02)	(0.010)
Late	135	10.3	17.6	48.3a	1.74a	1.21a	0.146a
		(1.8)	(0.7)	(2.2)	(0.05)	(0.01)	(0.008)

[1] N, number of entries; P_p, percent polymorphic loci; A_p, number of alleles per locus; A_{ep}, effective number of alleles per locus; H_{ep}, genetic diversity. See text for more complete definitions.

[2] Means followed by the same letter in a column are not significantly different at the 5% probability level.

Levels of significance: *, $p < 0.05$; **, $p < 0.01$; ***, $p < 0.001$; NS, not significant.

status were significantly different for the proportion of polymorphic loci and the number of alleles per locus while differences among categories of seed dispersal were significant for the proportion of polymorphic loci and genetic diversity. None of the genetic parameters was significantly different among the three successional stage categories.

Gymnosperms had higher percent polymorphic loci and mean number of alleles per locus than angiosperms (Table 3). Gymnosperms also had slightly more (6%) genetic diversity. Gymnosperms, thus, tend to have more loci with a common allele and several low frequency alleles in their populations than angiosperms.

Of the four geographic range categories, endemic species had the lowest levels of genetic diversity (Table 3). Genetic diversity of the endemic species was 25% of the value for the widespread species and 39% and 29% of the values for the narrowly and regionally distributed species, respectively. The more widespread species had more polymorphic loci, more alleles per polymorphic locus and allele frequencies were less skewed.

There were three significance classes for regional distribution: the first represented the boreal-temperate species; the second, temperate and temperate-tropical species; and the third, tropical species. Genetic diversity in populations of species from lower latitudes ranged from 61% to 71% of the genetic diversity in populations of the boreal-temperate trees.

There were no selfing trees in the population data set and there was only a single entry with a mixed mating, wind-pollinated breeding system. As a result, comparisons could only be made between species from the three remaining categories. The mixed mating, animal-pollinated species had low values for all genetic parameters. The genetic diversity value for this category was 23% of the value for the outcrossing wind-pollinated species and 21% of that of the outcrossing animal-pollinated species. The animal-pollinated outcrossed species had somewhat lower percent polymorphic loci and slightly higher genetic diversity indicating that species in this category have alleles with more equal frequencies. The number of alleles per polymorphic locus was virtually equal (2.53 vs. 2.58).

Tree species whose seeds are ingested by animals maintain significantly more genetic diversity at the population level than species with other forms of seed dispersal (Table 3). Trees with explosively dispersed seeds, with seeds dispersed by gravity, or with seeds dispersed by attachment to animals have much lower levels of genetic diversity (35—50% of the values of species with ingested seeds). Most of the differences in genetic diversity among categories are due to differences in the proportion of polymorphic loci.

Sample sizes in the two mode of reproduction categories are quite different. However, mean values for the four genetic parameters are significantly higher for species that reproduce both sexually and asexually. Genetic diversity for the sexual species is 57% of the mean value for species that utilize both mechanisms of reproduction. Significant differences for the four genetic parameters were not seen between species of different stages of succession.

Based on the GLM analyses, the seven traits explained 38.6% of the variation in population genetic diversity. Geographic range accounted for the majority of the variation (58.9%) among species explained by the model. Seed dispersal (12.9%), regional distribution (13.8%) and breeding system (9.5%) added significant but low proportions of the explained variation.

Variation among populations of woody plants

Relatively little variation occurred among categories of the seven traits for total genetic diversity at polymorphic loci (H_T, Table 4). Only differences among the categories of seed dispersal mechanism, mode of reproduction and successional status were significantly ($p < 0.05$) associated with total genetic diversity. Total genetic diversity (H_T) is dependent on the number of alleles and the evenness of allele frequencies at polymorphic loci. Thus, since H_T is a species-wide measure of genetic diversity at polymorphic loci, the higher levels of significance for the within species genetic diversity based on all loci (H_{es}; Table 2) must be due to differences in the proportion of polymorphic loci.

Differences in mean genetic diversity within and among populations indicate that woody species with different combinations of characteristics partition allozyme variation in different ways. The most significant differences in genetic diversity at polymorphic loci within populations (H_s, Table 4) occurred between species with different geographic ranges, seed dispersal mechanisms, modes of reproduction and successional status. Endemic species had a mean within population diversity (H_s) that was 49% of that of the widespread species and 56% of the value for regional species. Species whose seeds are dispersed by both gravity and animal attachment had mean genetic diversity values which were 53% of species with ingested seed and 74% of species with wind-dispersed seeds. Sexually reproducing species had a mean diversity within populations that was 69% of the mean of species with both sexual and asexual reproduction, while mid-successional species had a mean within population diversity value that was 83% of the value for late successional species (Table 4).

Significant differences in genetic diversity among populations were (G_{ST}, Table 4) seen for taxonomic status, geographic range and regional

Table 4. Distribution of allozyme variation among populations of long-lived woody species according to their attributes. Values are based on polymorphic loci only

Categories	N^1	Mean no. populations	Mean no. loci	H_T	H_s	G_{ST}
Taxonomic status				NS	NS	*
Gymnosperms	121	8.9	17.3	0.281a[2]	0.255a	0.073b
		(1.0)	(0.9)	(0.010)	(0.010)	(0.010)
Angiosperms	73	9.5	18.9	0.287a	0.249a	0.102a
		(2.4)	(0.8)	(0.016)	(0.014)	(0.012)
Geographic range				NS	***	***
Endemic	18	8.5	21.8	0.232a	0.150c	0.141a
		(2.1)	(2.0)	(0.051)	(0.029)	(0.049)
Narrow	40	5.3	21.1	0.267a	0.225b	0.124a
		(0.7)	(1.0)	(0.016)	(0.015)	(0.016)
Regional	127	11.6	16.8	0.292a	0.270a	0.065b
		(2.2)	(0.8)	(0.010)	(0.010)	(0.006)
Widespread	9	5.4	14.0	0.316a	0.306a	0.033b
		(1.3)	(1.7)	(0.016)	(0.015)	(0.009)
Regional distribution				NS	NS	**
Boreal-temperate	40	8.0	15.8	0.280a	0.268a	0.038b
		(1.2)	(1.6)	(0.015)	(0.014)	(0.006)
Temperate	125	12.2	17.7	0.287a	0.252a	0.092a
		(2.1)	(0.8)	(0.011)	(0.011)	(0.010)
Temperate-tropical	3	1.5	16.6	0.248a	0.216a	0.109ab
		(0.4)	(1.6)	(0.021)	(0.038)	(0.052)

Table 4. (Continued)

Categories	N^1	Mean no. populations	Mean no. loci	H_T	H_s	G_{ST}
Tropical	26	2.9	21.3	0.275a	0.235a	0.119a
		(0.4)	(1.4)	(0.031)	(0.023)	(0.025)
Breeding system				NS	NS	NS
Selfing	0	—	—	—	—	—
Mixed-animal	12	11.8	18.7	0.266a	0.179b	0.122a
		(3.2)	(2.7)	(0.076)	(0.042)	(0.038)
Mixed-wind	0	—	—	—	—	—
Outcrossing-animal	37	4.6	19.5	0.304a	0.268a	0.099a
		(0.8)	(1.0)	(0.018)	(0.017)	(0.017)
Outcrossing-wind	146	11.3	17.5	0.279a	0.254a	0.077a
		(2.0)	(0.8)	(0.009)	(0.009)	(0.009)
Seed dispersal				*	**	NS
Gravity	30	7.8	20.8	0.286b	0.231bc	0.131a
		(1.5)	(1.1)	(0.020)	(0.018)	(0.022)
Gravity-attached	12	7.3	16.8	0.203c	0.187c	0.099ab
		(2.0)	(1.6)	(0.031)	(0.028)	(0.024)
Attached	6	5.1	18.5	0.296abc	0.272abc	0.065ab
		(2.3)	(1.2)	(0.028)	(0.029)	(0.035)
Explosive	1	2.3	17.0	0.328abc	0.245abc	0.092ab
		(1.0)	(5.0)	—	—	—

Table 4. (Continued)

Categories	N[1]	Mean no. populations	Mean no. loci	H_T	H_s	G_{ST}
Ingested	14	21.0	14.0	0.371a	0.352a	0.051b
		(11.7)	(1.8)	(0.032)	(0.033)	(0.010)
Wind	131	8.4	18.3	0.280b	0.252b	0.076b
		(0.9)	(0.9)	(0.011)	(0.010)	(0.009)
<u>Mode of reproduction</u>				*	**	NS
Sexual	184	7.6	18.8	0.278b	0.247b	0.086a
		(0.7)	(0.6)	(0.009)	(0.008)	(0.008)
Sexual and asexual	10	45.2	8.7	0.376a	0.356a	0.051a
		(25.1)	(2.1)	(0.035)	(0.037)	(0.013)
<u>Successional status</u>				*	**	NS
Early	—	—	—	—	—	—
Mid	58	7.4	19.5	0.237b	0.221b	0.095a
		(0.81)	(1.31)	(0.013)	(0.012)	(0.020)
Late	137	10.3	17.6	0.295a	0.267a	0.080a
		(1.8)	(0.7)	(0.011)	(0.010)	(0.007)

[1] N, number of entries; H_T, total genetic diversity; H_s, genetic diversity within populations; G_{ST}, proportion of the total diversity among populations. See text for more complete definitions.

[2] Means followed by the same letter in a column are not significantly different at the 5% probability level.

Levels of significance: *, $p < 0.05$; **, $p < 0.01$; ***, $p < 0.001$; NS, not significant.

Standard errors are in parentheses.

distribution. The largest differences in among population diversity were associated with differences in the geographic range of species. Endemic species had a mean among population diversity more than four times greater than the value for widespread species and more than twice the value for regionally distributed species. Differences among species with different regional distributions were also large, with the genetic diversity among populations of boreal-temperate species averaging 30—40% of the other categories. Although differences were not significant, species with relatively limited gene movement (animal-pollinated species) have higher mean among population diversity than outcrossed, wind-pollinated species. A similar pattern is seen for seed dispersal; species with gravity dispersed seeds have more genetic diversity among populations than species with ingested, attached or wind-dispersed seeds.

The seven traits explained 16.1% of the heterogeneity in genetic diversity among populations. Geographic range accounted for 53% of the variation explained by the model, while seed dispersal and regional distribution explained an additional 16.7% and 12.5%, respectively.

Multivariate analyses

Multivariate analyses were performed on the seven species traits and within species genetic diversity (H_{es}), within population genetic diversity (H_{ep}) and among population diversity (G_{ST}). The seven traits were generally only loosely associated with one another. The highest correlations were between taxonomic status and breeding system (r = 0.65) and taxonomic status and seed dispersal mechanism (r = 0.69). A second group of significant correlations included taxonomic status and regional distribution (r = 0.40), regional distribution and breeding system (r = 0.38), breeding system and geographic range (r = 0.49), breeding system and seed dispersal (r = 0.45), and regional distribution and successional status (r = 0.32).

Variation in genetic diversity within species (H_{es}) was most highly correlated with geographic range (r = 0.37), mode of reproduction (r = 0.22) and breeding system (r = 0.15). In particular, high diversity is associated with large geographic ranges, sexual and asexual reproduction and outcrossing breeding systems.

Genetic diversity within populations (H_{ep}) was most highly correlated with geographic range (r = 0.45), breeding system (r = 0.28), regional distribution (r = 0.22) and mode of reproduction (r = 0.21). Based on the principal components analysis, species with high within population genetic diversity are widely distributed temperate or boreal gymnosperms with outcrossing breeding systems and widespread seed dispersal.

Variation among species in genetic diversity among populations was significantly correlated with geographic range ($r = 0.28$), regional distribution ($r = 0.23$) and seed dispersal ($r = 0.18$). The principal components analysis indicated that higher values of among population diversity were associated with angiosperm species, species from tropical regions, and endemic species with mixed mating, animal-pollinated breeding systems and gravity or gravity-attached seed dispersal mechanisms.

Correlations among genetic parameters

Associations among the genetic diversity measures were examined by calculating a correlation matrix based on individual species values. Genetic diversity within species based on all loci was very highly correlated with genetic diversity within populations ($r = 0.98$; $p < 0.0001$) and more moderately correlated with total and within population diversity based only on polymorphic loci ($r = 0.52$ and r = 0.73, respectively; $p < 0.0001$) but negatively correlated with among population diversity ($r = -0.24$; $p < 0.01$). Measures of within population diversity based on all loci were highly correlated with genetic diversity within populations based on polymorphic loci ($r = 0.68$; $p < 0.0001$) and more moderately correlated with total diversity (H_T) and among population diversity ($r = 0.46$; $p < 0.0001$, and $r = -0.28$; $p < 0.001$). Total genetic diversity at polymorphic loci was positively correlated with within population diversity ($r = 0.89$; $p < 0.0001$) and weakly correlated with variation among populations ($r = 0.13$; $p < 0.10$). Diversity within species based on all loci was negatively correlated with among population diversity ($r = -0.27$; $p < 0.0001$). The correlation between diversity within species based on all loci and total diversity at polymorphic loci is relatively low suggesting that much of the genetic diversity variation among species results from differences in the proportion of polymorphic loci rather than the number and frequency of alleles at polymorphic loci. Negative correlations between genetic diversity among populations and the two within population parameters (H_{ep} and H_s) indicate that differences among species in genetic diversity within populations are a function of genetic diversity within species and how this variation is partitioned among populations.

Discussion

Levels of variation maintained within species and populations are highly heterogeneous among woody plant taxa. Genetic diversity within species ranged from approximately 0.00 to 0.40. Species with high levels of genetic diversity included *Alseis blackiana* (Hamrick and Loveless 1989), *Picea glauca* (Tremblay and Simon 1989), *Robinia pseudoacacia* (Surles et al. 1989) and *Pinus sylvestris* (Kinloch et al. 1986). Species with exceptionally low levels of diversity were *Acacia mangium* (Moran et al. 1989), *Pinus resinosa* (Simon et al. 1986), *Pinus torreyana* (Ledig and Conkle 1983) and *Populus balsamea* (Farmer et al. 1988). Within population genetic diversity ranged from 0.00 to 0.35 (see examples above). Genetic diversity among populations nearly covered the possible range with some species having almost no variation among populations (e.g., *Acer saccharum* (Perry and Knowles 1989) and *Pinus banksiana* (Dancik and Yeh 1983) to a few having fifty percent or more of their genetic diversity at polymorphic loci among populations (e.g., *Acacia mangium* (Moran et al. 1989) and *Pinus torreyana* (Ledig and Conkle 1983). Most of this heterogeneity is at the species level as much less heterogeneity in these parameters is seen among genera (Table 5).

Even with such heterogeneity, certain generalizations can be made. First, long-lived woody plant species maintain, on average, higher levels of allozyme variation within their populations than other plant species. The average genetic diversity within populations of woody plants (0.148) is 31% higher than the average for all plant taxa (0.113) and 46% and 51% higher than the mean for annual (0.101) and perennial herbaceous species (0.098). Plants in general, and woody plants in particular, maintain more diversity within their populations than either invertebrates (H_{ep} = 0.100) or vertebrates (H_{ep} = 0.054; Nevo et al. 1984). Comparisons of genetic diversity within plant and animal species are not possible since, to our knowledge, reviews of within species diversity are not available in the animal literature.

Although the number of allozyme studies of woody plants has increased by more than 50 percent over the last two years there were only minor differences in the means of the genetic parameters produced by this review and those of Hamrick and Godt (1989). For example, mean genetic diversity within long-lived woody species was identical (0.177), while the within population value changed very little (0.148 vs. 0.149). In this study the mean genetic diversity among populations was somewhat higher (0.084 vs. 0.076). This is due to increased data from angiosperm species, which tend to have more among population diversity than gymnosperms.

Concern may be expressed that the number of loci and populations included in studies might affect estimates of genetic diversity. In response, we tested the relationships between measures of genetic diversity within and among species and populations and the number of populations and loci that were utilized. Correlations between the number of populations

Table 5. Genetic diversity within species and within and among populations for several genera of woody plants

Genera	N^1	Mean no. populations	Mean no. loci	H_{es}	H_{ep}	G_{ST}
Gymnosperms						
Abies	7	5.4	13.6	0.145	0.130	0.063
		(2.0)	(3.2)	(0.013)	(0.007)	(0.019)
Picea	28	9.4	12.2	0.219	0.218	0.055
		(2.6)	(1.6)	(0.020)	(0.020)	(0.011)
Pinus	93	7.8	19.9	0.157	0.136	0.065
		(0.9)	(1.3)	(0.010)	(0.009)	(0.008)
Pseudotsuga	11	15.5	15.5	0.201	0.163	0.074
		(9.1)	(3.0)	(0.026)	(0.012)	(0.029)
Angiosperms						
Acacia	13	2.6	21.4	0.125	0.096	0.206
		(1.0)	(1.8)	(0.034)	(0.019)	(0.099)
Eucalyptus	14	8.8	14.2	0.187	0.172	0.169
		(1.7)	(1.7)	(0.028)	(0.023)	(0.034)
Populus	10	9.0	24.0	0.161	0.154	0.041
		(2.5)	(3.8)	(0.035)	(0.067)	(0.019)
Prosopsis	13	1.2	17.5	0.186	0.098	0.072
		(0.1)	(1.2)	(0.041)	(0.012)	(0.008)
Quercus	28	5.5	17.5	0.186	0.109	0.107
		(2.0)	(0.7)	(0.015)	(0.017)	(0.028)
Tetramolopium	6	3.2	22.0	0.013	0.008	0.199
		(0.9)	(0.0)	(0.005)	(0.002)	(0.087)

[1] N, number of entries; H_{es}, genetic diversity within species; H_{ep}, genetic diversity within populations; G_{ST}, proportion of the total diversity among populations.

Standard errors are in parentheses.

analyzed and the three genetic diversity measures (H_{es}, H_{ep} and G_{ST}) were not significant and accounted for 0.6%, 0.1% and <0.1% respectively, of the variation in these parameters among species. Correlations between the number of loci analyzed and the three genetic diversity measures explained 2.6%, 0.1% and <0.1% of the variation among species for these parameters. The correlation between genetic diversity within species and the number of loci analyzed was significant ($p < 0.02$) while the other two correlations were not significant.

The second major generalization concerns the relationship between genetic diversity and species characteristics. Earlier reviews based on all seed plants found that long-lived, outcrossing, wind-pollinated species of the late stages of succession have higher levels of allozyme diversity within populations and less among population diversity than species with other combinations of traits (Hamrick et al. 1979; Brown 1979; Gottlieb 1981; Loveless and Hamrick 1984; Hamrick and Godt 1989). The results of this paper are generally the same: woody species with widespread distributions, outcrossing breeding systems, widely dispersed seeds, and with both sexual and asexual reproduction tend to have more genetic diversity within species and populations and less variation among populations than species with other combinations of traits. Our results differ from those of Ellstrand and Roose (1987) who found no differences between predominately or strictly clonally reproducing species and sexually reproducing species for genetic diversity within populations. Although the number of woody species in our sample that rely primarily on vegetative reproduction is low, it appears that woody species with asexual reproduction maintain more genetic diversity than those with sexual reproduction. The greater potential longevity of genotypes of species with asexual reproduction and the resistance to genetic drift that this longevity provides may partially explain these results.

The seven ecological and life history traits explained significant amounts of the interspecific variation in species, population and among population diversity for woody plant taxa. Over thirty percent of the variation among species was explained by the seven traits for genetic diversity within species and populations. A similar percent of the variation (24—28%) was explained when eight traits (life form was included) were considered for all plant taxa (Hamrick and Godt 1989). In contrast, only 16% of the variation in G_{ST} values among woody taxa was explained by the seven traits whereas 49% of the variation in diversity among populations was explained by eight traits when all plant taxa were included (Hamrick and Godt 1989). In the previous study breeding system and life form accounted for the majority of the variation in among population diversity when all plant taxa where included (Hamrick and Godt 1989). Since the current study was limited to woody species, life form was not included in the analyses. Also, woody species have a more limited range of breeding systems than all the seed plants. For example, no selfing woody species have been analyzed electrophoretically and only 12 of 195 entries in the G_{ST} analysis were classified as species with mixed mating systems. The limited range of breeding systems examined (75% of the entries were wind-pollinated, outcrossing species) allowed less discrimination in the analysis of interspecific variation in G_{ST} values.

Geographic range is the best predictor of levels of allozyme variation in long-lived woody species. Endemic woody species have the least genetic diversity whereas regionally distributed and widespread species maintain the most diversity. The breeding system of the species also provides some information concerning diversity within species, a result that is consistent with more comprehensive reviews (Hamrick and Godt 1989). Endemic species, and species with limited gene movement, should have smaller effective population sizes and should experience greater losses of genetic variation through drift. Thus, such species are expected to have fewer polymorphic loci and lower genetic diversities. More widespread species, on the other hand, may have historically consisted of large, continuous populations that were less susceptible to the random loss of genetic variation.

Genetic diversity within populations is also influenced by the geographic range of the species. Seed dispersal and breeding system have some additional predictive value, however. Widespread species have higher proportions of polymorphic loci and higher gene diversities than geographically limited species. Outcrossed species and species with animal-dispersed propagules also tend to have more within population diversity. Species with limited geographic ranges often occur in small isolated populations where gene flow is limited and, as a result, genetic variation within populations is reduced. Breeding systems that limit gene movement should also cause a reduction in genetic diversity within populations.

Geographic range is also the best predictor of genetic variation among populations. Endemic woody species have twice the genetic diversity among populations as more widespread species. Genetic structure in endemic species is probably highly influenced by the effects of genetic drift in small populations and population divergence due to limited gene flow among isolated populations. A graphic example is provided by the pines. Pine species that are distributed as scattered isolated populations have more genetic diversity among populations (e.g., *P. torreyana* (G_{ST} = 1.00, Ledig and Conkle 1983); *P. halepensis* (G_{ST} = 0.30, Scheller et al. 1985), and *P. muricata* (G_{ST} = 0.20, Millar 1983)) while more widespread and continuously distributed pines have less among-population diversity (e.g., *P. banksiana* (G_{ST} = 0.030, Dancik and Yeh 1983), *P. contorta* (G_{ST} = 0.036, Wheeler and Guries 1982), and *P. ponderosa* (G_{ST} = 0.015, Hamrick et al. 1989)).

Rather surprisingly, breeding systems play a relatively small role in predicting the level of genetic diversity among populations of woody plant species. This result contrasts markedly with that of Hamrick and Godt (1989). This difference may be due to the more limited range of breeding systems represented in this sample of woody taxa, as discussed previously. In addition, breeding systems were significantly correlated with taxonomic status (all conifers are outcrossing and wind-pollinated), regional distribution (most boreal-temperate trees analyzed are conifers), seed dispersal, and geographic range. As a result, breeding systems do not provide much additional predictive power for woody species. The association between geographic range and breeding systems was somewhat unexpected. Species with outcrossing breeding systems had more widespread distributions, while species with mixed mating systems were more geographically restricted. Island endemic species belonging to the genera *Dendroseris* and *Tetramolopium* represent many of the species in the mixed mating, animal-pollinated category and perhaps bias the data. If data from a larger number of mixed mating species were available, the apparent correlation between breeding system and geographic range might disappear.

Factors such as isolation, small populations and gene flow, all of which directly affect effective population sizes, may have a major influence on the levels of genetic diversity observed within and among populations of woody species. These same factors are also related to the total amount of genetic diversity maintained by species, as evidenced by the negative correlation between genetic diversity within and among populations. While the structuring of diversity within and among populations is expected to be related to effective population size and gene flow, it is less clear just how these factors might affect diversity within species. It could be argued that in species with limited gene exchange, new mutations have a lower probability of being incorporated into additional populations and as a result, are more likely to be lost. This would lead to fewer polymorphic loci, fewer alleles per locus and ultimately to lower genetic diversity in the species. This conclusion is supported by the significant negative correlations ($p < 0.001$) between diversity among populations and the proportion of polymorphic loci within species and populations ($r = -0.30$ and $r = -0.45$, respectively).

The question remains whether woody species maintain more genetic diversity because they often combine characteristics that ensure the preservation of genetic diversity (e.g., large continuous populations, large size, long lives, outcrossing breeding systems, and relatively long distance pollen and seed movement) or whether there is some inherent quality of woody plants that selects for increased levels of diversity (Hamrick 1978). Although we can not answer this question directly, the heterogeneity in genetic diversity among woody taxa suggests that there is nothing inherent in woody species that necessitates high levels of genetic diversity. Woody species are merely more likely than herbaceous species to combine those life history and population traits that act to preserve genetic diversity.

A goal of this review was to determine if genetic diversity within the species, and within and among populations was associated with the life history and ecological characteristics of woody species. We have shown that a significant amount of the genetic heterogeneity among species is associated with these traits. However, for all three measures of genetic diversity, life history and ecological traits explain less than forty percent of the variation among species. Other factors must, therefore, play major roles in determining the levels and distribution of genetic diversity. Certainly the evolutionary history of species must have a major influence on genetic diversity. Woody species that originated from a limited number of individuals or have spent a considerable proportion of their evolutionary history as small, isolated populations should have less variation at the species and population level and more variation among populations than species with a history of larger, more continuously distributed populations. Also, species that have recently experienced a series of severe population bottlenecks, but today are more widespread, might have less genetic diversity. Thus, while life-history and ecological characteristics of woody plant species play a significant role in shaping the present-day genetic structure of species and their populations, the evolutionary history of the species probably plays an equally significant role.

Acknowledgments

We are grateful to Jenna Hamrick, Lorrel Manning and Tracie Plaster for their assistance on this project and to Becky Hendren, Carla Ingram and Kim Aldridge for typing the manuscript. Portions of this work were supported by NSF Grants BSR8718803 and BSR8918420.

References

Brown, A. H. D. 1979. Enzyme polymorphism in plant populations. Theor. Pop. Biol. 15: 1—42.

Dancik, B. P. and Yeh, F. C. 1983. Allozyme variability and evolution of lodgepole pine (*Pinus contorta* var. *latifolia*) and jack pine (*Pinus banksiana*) in Alberta. Can. J. Genet. Cytol. 25: 57—64.

Ellstrand, N. C. and Roose, M. L. 1987. Patterns of genotypic diversity in clonal plant species. Am. J. Bot. 74: 123—131.

Farmer, R. E., Cheliak, W. M., Perry, D. J., Knowles, P., Barrett, J. and Pitel, J. A. 1988. Isozyme variation in balsam poplar along a latitudinal transect in northwestern Ontario. Can. J. For. Res. 18: 1078—1081.

Gottlieb, L. D. 1981. Electrophoretic evidence and plant populations. Prog. Phytochem. 7: 1—46.

Hamrick, J. L. 1976. Variation and selection in western montane species. II. Variation between and within stands of white fir on an elevational transect. Theor. Appl. Genet. 48: 27—34.

——. 1978. Genetic variation and longevity. pp. 84—113. In: Solbrig, O.T., Jain, S., Johnson, G. B. and P. H. Raven (Eds) Plant Population Biology. Columbia University Press, New York.

Hamrick, J. L. and Loveless, M. D. 1989. The genetic structure of tropical tree populations: Associations with reproductive biology. pp. 129—146. In: Bock, J. and Linhart, Y. B. (Eds) The Evolutionary Ecology of Plants. Westview Press, Boulder, Colorado.

Hamrick, J. L. and Godt, M. J. W. 1989. Allozyme diversity in plant species. pp. 43—63. In: Brown, A. H. D., Clegg, M. T., Kahler, A. L. and Weir, B. S. (Eds) Plant Population Genetics, Breeding and Genetic Resources. Sinauer Press, Sunderland, Mass.

Hamrick, J. L., Blanton, H. M., and Hamrick, K. J. 1989. Genetic structure of geographically marginal populations of ponderosa pine. Am. J. Bot. 76: 1559—1568.

Hamrick, J. L., Linhart, Y. B. and Mitton, J. B. 1979. Relationships between life history characteristics and electrophoretically-detectable genetic variation in plants. Ann. Rev. Ecol. Syst. 10: 173—200.

Kinloch, B. B. and Stonecypher, R. W. 1969. Genetic variation in susceptibility to fusiform rust in seedlings from a wild population of loblolly pine. Phytopathology 59: 1246—1255.

Kinloch, B. B., Westfall, R. D. and Forrest, G. I. 1986. Caledonian Scots pine: Origins and genetic structure. New Phytol. 104: 703—729.

Ledig, F. T. and Conkle, M. T. 1983. Gene diversity and genetic structure in a narrow endemic, Torrey pine (*Pinus torreyana* Parry ex Carr). Evolution 37: 79—85.

Libby, W. J., Stettler, R. F. and Setz, F. W. 1969. Forest genetics and forest tree breeding. Ann. Rev. Genet. 3: 469—494.

Loveless, M. D. and Hamrick, J. L. 1984. Ecological determinants of genetic structure in plant populations. Ann. Rev. Ecol. Syst. 15: 65—95.

Millar, C. I. 1983. A steep cline in *Pinus muricata*. Evolution 37: 311—319.

Moran, G. F., Muona, O. and Bell, J. C. 1989. *Acacia mangium*: A tropical forest tree of the coastal lowlands with low genetic diversity. Evolution 43: 231—235.

Nevo, E., Beiles, A. and Ben-Shlomo, R. 1984. The evolutionary significance of genetic diversity: Ecological, demographic and life history correlates. pp. 13—123. In: Mani, G. S. (Ed) Evolutionary Dynamics of Genetic Diversity. Lecture Notes in Biomathematics Vol. 53. Springer-Verlag, New York.

Nei, M. 1973. Analysis of gene diversity in subdivided populations. Proc. Natl. Acad. Sci. 70: 3321—3323.

Perry, D. J. and Knowles, P. 1989. Allozyme variation in sugar maple at the northern limits of its range in Ontario, Canada. Can. J. For. Res. 19: 509—514.

SAS Institute Inc. SAS/STAT TM Guide for Personal Computers. Version 6 Edition. 1987. SAS Institute Inc., Cary, N.C. 1028 pp.

Scheller, G., Conkle, M. T. and Griswald, L. 1985. Local differentiation among Mediterranean populations of Aleppo pine in their isozymes. Silv. Genet. 35: 11—19.

Simon, J. P., Bergeron, Y. and Gagnon, D. 1986. Isozyme uniformity in populations of red pine (*Pinus resinosa*) in the Abitibi Region, Quebec. Can. J. For. Res. 16: 1133—1135.

Stonecypher, R. W., Zobel, B. J. and Blair, R. 1973. Inheritance patterns of loblolly pine from a nonselected natural population. N.C. Agric. Exper. Station Tech. Bull. No. 220. 60 pp.

Surles, S. E., Hamrick, J. L. and Bongarten, B. C. 1989. Allozyme variation in black locust (*Robinia pseudoacacia*). Can. J. For. Res. 16: 471—479.

Tremblay, M. and Simon, J. P. 1989. Genetic structure of marginal populations of white spruce (*Picea glauca*) at its northern limits of distribution in northern Quebec. Can. J. For. Res. 19: 1371—1379.

Weir, B. S. 1989. Sampling properties of gene diversity. pp. 23—42. In: Brown, A.D.H., Clegg, M. T., Kahler, A. L. and B. S. Weir (Eds) Plant Population Genetics, Breeding and Genetic Resources. Sinauer Press, Sunderland, Mass.

Wells, O. O. and Snyder E. B. 1976. Longleaf pine half-sib progeny test. For. Sci. 22: 404—406.

Wheeler, N. C. and Guries, R. P. 1982. Population structure, genetic diversity and morphological variation in *Pinus contorta* Dougl. Can. J. For. Res. 12: 595—606.

New Forests **6**: 125–158, 1992.

Review paper

Biochemical and molecular genetic markers in biosystematic studies of forest trees

S. H. STRAUSS,[1] J. BOUSQUET,[1,2] V. D. HIPKINS[1] and Y.-P. HONG[1]

[1] *Department of Forest Science, College of Forestry, Peavy Hall 154, Oregon State University, Corvallis, OR 97331–5705, USA;* ([2] *Present address: Centre de Recherche en Biologie Forestière, Faculté de Foresterie et de Géomatique, Université Laval, Ste-Foy (Québec), Canada G1K 7P4)*

Received 25 October 1990; accepted 11 March 1991

Key words: phylogeny, allozyme, conifer, clustering, terpene, DNA, RFLP

Application. Evolutionary history may affect the genetic architecture of adaptation. It is therefore important to separate legacy from current adaptedness when defining seed transfer rules. Biochemical and molecular genetic markers are the most powerful tools available for understanding evolutionary relationships and biogeographic histories.

Abstract. Biochemical and molecular markers have proven to be powerful tools for discerning biosystematic, biogeographic, and phylogenetic relationships. Biosystematic information can be important for guiding traditional breeding programs, gene transfer, interspecific hybridization, and gene conservation. A phylogenetic framework is usually necessary, but frequently ignored, for making valid statistical tests in studies of adaptive evolution. Several studies have indicated a strong correlation between biochemical "races" and traits important to growth and adaptation, suggesting that evolutionary legacy may affect genetic architecture of fitness traits — with consequences for seed transfer, breeding strategies, and tolerance of climate change. A number of methods for phylogenetic analysis exist, but differ in their assumptions. Use of an inappropriate method — such as a method that assumes constant rates of evolution when rates in fact vary — can lead to incorrect phylogenies. Because of their complexity, phylogenetic topologies are often difficult to determine unambiguously; estimates of statistical confidence should therefore accompany phylogenetic trees if they are to be regarded as providing new knowledge, or strong confirmation, of relationships. Molecular genetic markers are more expensive than biochemical markers such as allozymes and terpenes, but they provide increased accuracy and expanded scope of biosystematic inference, and facilitate statistical analyses of phylogenetic trees.

Abbreviations. BMGMs — biochemical and molecular genetic markers, cpDNA — chloroplast DNA, OTU — operational taxonomic unit, *rbc*L — chloroplast gene encoding ribulose 1,5-bisphosphate carboxylase large subunit, *rbc*S — nuclear gene encoding ribulose 1,5-bisphosphate carboxylase small subunit, rDNA — ribosomal DNA, RFLP — restriction fragment length polymorphism, UPGMA — unweighted pair group method of clustering using averages

This is paper 2708 of the Forest Research Laboratory, Oregon State University.

Introduction

Goals and scope of paper

BMGMs have been widely used to study to the evolution of forest trees. The goals of this paper are to review work to date in this area, focusing on:

- theory and methods for analysis;
- contributions made by BMGMs to knowledge of relationships; and
- challenges for future work.

Undertaking a literature review of this area, particularly for conifers, is a major task; nearly every survey of genetic variability has included some consideration of relationships — usually via calculation of genetic distances — among the taxa studied. To make our job manageable, we focus primarily on those studies for which biosystematics was a main goal. For the most part, this corresponded to the production of an explicit hypothesis of relationships as depicted in some kind of dendrogram or multivariate groupings. We also focus mainly on allozyme and DNA markers, providing only a brief review of terpenoids and other secondary biochemicals. We proceed by:

- defining terms;
- discussing the importance of biosystematic information;
- contrasting the various biosystematic characters, emphasizing comparison of BMGMs to other traits;
- surveying quantitative methods for biosystematic analysis, emphasizing construction of phylogenetic trees; and
- reviewing biosystematic studies of forest trees that employed BMGMs, first considering studies of gymnosperms and then angiosperms.

Defining terms and concepts

Biochemical and molecular markers

As we argue below when considering the advantages and disadvantages of BMGMs, there is an important distinction between the nucleotide sequence information in the hereditary material, as reflected in DNA or unedited, unprocessed RNA sequences, and the expression of that material, as reflected in methylation of DNA, splicing of RNA, protein sequences, and cellular biochemicals other than nucleic acids and proteins. We define molecular genetic markers as those derived from direct analysis of genetic polymorphism in DNA sequences. Biochemical markers are those derived from study of the chemical products of gene expression, such as protein sequences or net charges, and composition of secondary chemicals such as terpenoids. Because of the confusion surrounding use of the term "molecular," which has frequently (but inconsistently) been used to refer to DNA but not to molecules such as enzymes, we suggest adoption of the term "molecular *genetic* marker," and abandonment of "molecular marker," when referring to direct analysis of genetic polymorphism in DNA sequences. "Molecular genetic marker" connotes study of DNA, whereas "molecular marker" connotes study of any cellular molecule.

Biosystematics, biogeography, and phylogenetics
These terms have been used in a variety of ways, and overlap in their meanings. Taken as a whole, we consider them to refer to *the migrational history and evolutionary relationships of populations and species.* Taken individually, the broadest of the three terms is biosystematics, which has been defined as the study of organismic diversity in relation to evolutionary relationships among taxa such as populations or species (Wiley 1981). In this paper, we use biosystematics as a shorthand term for referring to the entire area of biosystematics, biogeography, and phylogenetics.

Importance of biosystematic studies

No scientist would question the importance of taxonomy. Without a system of nomenclature, identification, and classification, science cannot proceed. Because of the need for an expedient and practical system, the emphasis in taxonomy is on identifying and classifying, often with no assumptions regarding phylogenetic relationships. Many biologists, however, confuse taxonomy, which is usually highly subjective and pragmatic in nature, with biosystematics, which strives to be an objective and quantitative evolutionary science. From a biosystematist's point of view, most taxonomic classifications can be viewed as hypotheses of evolutionary relationships awaiting testing. But biosystematists seek not only to test the naturalness of taxonomic classifications, but to embellish them by adding information about ancestor/descendent relationships among taxa, relative amounts of divergence, and times of origin. Such rigor is highly desirable if phylogenetic information is to be useful for plant improvement, and if it is to serve as a framework for studying the evolution of adaptive characters. It is in providing evolutionary rigor that BMGMs can make a large contribution.

Why should scientists concerned with plant improvement be interested in biosystematics? Plant breeders wish to have as accurate information as possible about relationships as a guide for planning intraspecific and interspecific hybridization programs. Heterosis is usually greatest in wide, but not too distant, crosses (Falconer 1981). For introgressive hybridization, breeders desire species or populations with unique adaptive or economic qualities that are as little diverged from the crop species as possible. The probability of meiotic problems such as segregation distortion, and developmental problems such as embryo abortion, increases with time of evolutionary divergence (e.g., Saylor and Smith 1966).

Plant biotechnology can also benefit from biosystematic knowledge. Programs of somatic hybridization are likely to be successful where the species to be crossed are as closely related as possible — and therefore likely to have compatible organelles, and nuclear genomes that undergo mitosis and meiosis normally. Gene transfer is likely to be successful when inserted genes and promoters are as similar as possible to native ones. This will become increasingly important as the field of genetic engineering matures, and today's predominantly constitutive promoters are replaced by ones designed to direct complex patterns of gene expression in response to environmental and developmental cues. It is well known that many genes are not expressed, or are expressed improperly, when transferred between monocots and dicots (e.g., Bruce et al. 1989). Problems have also been encountered when moving genes between plant genera and families (e.g., Benfey and Chua 1989). Developing accurate sets of "rules" to guide transfer of various classes of genes will be aided by the increasingly clear picture of plant phylogeny provided by BMGMs.

Genetic conservation can be most efficient when informed by knowledge of relationships and levels of genetic variability (Ledig 1988; Millar and Westfall 1992, this issue, pp. 347—371). Efficiency is critical because conservation programs are typically expensive and budgets low. Efforts to conserve germplasm for potential use in presently important species should focus on the most closely related species that possess contrasting adaptive and economic traits, and that have most genetic variability. Within species, efforts should be directed at populations with most BMGM variability — which are often identified in phylogenetic analyses as ancestral (e.g., Conkle et al. 1988; Millar et al. 1988).

Phylogenetic knowledge should serve as the framework for studying the evolution of adaptation at all its levels — ecological through molecular. It is impossible to understand how organisms function, and why they function in one way rather than another, without knowing their history. This point was well developed for ecology by Wanntorp et al. (1990), who discussed several cases of how phylogenetic analyses have informed ecological studies. One example concerned an analysis of why some species of north-temperate oaks retain their leaves after leaf senescence. They pointed out that it was not until the phylogeny of the oaks in question was clarified could a meaningful hypothesis could be formulated. The north-temperate oaks had evolved from subtropical evergreen species; thus, the question should not have been why some oaks retain their leaves, but why evolution of the newly acquired trait of seasonal leaf abscission had not proceeded to completion.

Many ecological and molecular biologists consider phylogenetic information as relatively unimportant; interesting but not crucial to their studies of gene, character, or adaptive evolution. Without a phylogenetic framework, however, most such studies lack a credible means for conducting statistical tests. Felsenstein (1985a) discussed this problem at length, pointing out that because of common history, species or populations cannot be assumed to be statistically independent observations. This means, for example, that the common practice of statistically examining the correlation of two traits (e.g., growth and frost tolerance) in a provenance test is invalid unless racial phylogenetic structure of the species is accounted for. It is also a problem when one tries to test the correlation between genetic traits such as allozyme diversity and species life history attributes, and when one tests genetic correlations among traits in newly formed populations that are composites of germplasm from several distinct areas. To conduct valid statistical tests, one must either be able to justifiably assume that the traits in question are not constrained by their phylogenetic history, or construct tests with reference to a phylogenetic tree. Felsenstein (1985a) and Burt (1989) discussed how a number of statistical contrasts can be extracted by using a phylogenetic tree as a framework. An example of a phylogeny-guided test of an adaptive hypothesis is provided by Sessions and Larson's (1987) study of the relationship between genome size and rate of development in salamanders. BMGMs can help to provide more complete, detailed, and reliable phylogenies, which will help forest biologists to choose the most appropriate species for comparative analysis, and enable them to conduct valid statistical tests.

Comparison of characters for biosystematic analysis

The relative merits of the various characters used in biosystematic analysis have been widely discussed (e.g., Sytsma 1990). Compared to morphological characters, BMGMs are relatively difficult to assay. They are therefore impractical for routine taxonomic purposes such as species identification. However, they have been readily adopted by evolutionary systematists for a number of reasons. First, BMGMs are directly related to genotype, whereas morphological traits are often strongly subject to environmental influences, thus controlled experiments and statistical analyses are required to estimate genetic effects that are unconfounded with environment. Second, because morphological traits are often key parts of the external structure of plants, they can be subject to the direct and variable action of natural selection, and thus are more prone to convergence and variability in evolutionary rate than BMGMs. This is well illustrated by seed and seed-cone structure in pines, which provide major taxonomic characters, yet show much morphological variability of importance to adaptation to fire, seed dispersal, predation, and other factors. Excessive reliance on them has resulted in some major errors in pine taxonomy [discussion in Strauss and Doerksen 1990; also see discussion of cone vs. allozyme analyses by Wheeler and Guries (1982a) and Wheeler et al. (1983) below]. In contrast, BMGMs are probably much less subject to natural selection (Kimura 1983) and thus better indices of the length of time that taxa have been separated. Third, BMGMs — either in the form of gene frequencies or nucleotide substitutions — provide a natural metric for calculation of evolutionary divergence. Morphologic traits, in contrast, are severely constrained by natural selection and ontogenetic programs of development (Gould and Lewontin 1979), and will thus rarely have a simple relationship between changes in their form and time since divergence. Moreover, when studying large evolutionary distances it is frequently difficult to distinguish analogous from evolutionarily homologous morphological structures. Homologous genes, on the other hand, can be readily identified across vast phylogenetic distances. Finally, although BMGMs are not free from convergent evolution, the large numbers of characters available, and knowledge of the mechanisms of molecular evolution, help to overshadow the effects of convergence and allow adjustments that minimize its impacts, respectively. An example of the latter is the practice of excluding or down-weighting transitions, which are subject to rapid evolution and therefore to more frequent homoplasy, relative to transversions (Williams and Fitch 1990).

Prior to the arrival of BMGMs, a number of other traits were used that provided substantial improvement over morphological traits for biosystematic studies. Cytological traits such as chromosome arm lengths and banding patterns are probably less subject to the direct action of natural selection than morphology; however, they may reflect strong architectural constraints imposed by the need for proper mitotic function and coordinated gene expression (Zuckerkandl 1986; Holmquist 1989). Likewise, highly repetitive DNA may be subject to architectural constraints and major genome turnover events such as invasions by transposable elements

(Schmid and Jelinek 1982; Evans et al. 1983; Rose and Doolittle 1983), impairing its value as an indicator of divergence time. Crossability has been useful for making rough assessments of relationships; however, it can be highly erratic (Sytsma 1990). In pines, interpopulation or interracial crosses can be less successful than interspecific or intersubsectional crosses (reviewed by Critchfield 1975).

DNA sequence information is what is actually transmitted among generations; it is therefore potentially the most accurate character for analysis of relationships (Zuckerkandl and Pauling 1965). Biochemical traits, as products of gene expression, are not transmitted, and as part of the phenotype, are more directly subject to natural selection. Terpenoid composition, while strongly inherited, probably plays a role in resistance to herbivory (Hanover 1992, this volume, pp. 159—178). Allozymes have also been argued to have an impact on fitness (Bush and Smouse 1992, this volume, pp. 179—196). By focusing on DNA sequences that are extremely weakly related to fitness (e.g., third codon positions and non-coding DNA), evolutionary inferences may be nearly free of the constraints imposed by natural selection. Moreover, even among nonsynonymous nucleotide changes, many evolutionary events are likely to be only weakly selected — for example, amino acid substitutions that do not alter hydrophobicity or charge, and those in specific regions of the molecule, usually away from the active site and key regulatory regions. Such changes are likely to be motivated primarily by stochastic rather than selective forces.

Other advantages of molecular genetic markers compared to biochemical markers are:

- The greater variety of scales on which evolutionary processes can be studied — which range from DNA fingerprints and large numbers of RFLPs that can differentiate trees uniquely from all others — to coding regions of ribosomal RNA and chloroplast genes, at which given base substitutions may occur only once every 10 billion years (Zurawski and Clegg 1987).
- The ability to readily interpret genotypes from phenotypes. A substantial fraction of allozyme polymorphisms are unrecognized by conventional starch gel electrophoresis (Selander and Whittam 1983; Lewontin 1985). For terpenoids, only genetic changes that substantially change gene expression or biosynthetic enzyme activity will be detected; thus products of many different mutational events will be confounded or undetected. The net result is that phenetic similarity will often be a poor index of genetic similarity, leading to incorrect assessments of homologies. Such errors can greatly weaken and distort

phylogenetic analyses — especially where few genes are studied. In contrast, errors of interpreting homology should be far lower with RFLP analyses and DNA sequencing.

— Analysis of DNA has permitted extensive study of organelle genetic variation, which can provide new insights into phylogenetic histories and genetic architecture compared to those provided by nuclear genes (e.g., Avise et al. 1987). Apart from rare mutation hotspots, the low genetic polymorphism of chloroplast DNA within species facilitates study of interspecific phylogeny; large intraspecific surveys are not required to characterize allele frequencies (reviewed by Strauss et al. 1989).

— Because of the great amount of information that can be provided by study of one or a few genes, a larger number of homologous characters (nucleotides or restriction sites) can be identified with molecular genetic studies than are possible with biochemical traits — which are limited in number (e.g., allozyme loci) and often strongly genetically interdependent (e.g., terpene concentrations). And finally,

— the ability to infer the mutational basis of genetic changes with molecular genetic studies facilitates the use of statistical methods for assessing reliability of phylogenetic trees (discussed below). Such methods require independence of characters (e.g., bootstrap) and sometimes knowledge of the nature of mutational processes (e.g., maximum likelihood) — conditions which are generally more difficult to meet with biochemical characters. Moreover, as described above, knowledge of the nature of mutations can be critical for effectively dealing with convergence.

For the reasons given above, molecular genetic markers are more accurate and sensitive analytic tools than biochemical markers. There are, however, reasons why biochemical markers may be preferable for biosystematic analyses, at least at present. First, molecular markers are still relatively new for forest trees; the technology is not widely available, and its effectiveness incompletely proven. This is particularly true for studies within and among closely related species. At present, the commonly employed DNA markers — ribosomal and chloroplast DNA — code for highly conserved sequences. With the possible exception of ribosomal DNA spacer length variability, allozymes are more polymorphic, and have proven effectiveness for elucidating relationships at these lower systematic levels (e.g., Steinhoff et al. 1983; Wheeler et al. 1983; Millar et al. 1988; Moran et al. 1988; Li and Adams 1989; Lagercrantz and Ryman 1990; Niebling and Conkle 1990). With the ongoing development of single-copy nuclear DNA probes for RFLP mapping, however, highly polymorphic markers will soon become widely available (Neale 1992, this volume, pp. 391—407). Second, many biosystematic problems may be sufficiently simple, or require only approximate answers, that the additional power of molecular genetic markers is not required. And, third, the added cost of laboratory set up and supplies for molecular analyses is a substantial deterrent to their use (Clegg 1989). Costs should decline substantially, however, as the technology matures and the necessary laboratory hardware becomes more commonplace.

Quantitative methods

Along with the development of powerful biochemical and molecular methods over the past 20 years, numerical methods of biosystematic inference have also been revolutionized. We will briefly describe the key methods according to the different philosophies of analysis — multivariate, phenetic, cladistic, and maximum likelihood — focusing on their assumptions and limitations.

Multivariate methods

Multivariate methods such as principal components analysis and canonical analysis have frequently been used to discern geographical trends or subspecific structure within species (Westfall 1992, this volume, pp. 279—309). These methods can both reduce the dimensionality of large data sets, simplifying analysis, and be used to construct composite variables that maximize correlation or discrimination among groups. Analyses can start from either continuous data such as allele frequencies, common with principal components analysis, or from discrete data, such as allelic scores for individual genotypes (Smouse and Neel 1977). When multivariate distances such as Mahalanobis distances are calculated among OTUs, the original multidimensional space delineated by the allelic variables is conserved. With clustering methods, however, the averaging procedures used collapse the data into a single dimension (see below).

Multivariate methods have generally been preferred to clustering methods because of their superior ability to take advantage of correlations among alleles or loci. This would help to detect differences among populations that are very similar on a single locus basis. To our knowledge, there are no simulation studies of the relative efficiency of multivariate vs. clustering methods for detecting genetic structure. In the few studies where both methods were used, however, essentially the same structure was discerned among local populations (Kinloch et al. 1986; Bousquet et al.

1987a), distant populations from a rangewide sample (Bousquet et al. 1987b; Lagercrantz and Ryman 1990), and subspecies (Bousquet et al. 1990). This is probably a consequence of the paucity of correlations among alleles at different loci (linkage disequilibrium) in forest tree species (Epperson 1992, this volume, pp. 257—278). Multivariate methods do not attempt to reconstruct phylogenetic relationships among OTUs, so their utility is generally limited to population genetic and taxonomic studies.

Phenetic methods

With phenetic, or distance, methods, a phenogram is obtained by considering pairwise similarities or distances among OTUs. Where the genetic basis of the phenotypes studied is unknown, a number of metric and nonmetric pairwise indices can be estimated (Sneath and Sokal 1973); however, the derived trees cannot be interpreted in terms other than of strict phenetic resemblance.

When the genetic basis of the observed phenotypes is known, a variety of genetic identities and distances can be calculated by following various evolutionary models. Genetic distances are usually measured as gene substitutions per locus for biochemical markers such as allozymes, and as nucleotide substitutions per site for molecular markers such as DNA or amino acid sequences.

Nei (1987) reviewed a number of methods for calculating genetic distances that differ in their assumptions about homogeneity of evolutionary rates or proportionality to time since divergence. The most common genetic distances used in allozyme studies of trees are Nei's (1972, 1978), Rogers' (1972), and Cavalli-Sforza and Edwards' (1967) distances, and these differ substantially from one another with respect to a number of assumptions. Few workers have explicitly assessed the appropriateness of the various assumptions for their data sets, however, before proceeding with phenetic analyses. Consequently, the topology of derived trees can vary significantly among distances for a given data set (Guttman and Weigt 1989).

Studies of both DNA sequences and restriction sites (Nei and Li 1979; Nei and Tajima 1983) can be used to estimate nucleotide similarity among OTUs. The observed number of nucleotide substitutions is an underestimate of the actual amount of sequence divergence, however, especially when sequences are highly diverged. This is because a number of sites will have mutated more than once, some returning to their original state. The accumulated number of nucleotide substitutions per site can be estimated from the matrix of observed nucleotide similarity by following a variety of mathematical models. The methods of Jukes and Cantor (1969) and

Tajima and Nei (1984) correct for multiple substitutions at the same site. The method of Kimura (1980) additionally takes into account differences in probabilities of transitions and transversions. The methods of Li et al. (1985) and Nei and Gojobori (1986) consider synonymous and non-synonymous nucleotide substitutions separately (the frequency of amino-acid-altering nonsynonymous substitutions is generally far less than that of synonymous substitutions). When sequences contain insertions or deletions, they must be correctly aligned before similarity measures are calculated. A variety of strategies exist, most of which optimize a weighted-sum function of correctly paired residues, with various penalties for insertion of gaps (reviewed by Doolittle 1990).

Phenetic methods of clustering can be distinguished on the basis of whether they assume homogeneity of evolutionary rates. Among methods that assume homogeneity, UPGMA is the most widely used, although other weighting strategies also exist [e.g., weighted pair group methods (WPGMA) and methods based on single or complete linkage (UPGMS, UPGMC) (reviewed by Sneath and Sokal (1973))]. With all of these methods, the tips of the external branches of the dendrogram are all contemporary (i.e., extant species), and the tree is rooted at the midpoint of the two most divergent taxa or subclusters. Other methods such as multidimensional scaling have also been used with genetic distances, but no direct information about ancestry can be inferred unless other methods are simultaneously used (Yacine and Lumaret 1989).

Homogeneity of substitution rates can usually be safely assumed when comparing conspecific populations of temperate tree species that are linked by gene flow. This assumption may not hold, however, when isolated populations, subspecies, species, and genera are studied. Heterogeneity of substitution rates may cause OTUs to be grouped on the basis of similarity of speed of evolution rather than on genetic identity. For example, the northern populations of Bishop pine (*P. muricata* D. Don) have accumulated cpDNA mutations three- to four-fold faster than its southern populations or Monterey pine (*P. radiata* D. Don). Use of a phylogenetic method that accommodates rate heterogeneity (neighbor-joining) correctly grouped all populations of Bishop pine together, whereas one that assumes homogeneity (UPGMA) forced the northern populations to branch off independently (Fig. 1).

Methods such as that of Fitch and Margoliash (1967), Li (1981), the Distance Wagner procedure (Farris 1970, 1972), and the neighbor-joining method (Saitou and Nei 1987) do not assume homogeneity of substitution rates; therefore the length of branches emerging from particular nodes are free to vary independently from one another. The resulting dendrograms are unrooted, and are often termed networks. These methods can be

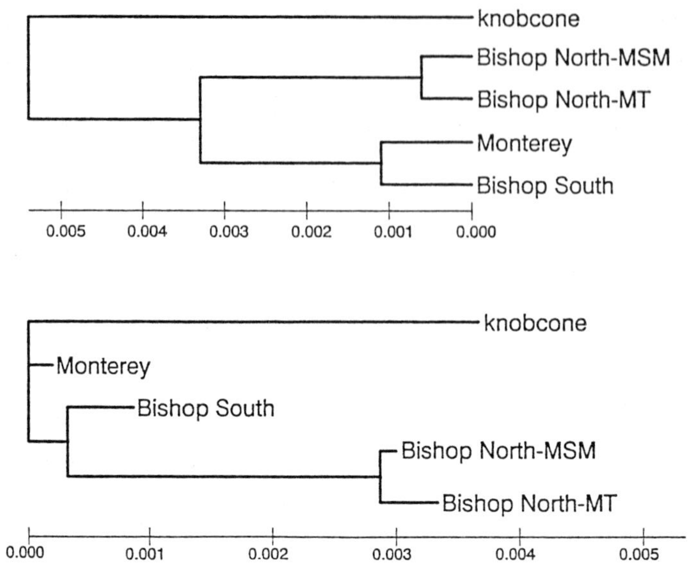

Fig. 1. How heterogeneity of substitution rate can cause incorrect phylogenetic reconstructions with methods that assume homogeneity. Top: UPGMA, which assumes rate homogeneity, caused the rapidly evolving northern populations of Bishop pine to branch off early — separating it from the southern populations of Bishop pine, and making Bishop appear polyphyletic. Bottom: The neighbor-joining method accomodates rate heterogeneity and correctly placed all the populations of Bishop pine together. Scale is number of substitutions per site estimated from chloroplast DNA restriction site mutations according to Nei and Miller (1990) (Hong and Strauss, unpub. data).

further divided into stepwise clustering methods (such as neighbor-joining or Farris' methods) where local topological relationships are first examined and the best tree constructed step by step, or exhaustive search methods (such as Fitch and Margoliash's) where a large number of trees are examined and the best one is chosen on the basis of a goodness-of-fit criterion such as weighted least squares. Several studies have used simulations to compare the robustness of various clustering algorithms under variable evolutionary rates and other conditions (Sourdis and Nei 1988; Saitou and Imanishi 1989; Jin and Nei 1990; Pamilo 1990), facilitating rational choices of the methods appropriate for given kinds of data.

Cladistic methods

In contrast to distance-based methods, cladistic methods attempt to infer phylogenies and build trees by analyzing the gains and losses of individual characters. Maximum parsimony (Fitch 1977) is the most commonly used method, and can accommodate any kind of discrete character. Examples include allozymes, restriction sites, and amino acid or nucleic acid sequences. The method does not assume homogeneity of substitution rates, although it has been shown to be inconsistent when rates vary greatly (Felsenstein 1981). Once an initial tree is estimated, various search procedures involving local or global branch swapping are used to find the shortest (most parsimonious) tree — the one which minimizes the total number of evolutionary events (steps) necessary to explain the observed character arrays of the OTUs. Cladograms can be compared on the basis of homoplasy — the need to invoke multiple independent origins of the same character — present in the trees, for which several indices are available (Archie 1989).

Various weighted parsimony techniques applicable to DNA sequences have been developed in recent years (Williams and Fitch 1990), and range from character weighting by the reciprocal of substitution frequencies at first, second and third codon positions, to weighting of character state *change* where transitions and transversions are differently weighted — e.g., in the evolutionary parsimony method of Lake (1987) or the more recent compositional statistics method of Sidow and Wilson (1990).

Maximum likelihood methods

With maximum likelihood, a probabilistic model of substitution is compared to the data and the tree chosen such that the probability that a particular phylogeny could have occurred is maximized (Felsenstein 1981). The method can be applied to restriction site data as well as DNA sequences, and provides unrooted trees. Because the probability of a given set of data arising on a given tree is computed site by site, and the product of the probabilities across sites is then taken, a very large number of computations are involved and most readily available programs cannot accommodate more than a few taxa.

Statistical testing of phylogenetic trees

Because of the complexity of the phylogenetic hypotheses represented by dendrograms, it is usually difficult to specify many aspects of a phylogeny with a high level of confidence. Assessing the statistical significance of a given branching structure, as well as the variance of branch lengths, is therefore highly desirable. Quantification of levels of uncertainty is especially crucial in genetic studies for conservation programs, since decisions regarding gene and species extinctions may be guided by knowledge of relationships. Several statistical methods have been developed in recent years; however, the field is still in its infancy and methods are in need of substantial improvement. Additional computer programs need to be developed that handle a variety of biochemical and molecular data.

As discussed above, detection of heterogeneity of evolutionary rates is important for making decisions about the appropriate method of phylogenetic analysis. A commonly used means of statistically assessing rate heterogeneity is via relative rate tests. They proceed by comparing nucleotide similarities of pairs of taxa to a reference taxon, usually from a closely related sister group (Fitch 1976; Wu and Li 1985; Li and Tanimura 1987).

One method for examining the reliability of a topology calculates standard errors of branch lengths. This has been adapted for UPGMA dendrograms derived from molecular data (Nei et al. 1985), phenetic methods that do not assume homogeneity of substitution rates (Li 1989), and maximum likelihood methods (Ritland and Clegg 1987).

Resampling procedures have been widely used to place confidence limits on phylogenetic topologies. Because they do not rely on particular statistical distributions, they can be applied to any data set where the observations can be assumed to be independently drawn. The jackknife is a resampling method in which observations are systematically dropped from the data set and the resulting variance in phylogenetic parameters is examined. Mueller and Ayala (1982) randomly dropped allozyme loci one at a time to study the variance in branch lengths under UPGMA. A closely related method, the bootstrap, consists of random sampling with replacement. It is usually applied to the character array, but could also be applied to OTUs. For phylogenetic purposes, with each sample of characters a new topology is calculated, and then the sampling process and topology construction is begun anew. The resampling continues until a reasonable level of confidence is attained (usually 50—100 replicates). For parsimony analyses, a common way to summarize the results is to record the number of trees in which a given group of species is found to be monophyletic (Felsenstein 1985b). The fraction of trees in which given groups appeared are indicated at the appropriate nodes of the most frequently observed tree. Bootstrap programs for parsimony analysis are available on the PHYLIP (Felsenstein 1985b) and PAUP (Swofford 1990) computer programs. The bootstrap has also been adapted to UPGMA analyses of sequence data (Ritland and Clegg 1987).

Biosystematic studies of gymnosperms

Terpenoids

Terpenoids have been, and continue to be, valuable biosystematic markers.

Other plant secondary compounds such as phenolics (e.g., Erdtman et al. 1966; Forrest 1975; Thielges 1969; Lebreton 1990) and resin acids (Zinkel 1977) have been infrequently used. Because terpenoid composition is highly genetically variable yet does not often appear to be strongly affected by natural selection (Hanover 1992, this volume, pp. 159—178), it can be a good indicator of evolutionary history. With the exception of the work of Kinloch et al. (1986) and Baradat (1992, this volume), however, terpenoids have not to our knowledge been interpreted as Mendelian traits for biosystematic studies; thus, the detailed genetic analyses and interpretations possible with allozymes and DNA markers are precluded. Much of the work with terpenoids was conducted prior to the widespread use of numerical methods for production of phylogenetic trees. Many studies therefore have not gone further than to present maps or matrices of distances.

Terpenoids are useful at both intra- and interspecific levels. Some of the earliest intraspecific work recognized distinct monoterpene races in Bishop pine (*P. muricata* D. Don) (Forde and Blight 1964; Mirov et al. 1966). Other key studies of intraspecific structure included that of Douglas-fir's (*Pseudotsuga menziesii* (Mirb.) Franco) chemical races by Zavarin and Snajberk (1973) and von Rudloff and Rehfeldt (1980); Lester's (1974) study of eastern and western races of balsam fir (*Abies balsamea* Mill.); Arbez et al.'s (1974) analysis of provenances of Austrian pine (*Pinus nigra* Arnold); Forrest (1980, 1982) and Kinloch et al.'s (1986) studies of Scotch pine's (*P. sylvestris* L.) refugia and migration history; and Smith's (1977; reviewed by Conkle and Critchfield 1988) studies of xylem monoterpene races of ponderosa pine (*Pinus ponderosa* Dougl.). Critchfield's (1984) use of terpenoid data substantially aided his interpretations of migrational history for white spruce (*Picea glauca* Voss), balsam fir, subalpine fir (*Abies lasiocarpa* Nutt.), and Douglas-fir.

Interspecific studies based on terpenoids have been less common. Adams and Simmons (1987) used terpene analysis to resolve relationships among species and populations of *Callitris* (Cupressaceae) from Australia, including the identification of natural hybrids. Pollack and Dancik (1985) and von Rudloff and Lapp (1987) used terpenes to identify introgression between jack (*P. banksiana* Lamb.) and lodgepole pines (*P. contorta* Loud.). Terpenoids and phenolics have been used to help define subgeneric groupings of species within pines (Duffield 1952; Erdtman et al. 1966).

Allozymes-introgression and genetic diversity

Allozymes have been effective for defining zones of natural hybridization and introgression. One of the first such studies was that of Sitka spruce (*Picea sitchensis* Carr.) and white spruce (*P. glauca*) by Copes and Beckwith (1977) — a species pair that was later analyzed by Yeh and Arnott (1986). Jacobs et al. (1984) found that *Abies balsamea* var. *phanerolepis* was unlikely to be the result of hybridization between balsam fir and fraser fir (*Abies fraseri* Poir.), as had been hypothesized. Schiller et al. (1986) found that introgression of genes from *Pinus brutia* Ten. into Aleppo pine (*P. halpensis* Mill.) was responsible for the elevated genetic diversity of Aleppo pine's eastern European race. Wheeler and Guries (1987) conducted an in depth analysis of introgression in jack and lodgepole pines. They subjected allozyme and morphological traits to classical and multivariate analyses to distinguish hybrids. All traits allowed clear separation of hybrids from pure species, and indicated that a number of localized populations in Canada have undergone multiple rounds of introgressive hybridization. In a survey of the mainland populations of Monterey pine (*P. radiata* D. Don.), Plessas and Strauss (1986) found that introgression of genes from knobcone pine (*P. attenuata* Lemm.) into Monterey pine was taking place in a stand at Año Nuevo where the two species were in admixture. Millar (1983) used a distinctive allele at the GOT locus (glutamate-oxaloacetate transaminase) to monitor introgression between the distinct "blue and green" races of Bishop pine; additional loci were found to usefully differentiate between the races in later studies (Millar 1989).

A difficulty with allozymes for studying introgression is that species-specific alleles are rarely fixed in closely related species. Studies therefore usually require analyses of multiple loci, large samples of trees to estimate gene frequencies, and sophisticated statistical methods. In contrast, species-specific cpDNA RFLP variants appear to be common (e.g., Wagner et al. 1987; Szmidt et al. 1988a) — allowing simple and precise estimates of introgression to be made. In pines, combining paternally inherited cpDNA (Strauss et al. 1989) with maternally inherited mitochondrial DNA (Neale and Sederoff 1989) could provide an opportunity for an incisive description of the dynamics of introgression.

Biochemical markers have been useful for monitoring changes in genetic diversity caused by historical events — especially Pleistocene climatic changes (reviewed by Critchfield 1984). Severe reductions of species-level genetic diversity have been attributed to reductions of population size in red pine (*P. resinosa* Ait.) (Fowler and Morris 1977), western red cedar (*Thuja plicata* Donn) (Yeh 1988), and Torrey pine (*P. torreyana* Parry) (Ledig and Conkle 1983). Substantial reductions of genetic diversity have been observed in sections of the ranges of western white pine (*P. monticola* Dougl.) (Steinhoff et al. 1983), Douglas-fir (Li and Adams 1989), and Norway spruce (*Picea abies* Karst.) (Lagercrantz and Ryman 1990).

Allozymes-intraspecific relationships

Allozymes have been useful for studying intraspecific relationships and biogeographic history. Wheeler and Guries (1982a) found that genetic distances among subspecies and populations of lodgepole pine largely agreed with expectations from an earlier taxonomic treatment of the species. A population in the Cascade Range of southern Oregon allozymically allied to subspecies *latifolia*, however, had been incorrectly classified as subspecies *murryana* on the basis of seed-cone characters. On the basis of differentiation among populations and the distribution of rare alleles, Wheeler and Guries (1982b) identified a number of refugia in which lodgepole pine persisted during the Pleistocene. Study of four loci indicated that subspecies of Austrian pine clustered into three major groups, which, however, often ran contrary to geographic proximity (Nikolic and Tucic 1983). Harry (1984) found that regional samples of incense cedar (*Calocedrus decurrens* (Torr.) Florin) taken from a variety of aspects and elevations clustered together, whereas samples from similar aspects or elevations in different regions did not. Although morphological traits indicated otherwise, the UPGMA cluster analysis of Gullberg et al. (1985) provided no evidence for a northern and southern subspecies of Scotch pine in Sweden.

Fins and Seeb (1986) found that the Payette population of western larch (*Larix occidentalis* Nutt.) was highly diverged from populations in northern Idaho, eastern Washington, and western Montana. Furnier and Adams (1986) found that Jeffrey pine (*P. jeffreyi* Grev. & Balf.) from Klamath and Sierran/Southern Californian regions were distinct from one another. The single Sierran population that occurred on an ultramafic soil was most similar to the Klamath populations, which also occur primarily on ultramafic soils, suggesting genetic adaptation to such soils. Moran and Bell (1987) and Moran et al. (1988) found that Australian breeding populations of Monterey pine were derived primarily from Año Nuevo and Monterey, that the Cedros Island population was highly differentiated from the other four populations, and that Monterey and Año Nuevo were the most closely related of the three mainland populations. The latter result differed from findings of Plessas and Strauss (1986), and varied according to the clustering method used by Millar et al. (1988) — indicative of the difficulty of determining phylogenetic relationships among recently diverged populations. Differences in allozymes among eastern and western varieties of ponderosa pine, in concert with data from some other traits, prompted Conkle and Critchfield (1988) to suggest that the species may have been derived from two distinct progenitor lines — one from the west coast of Mexico and another from the country's more central montane region. Interestingly, geographic patterns for terpenoids had a number of significant differences from those for allozymes — particularly with respect to differentiation of the North Plateau and Northern Rocky Mountain varieties. Millar et al. (1988) found that the coastal species in Subsection Oocarpae — Bishop and Monterey pines — had migrated from south to north along the California coast, whereas the inland species — knobcone pine — first migrated northward along the Sierra Nevada, then through the Klamath region, and finally southward along the California coast. Millar and Marshall (1990) found that the genetic distance among inland and coastal regions in Port Orford cedar (*Chamaecyparis lawsoniana* (A. Murr.) Parl.) was three times higher than that among disjunct populations within regions.

Allozymes have been used to demonstrate the existence of "transient" races within widespread species that formed as a consequence of multiple Pleistocene refugia (reviewed by Critchfield 1984). Steinhoff et al. (1983) found that western white pine was differentiated into a northern and a southern race with a transition zone in Northern California-Southern Oregon. The races also differed in amount of genetic diversity, with more diversity maintained in the southern race. Kinloch et al. (1986) found that two distinct biochemical "races" of Scotch pine occur in Scotland, a result that suggests the existence of at least two Pleistocene refugia. One of the races, Webster Ross, also had less genetic diversity than most other Scottish populations. Li and Adams (1989) found that Douglas-fir was composed of three major groups; the southern Rocky Mountain group had less than half the genetic diversity present in the other two groups. Lagercrantz and Ryman (1990) showed that Norway spruce clusters into four groups, one of which has substantially reduced genetic diversity. From their data, they were able to infer locations of Pleistocene refugia and migration routes.

Several studies have used multivariate techniques to help detect genetic similarities within species. Yeh et al. (1985) found that two canonical vectors accounted for 38% of the allozyme variation in lodgepole pine (var. *latifolia*) from the Yukon and British Columbia, and that the first function provided good separation of the northern and southern populations. In a similar study, Yeh et al. (1986) found that two canonical discriminant functions accounted for more than half of the allozyme variation in black spruce (*Picea mariana* (Mill.) B.S.P.) from Newfoundland; however, the functions had no simple relationship to geography. O'Reilly et al. (1985) found that 70% of black spruce trees could be correctly classified as to whether they originated from upland or lowland sites on the basis of discriminant analysis of allozyme frequencies. Merkle et al. (1988) applied canonical analyses in an attempt to detect allozyme differentiation associated with environment and breeding zones in Douglas-fir from southwest Oregon. Univariate approaches had been unsuccessful in detecting much differentiation among seed zones (Merkle and Adams 1987) or among aspects and elevations (Moran and Adams 1989). Merkle et al. (1988) found that canonical discriminant analysis was a substantial improvement over univariate analysis for detecting differences among breeding zones. Less than 10% of the genetic variability could be accounted for, however, when the canonical scores were regressed against location variables related to latitude, distance from the ocean, and elevation. Moreover, the level of discrimination among breeding zones was too small to be useful in breeding programs. Two significant vectors from canonical correlation analysis with location variables explained 39% of the genetic variability; however, the vectors had no simple relationship to geography. Multivariate techniques aided discrimination of biochemical races in Scotch pine (Kinloch et al. 1986) and Norway spruce (Lagercrantz and Ryman 1990); nonetheless, the groupings observed were largely the same as observed with cluster analysis.

Significance of evolutionary footprints

As discussed above, a number of studies have used allozymes to detect, or to better define, what Critchfield (1984) called "transient races." Lagercrantz and Ryman (1990), however, went a step further than most studies by analyzing the relationship of transient races to growth and adaptive traits. They had studied Norway spruce collections that were also present in a series of provenance tests, where data on height growth, shoot phenology, and seed weight had been taken. Using principal components analyses, they showed that a very strong correlation existed between the allozyme and morphological data sets. This correlation prompted them to suggest that the historical factors that altered allozyme frequencies also left a lasting imprint on growth and morphological traits, and that tree improvement programs need to take account of this. Most other studies that discerned transient races have also reported, though did not quantitate, associations with growth and adaptive traits (Steinhoff et al. 1983; Kinloch et al. 1986; Conkle and Critchfield 1988; Li and Adams 1989). Most workers seem to have implicitly assumed that such correlations are a result of natural selection in the different environments that these races presently occupy. The results of Lagercrantz and Ryman (1990) do not prove otherwise, but they seriously call this assumption into question.

Other studies are needed that carefully examine this question. Such work should be careful to distinguish between traits of significance for adaptation to the current environment, such as shoot phenology, from traits related to migration ability, such as seed weight (Cwynar and McDonald 1987). Unfortunately, Lagercrantz and Ryman (1990) did not distinguish seed weight from height growth and shoot phenology in their work. Another experimental approach would be to use a combination of biochemical marker and common garden studies to compare the growth of transient races that exist in sympatry or close proximity. If an "evolutionary footprint" persists, then despite coexistence in a similar natural environment, different patterns of growth and adaptation in the common garden will be observed. The work of Sorensen et al. (1990) comes closest to such a study. They analyzed a complex of true firs (*Abies*) in western North America that extends over a wide variety of latitudes and encompasses two to three putative races or sibling species. By comparing a variety of traits to environmental indices, particularly latitude, and traits to one another, they hoped to identify whether natural selection or racial history was the main influence on trait genetic architecture. They reported good agreement among traits as diverse as shoot phenology and β-phellandrene concentration, and that division of the complex into three regions (corresponding to putative racial divisions) with step clines between them accounted for essentially all of the latitudinal variation. Although the effects of natural selection were also evident (e.g., among locales within latitudes), these results appear to provide the strongest example to date that evolutionary history might overshadow, or constrain, adaptation to current environments.

If evolutionary footprints do indeed affect growth and adaptive traits, then the current emphasis on use of local seed in most tree improvement programs may be misplaced; greater gains — at least in the medium to long term — may be obtained by emphasizing recovery from past directional selection and drift, which would imply starting from a highly diverse, rather than a local, base population. Observed patterns of genecological adaptation to elevation and aspect (e.g., Rehfeldt 1984) may represent "the best that a species can do" with what's left over after evolution in past environments, rather than an expression of the potential fine tuning to the environment that tree species are capable of. Western white pine may be an extreme example: its northern populations lost much genetic variability during their residence in a northern Pleistocene refugium (Steinhoff et al. 1983), and are notable for having almost no differentiation of growth and adaptive traits according to latitude and elevation (Rehfeldt 1984). Studies of how past climatic challenges have affected the genetic structure of forest tree species should be useful in designing strategies to help them cope with anthropogenic climate changes.

Allozymes-interspecific relationships

Allozymes have shed light on interspecific relationships in conifers. One of the first such studies was that of Wheeler et al. (1983), who analyzed pines in Subsection Contortae, which includes jack, lodgepole, Virginia (*P. virginiana* Mill.) and sand pines (*P. clausa* (Chapm.) Vasey). Surprisingly, they found that differentiation between the southeastern species — sand and Virginia pines — were equivalent to that among races of lodgepole pine. They also found that the southeastern species were more similar to lodgepole than to jack pine, a pattern opposite to their geographic proximities. This work also provided a nice contrast between morphological and biochemical traits for biosystematic analysis. Phylogenetic trees constructed from seed cone measurements suggested a polyphyletic origin of lodgepole pine — whereas the dendrograms generated from allozymes maintained a monophyletic origin of each species. The cone analyses were very probably a result of convergent morphological evolution.

Other interspecific studies include that of Wellendorf and Simonsen (1979), where cluster analysis of small samples of trees from 19 species indicated that affinities among species of spruces often differed markedly from expectations based on taxonomy or geographic proximity. Jacobs et al. (1984) found that two species and a variety of *Abies* from the northeastern US (*balsamea*, *balsamea* var. *phenerolepis*, and *fraseri*) are very weakly differentiated, and probably represent geographically disjunct members of single species. Conkle et al. (1988) showed that Aleppo pine was derived from the *P. brutia* complex by fixation of a subset of alleles still polymorphic in the complex. Millar et al. (1988) studied the closely related Californian species Bishop, Monterey, and knobcone pines, but found inconsistent branching patterns between midpoint-rooted UPGMA and Distance Wagner trees. Karalamangala and Nickrent (1989) studied several Mexican pines, and found that *P. leiophylla* was distinct, as predicted from morphology. Niebling and Conkle (1990) found that the North Plateau race of ponderosa pine was genetically more similar to Washoe pine (*P. washoensis* Mason & Stockwell) than to the Pacific race or Rocky Mountain variety of ponderosa pine. Midpoint-rooted UPGMA and Distance Wagner trees, however, gave conflicting pictures of Washoe pine's racial origins. Wang et al. (1990) found that *P. densata* from China was intermediate between *P. yunanensis* and *P. tabulaeformis*, a position that supports the hypothesis of its hybrid origin.

Molecular genetic studies

Analyses of DNA sequences have largely been based on restriction enzyme studies of cpDNA (Strauss et al. 1989). Sigurgeirsson and Szmidt

(1988) used UPGMA to analyze the percentage of cpDNA fragments in common among nine species of spruces after each was cut with six restriction enzymes. Sitka spruce (*P. sitchensis* (Bong.) Carr. and Brewer spruce (*P. breweriana* S. Watson) were surprisingly diverged from one another, despite their placement in the same Section. Results, however, were in good general accordance with morphologically based taxonomies. Szmidt et al. (1988b) also used UPGMA to analyze the percentage of cpDNA fragments from four restriction enzymes that were in common among 20 species of pines from nine Subsections. Results showed clear differentiation of the subgenera, and in agreement with the findings of Strauss and Doerksen (1990), bristlecone pine and lacebark pine (*P. bungeana* Zucc.) of Section Parrya were the most ancestral species in Subgenus Strobus, and Section Strobus was the most recent. Wang and Szmidt (1990) found that *P. densata* had cpDNA of both *P. yunanensis* and *P. tabulaeformis*, supporting the hypothesized hybrid origin of the species, and agreeing with allozyme analyses (Wang et al. 1990). Parsimony analysis and bootstrapping of restriction site mutations in chloroplast, mitochondrial, and nuclear genes have been employed to study phylogenetic relationships in *Pinus* (Strauss and Doerksen 1990) and *Pseudotsuga* (Strauss et al. 1990). Among the 18 species of pines from 14 Subsections studied, some of the statistically significant findings were that Subsection Parrya — which included bristlecone, pinyon (*P. edulis* Engelm.), and chilgoza pines (*P. gerardiana* Wall.) — is the most ancestral of extant taxa in Subgenus Strobus; *P. leiophylla* has been incorrectly placed in Section Ternatae; and that the hard pines are strongly differentiated into Sections Pinus and Ternatae. In *Pseudotsuga*, the North American and Asian groups of species were strongly coherent and statistically significant units; five of six interpretable mutations that were shared with an outgroup, western larch, indicated that the genus is likely to have evolved in North America and then migrated to Asia. More recently, we have begun to sequence the chloroplast gene *rbc*L in a number of tree species. Results indicate a general slowdown of evolutionary rates in conifers and long-lived, dicotyledonous tree species in comparison with herbaceous flowering plants. This result parallels that found for the palm family (Wilson et al. 1990), suggesting that long-lived plant species have an intrinsically slower chloroplast "molecular clock" than short-lived species. Zimmer et al. (1989) used nuclear ribosomal RNA sequences to study higher-order phylogenetic relationships among land plants. Seed plants were found to be monophyletic, and gymnosperms paraphyletic within the seed plant clade. *Ginkgo*, conifers, and cycads were allied within the Gymnosperms.

Other studies

Immunological analyses of seed proteins were used to study relationships of genera in the Pinaceae (Price et al. 1987) and Taxodiaceae (Price and Lowenstein 1989). The Pinaceae clustered into two major groups (Abietoid and Pinoid), and the Cupressaceae and Sciadopityaceae were found to be distinct but subordinate to the Taxodiaceae. Martin and Dowd (1986) included four gymnosperms (*Metasequoia, Ginkgo, Ephedra*, and *Macrozamia*) and 15 monocots in their parsimony analysis of the first 40 N-terminal amino acids of the small subunit of *rbc*S. The gymnosperms were distinct from the monocots, and as predicted from traditional phylogenetic schemes, *Ephedra* was most similar to them.

Biosystematic studies of angiosperms

In the perennial angiosperms, BMGMs have only recently been used to study population genetics and biosystematics. To date, only allozyme markers have been used at the intraspecific level, and few of the studies have specifically addressed biosystematic questions. Furthermore, the study ranges have often been too small to allow geographical structure related to migration history to be detected. Nonetheless, the few biosystematic studies conducted to date have contributed significant insights.

Allozymes-intraspecific relationships

In a survey of widely separated populations of *Casuarina cunninghamiana* Miq., Moran et al. (1989) were able to confirm races and identify the center and origin of distribution of the species. In a regional study of the same species (Moore and Moran 1989), the authors could distinguish among drainage systems by using a phenetic procedure (Wagner network), and detected substantial heterogeneity of evolutionary rates. Isolation by distance and recent divergence of clustering populations were suggested as explanations of the pattern of differentiation observed. In a rangewide study of *Alnus crispa* (Ait.) Pursh, phenetic analysis of distances between populations showed that northern ones were recently derived from more central ones (Bousquet et al. 1987b). Studies on *Acer saccharum* Marsh. (Perry and Knowles 1989), *Eucalyptus diversicolor* F. Muell. (Coates and Sokolowski 1989), *Fagus silvatica* L. (Comps et al. 1987), *Liriodendron tulipifera* L. (Brotschol et al. 1986) and *Populus deltoides* Marsh. (Rajora 1989) suggested relationships among the populations and clones studied. Isolation by distance appeared to be responsible for the observed affinities among populations of *Camellia japonica* L. (Wendel and Parks 1985), *Populus trichocarpa* Hook. (Weber and Stettler 1981), and *Quercus ilex* L. (Yacine and Lumaret 1989). An analysis of populations of black locust (*Robinia pseudoacacia* L.) (Surles et al. 1989) indicated that the species may have migrated into its present-day range much more recently than did the majority of eastern American forest tree species.

Allozyme markers have also been used to assess subspecific classifications. Phenetic analysis of frequencies and cladistic analysis of shared allelic states showed that subsp. *caesia* of *Eucalyptus caesia* Benth. is paraphyletic while subsp. *magna* was monophyletic and probably derived from one of the identified populations of subsp. *caesia* (Hopper and Burgman 1983; Moran and Hopper 1983). Using phenetic and multivariate analyses, Bousquet et al. (1990) confirmed a subspecific morphological classification and identified natural hybridization between subspecies of *Alnus viridis* (Villars) Lamark & DeCandolle.

Allozymes-interspecific relationships

Allozymes have been useful for characterizing closely related species and interspecific hybridization in *Populus* (Rajora and Zsuffa 1989; Rajora 1990) and *Alnus* (Bousquet et al. 1988, 1990). Several studies focused on the genus *Quercus*. Manos and Fairbrothers (1987) analyzed six red oak species of subgenus *Erythrobalamus*, and showed that despite the existence of far more variation among species than that among conspecific populations, few alleles could be found that distinguished the species. An exception was *Q. palustris* Muench., which was strongly differentiated from the others. Schnabel and Hamrick (1990) used phenetic and discriminant analyses of several populations of *Q. macrocarpa* Michx. and *Q. gambelii* Nutt. They found unexpectedly high genetic similarity between the two species and intermediacy of some populations on their dendrogram. Gene flow during past hybridization events was proposed to account for these observations. Guttman and Weigt (1989) sampled 18 species in the two subgenera — *Quercus* and *Erythrobalamus* — and analyzed several types of genetic distances among species. They also examined a variety of phenetic methods that differed in relation to their assumptions of homogeneity of substitution rates. Differentiation between subgenera was larger than within subgenera, and subgenus Quercus was more differentiated than Erythrobalamus. They invoked interspecific hybridization and introgression to explain the small genetic divergence observed among species within each subgenus, and identified substantial heterogeneity of evolutionary rates.

Sytsma and Schaal used both allozymes (1985a) and restriction fragment analysis (1985b) to study relationships among populations and species in the shrubby genus *Lisianthius*. For allozymes, ancestral relationships were inferred by using both phenetic analysis of allele frequencies and cladistic (parsimony) analysis of shared allelic states. The trees obtained by the different approaches were generally congruent, and the modern derivative of the ancestral taxon of the group was identified. Heterogeneity of substitution rates was apparent among lineages, and founder effects were invoked as the major evolutionary force responsible for the biogeographic and genetic patterns observed.

Molecular genetic studies

Within genera, studies have largely relied on RFLPs to assess interspecific relationships. In *Populus*, Sytsma (1990) reported that cpDNA RFLPs could distinguish the European black poplar from the American cottonwoods, but not from the European white poplar. Nuclear rDNA RFLPs, however, could distinguish black poplar from both the cottonwoods and white poplar, suggesting that hybridization with white poplar as a maternal parent occurred as part of the origin of black poplar. In American cottonwoods, random nuclear DNA as well as cpDNA RFLPs have been used to identify unidirectional introgression (Keim et al. 1989). In *Alnus*, nuclear rDNA RFLPs were used to verify taxonomic relationships among species (Bousquet et al. 1989). In the subtropical genus *Lisianthius*, Sytsma and Schaal (1985b) used parsimony analysis of rDNA and cpDNA RFLPs to study phylogenetic relationships, and found good agreement with a concurrent allozyme study of the species complex, as discussed above (Sytsma and Schaal 1985a).

Recently, Wilson et al. (1990) performed a restriction fragment analysis of cpDNA from 22 species belonging to 18 genera in the palm family. They were able to reconstruct the phylogeny of the family, and found a much slower rate of substitution than had been reported in annual plant species. Partial DNA sequences of *rbc*L were obtained for two palm species and confirmed the slow rate of evolution inferred from restriction site analysis (Wilson et al. 1990). A partial *rbc*L sequence has also been obtained from a 17- to 20-million-year-old fossil leaf of *Magnolia* and included in a phylogenetic analysis of the genus (Golenberg et al. 1990). Phenetic analysis showed the fossil sequence to be very similar to contemporary *Magnolia* and *Liriodendron* sequences.

Other studies

Martin and Dowd (1988) used partial amino acid sequences of *rbc*S to assess phylogenetic relationships in the genus *Nothofagus*. Using parsimony methods, they found substitution rates to be highly heterogeneous. With the same strategies, partial phylogenetic analysis of the families Magnoliaceae, Winteraceae, Fagaceae and Proteaceae showed strong congruence with morphological phylogenies (Martin and Dowd 1984a, b, 1988). From intergeneric comparisons of taxa from different continents in the families Fagaceae, Betulaceae, Winteraceae, and Proteaceae, Martin and Dowd (1988) were also able to calibrate a molecular clock by using partial *rbc*S amino acid sequences. They estimated divergence times and compared them to continental separation dates derived from geologic studies. Inferred times of species divergence were much more recent than predicted from dates of continental separation.

Conclusions and directions for future work

The last 10—20 years have brought about major changes in biosystematics. Biochemical and molecular methods of analysis have largely replaced morphology and artificial hybridization as the traits of choice for phylogenetic study. Advanced numerical and statistical methods have revolutionized the inferences that can be drawn from biosystematic datasets. These advances have brought biosystematics — formerly a highly subjective field of study — into the mainstream of evolutionary science.

The impacts of this metodological revolution have been strongly felt by forest geneticists. The economic importance of forest trees, combined with an abundance of natural populations that show complex histories and adaptations to environment, have prompted numerous studies of forest tree evolution. Our rich knowledge of forest trees — particularly of temperate zone conifers — is the envy of many plant biologists.

A number of challenges exist, however, if we are to take full advantage of the opportunities provided by forest trees and the new methods of analysis. We emphasized the need for statistical testing of phylogenetic hypotheses, selection of appropriate methods of dendrogram construction, phylogeny-guided statistical tests of adaptive hypotheses, and the need to understand evolutionary footprints. We believe that the greatest challenge, however, will be to find the means to fully explore the power of molecular genetic techniques. Though expensive, by providing huge numbers of polymorphisms for intraspecific analysis, and highly conserved gene sequences for analysis at the interspecific level and above, they promise to provide a quantum leap in understanding of biosystematic relationships at all levels of biological organization.

References

Adams, R. and Simmons, D. 1987. A chemosystematic study of *Callitris* (Cupressaceae) in south-eastern Australia using volatile oils. Aust. For. Res. 17: 113—125.

Arbez, M., Bernard-Dagan, C. and Fillon, C. 1974. Intraspecific variability of *Pinus nigra* monoterpenes: analysis of the first results. Ann. Sci. For. 31: 57—70.

Archie, J. W. 1989. Homoplasy excess ratios: new indices for measuring levels of homoplasy in phylogenetic systematics and a critique of the consistency index. Syst. Zool. 38: 253—269.

Avise, J. C., Arnold, J., Ball, R. M., Bermingham, E., Lamb, T., Neigel, J. E., Reeb, C. A. and Saunders, N. C. 1987. Intraspecific phylogeography: the mitochondrial DNA bridge between population genetics and systematics. Annu. Rev. Ecol. Syst. 18: 489—522.

Benfey, P. N. and Chua, N.-H. 1989. Regulated genes in transgenic plants. Science 244: 174—181.

Bousquet, J., Cheliak, W. M. and Lalonde, M. 1987a. Genetic differentiation among 22 mature populations of green alder (*Alnus crispa*) in central Quebec. Can. J. For. Res. 17: 219—227.

——. 1987b. Genetic diversity within and among 11 juvenile populations of green alder (*Alnus crispa*) in Canada. Physiol. Plant. 70: 311—318.

——. 1988. Allozyme variation within and among mature populations of speckled alder (*Alnus rugosa*) and relationships with green alder (*A. crispa*). Am. J. Bot. 75: 1678—1686.

Bousquet, J., Cheliak, W. M., Wang, J. and Lalonde, M. 1990. Genetic divergence and introgressive hybridization between *Alnus sinuata* and *A. crispa* (Betulaceae). Plant Syst. Evol. 170: 107—124.

Bousquet, J., Girouard, E., Strobeck, C., Dancik, B. P. and Lalonde, M. 1989. Restriction fragment polymorphisms in the rDNA region among seven species of *Alnus* and *Betula papyrifera*. Plant Soil 118: 231—240.

Brotschol, J. V., Roberds, J. H. and Namkoong, G. 1986. Allozyme variation among North Carolina populations of *Liriodendron tulipifera* L. Silvae Genet. 35: 131—138.

Bruce, W. B., Christensen, A. H., Klein, T., Fromm, M. and Quail, P. H. 1989. Photoregulation of a phytochrome gene from oat transferred into rice by particle bombardment. Proc. Natl. Acad. Sci. USA 86: 9692—9696.

Burt, A. 1989. Comparative methods using phylogenetically independent contrasts, pp. 33—53. In: Harvey, P. H. and Partridge, L. (Eds) Oxford Surveys in Evolutionary Biology, Vol. 6. Oxford Univ. Press, England.

Bush, R. M. and Smouse, P. E. 1992. Evidence for the adaptive significance of allozymes in forest trees. This issue, pp. 179—196.

Cavalli-Sforza, L. L. and Edwards, A. W. F. 1967. Phylogenetic analyses: models and estimation procedures. Am. J. Human Genet. 19: 233—257.

Clegg, M. T. 1989. Molecular diversity in plant populations, pp. 98—115. In: Brown, A. H. D., Clegg, M. T., Kahler, A. L. and Weir, B. S. (Eds) Plant Population Genetics, Breeding, and Genetic Resources. Sinauer Associates, Inc. Publishers, Sunderland, Massachusetts.

Coates, D. J. and Sokolowski, R. E. 1989. Geographic patterns of genetic diversity in Karri (*Eucalyptus diversicolor* F. Muell.). Aust. J. Bot. 37: 145—156.

Comps, B., Barriere, G., Merzeau, D. and Letouzey, J. 1987. La variabilité alloenzymatique des hêtraies dans les sous-domaines medio- et eu-atlantiques d'Europe. Can. J. For. Res. 17: 1043—1049.

Conkle, M. T. and Critchfield, W. B. 1988. Genetic variation and hybridization of ponderosa pine, pp. 27—43. In: Ponderosa Pine: The Species and its Management. Washington State Univ. Cooperative Extension, Pullman, WA.

Conkle, M. T., Schiller, G. and Grunwald, C. 1988. Electrophoretic analysis of diversity and phylogeny of *Pinus brutia* and closely related taxa. Syst. Bot. 13: 411—424.

Copes, D. L. and Beckwith, R. C. 1977. Isoenzyme identification of *Picea glauca, P. sitchensis,* and *P. lutzii* populations. Bot. Gaz. 138: 512—521.

Critchfield, W. B. 1975. Interspecific hybridization in *Pinus*: a summary review, pp. 99—105 In: Fowler, D.P. and Yeatman, C. W. (Eds) Proc. 14th Meet., Canad. Tree Improvement Assoc., Part 2.

——. 1984. Impact of the Pleistocene on the genetic structure of North American conifers, pp. 70—118. In: Lanner, R. M. (Ed) Proc. 8th North American Forest Biology Workshop, Dept. of Forest Resources, Utah State Univ., Logan, UT.

Cwynar, L. C. and MacDonald, G. M. 1987. Geographical variation of lodgepole pine in relation to population history. Am. Nat. 129: 463—469.

Doolittle, R. F. (Ed) 1990. Methods in Enzymology, Vol. 183. Molecular Evolution: Computer Analysis of Protein and Nucleic Acid Sequences. Academic Press, San Diego, California, 736 pp.

Duffield, J. W. 1952. Relationships and species hybridization in the genus *Pinus*. Z. Forstgenetik 1(4): 93—97.

Epperson, B. 1992. Spatial structure of genetic variation within forest trees. This issue, pp. 257—278.

Erdtman, H., Kimland, B. and Norin, T. 1966. Pine phenolics and pine classification. Bot. Mag. Tokyo 79: 499—505.

Evans, I. J., James, A. M. and Barnes, S. R. 1983. Organization and evolution of repeated DNA sequences in closely related plant genomes. J. Mol. Biol. 170: 803—826.

Falconer, D. S. 1981. Introduction to Quantitative Genetics, Second Ed. Longman, NY.

Farris, J. S. 1970. Methods for computing Wagner trees. Syst. Zool. 19: 83—92.

——. 1972. Estimating phylogenetic trees from distance matrices. Am. Nat. 106: 645—668.

Felsenstein, J. 1981. Evolutionary trees from DNA sequences: a maximum likelihood approach. J. Mol. Evol. 17: 368—376.

——. 1985a. Phylogenies and the comparative method. Am. Nat. 125: 1—15.

——. 1985b. Confidence limits on phylogenies: an approach using the bootstrap. Evolution 39: 783—791.

——. 1988. Phylogenies from molecular sequences: inference and reliability. Annu. Rev. Genet. 22: 521—565.

Fins, L. and Seeb, L. W. 1986. Genetic variation in allozymes of western larch. Can. J. For. Res. 16: 1013—1018.

Fitch, W. M. 1976. Molecular evolutionary clocks, pp. 160—178. In: Ayala, F. J. (Eds) Molecular evolution. Sinauer Associates, Inc., Publishers, Sunderland, Massachusetts.

Fitch, W. M. 1977. On the problem of discovering the most parsimonious tree. Am. Nat. 111: 223—257.

Fitch, W. M. and Margoliash, E. 1967. Construction of phylogenetic trees. Science 155: 279—284.

Forde, M. B. and Blight, M. M. 1964. Geographical variation in the turpentine of Bishop pine. New Zeal. J. Bot. 2: 44—52

Forrest, G. I. 1975. Polyphenol variation in Sitka spruce. Can. J. For. Res. 5: 26—37.

——. 1980. Genotypic variation among native Scots pine populations in Scotland based on monoterpene analysis. Forestry 53: 101—128.

——. 1982. Relationship of some European Scots pine populations to native Scottish woodlands based on monoterpene analysis. Forestry 55: 19—37.

Fowler, D. P. and Morris, R. W. 1977. Genetic diversity in red pine: evidence for low genic heterozygosity. Can. J. For. Res. 7: 343—347.

Furnier, G. R. and Adams, W. T. 1986. Geographic patterns of allozyme variation in Jeffrey pine. Am. J. Bot. 73: 1009—1015.

Golenberg, E. M., Giannasi, D. E., Clegg, M. T., Smiley, C. J., Durbin, M., Henderson, D. and Zurawski, G. 1990. Chloroplast DNA sequence from a Miocene *Magnolia* species. Nature 344: 656—658.

Gould, S. J. and Lewontin, R. C. 1979. The spandrels of San Marco and the Panglossian paradigm: a critique of the adaptationist programme. Proc. Roy. Soc. Lond. B 205: 581—598.

Gullberg, U., Yazdani, R., Rudin, D. and Ryman, N. 1985. Allozyme variation in Scots pine (*Pinus sylvestris* L.) in Sweden. Silvae Genet. 34: 193—201.

Guttman, S. I. and Weigt, L. A. 1989. Electrophoretic evidence of relationships among *Quercus* (oaks) of eastern North America. Can. J. Bot. 67: 339—351.

Hanover, J. W. 1992. Tree fitness as a function of terpene composition. This issue, pp. 159—178.

Harry, D. E. 1984. Genetic structure of incense-cedar (*Calocedrus decurrens*) populations. Ph.D. thesis. Univ. of California, Berkeley, 163 pp.

Holmquist, G. P. 1989. Evolution of chromosome bands: molecular ecology of noncoding DNA. J. Mol. Evol. 28: 469—486.

Hopper, S. D. and Burgman, M. A. 1983. Cladistic and phenetic analyses of phylogenetic relationships among populations of *Eucalyptus caesia*. Aust. J. Bot. 31: 35—49.

Jacobs, B. F., Werth, C. R. and Guttman, S. I. 1984. Genetic relationships in *Abies* (fir) of eastern United States: an electrophoretic study. Can. J. Bot. 62: 609—616.

Jin, L. and Nei, M. 1990. Limitations of the evolutionary parsimony method of phylogenetic analysis. Mol. Biol. Evol. 7: 82—102.

Jukes, T. H. and Cantor, C. R. 1969. Evolution of protein molecules, pp. 121—132. In: Munro, H. N. (Ed) Mammalian Protein Metabolism. Academic Press, NY.

Karalamangala, R. R. and Nickrent, D. L. 1989. An electrophoretic study of representatives of subgenus Diploxylon of *Pinus*. Can. J. Bot. 67: 1750—1759.

Keim, P., Paige, K. N., Whitham, T. G. and Lark, K. G. 1989. Genetic analysis of an interspecific hybrid swarm of *Populus*: Occurrence of unidirectional introgression. Genetics 123: 557—565.

Kimura, M. 1980. A simple method for estimating evolutionary rate of base substitutions through comparative studies of nucleotide sequences. J. Mol. Evol. 16: 111—120.

——. 1983. The Neutral Theory of Molecular Evolution. Cambridge Univ. Press, Cambridge, England.

Kinloch, B. B., Westfall, R. D. and Forrest, G. I. 1986. Caledonian Scots pine: origins and genetic structure. New Phytol. 104: 703—729.

Lagercrantz, U. and Ryman, N. 1990. Genetic structure of Norway spruce (*Picea abies*): Concordance of morphological and allozymic variation. Evolution 44: 38—53.

Lake, J. A. 1987. A rate-independent technique for analysis of nucleic acid sequences: evolutionary parsimony. Mol. Biol. Evol. 4: 167—191.

Lebreton, P. 1990. La chimiotaxonomie des Gymnospermes. Bull. Soc. Bot. France 137, Lett. Bot. (1) 35—46.

Ledig, F. T. 1988. The conservation of diversity in forest trees. BioScience 38: 471—479.

Ledig, F. T. and Conkle, M. T. 1983. Gene diversity and genetic structure in a narrow endemic, Torrey pine (*Pinus torreyana* Parry ex Carr.). Evolution 37: 79—85.

Lester, D. T. 1974. Geographic variation in leaf and twig monoterpenes of balsam fir. Can. J. For. Res. 4: 55—60.

Lewontin, R. C. 1985. Population genetics. Annu. Rev. Genet. 19: 81—102.

Li, P. and Adams, W. T. 1989. Range-wide patterns of allozyme variation in Douglas-fir (*Pseudotsuga menziesii*). Can. J. For. Res. 19: 149—161.

——. 1981. Simple method for constructing phylogenetic trees from distance matrices. Proc. Natl. Acad. Sci. USA. 78: 1085—1089.

——. 1989. A statistical test of phylogenies estimated from sequence data. Mol. Biol. Evol. 6: 424—435.

Li, W.-H. and Tanimura, M. 1987. The molecular clock runs more slowly in man than in apes and monkeys. Nature 326: 93—96.

Li, W.-H., Wu, C.-I. and Luo, C.-C. 1985. A new method for estimating synonymous and nonsynonymous rates of nucleotide substituion considering the relative likelihood of nucleotide and codon changes. Mol. Biol. Evol. 2: 150—174.

Manos, P. S. and Fairbrothers, D. E. 1987. Allozyme variation in populations of six northeastern American red oaks (Fagaceae: *Quercus* subg. Erythrobalanus). Syst. Bot. 12: 365—373.

Martin, P. G. and Dowd, J. M. 1984a. The study of plant phylogeny using amino acid sequences of ribulose-1,5-bisphosphate carboxylase. IV. Proteaceae and Fagaceae and the rate of evolution of the small subunit. Aust. J. Bot. 32: 291—299.

——. 1984b. The study of plant phylogeny using amino acid sequences of ribulose-1,5-bisphosphate carboxylase. V. Magnoliaceae, Polygonaceae and the concept of primitiveness. Aust. J. Bot. 32: 301—309.

——. 1986. A phylogenetic tree for some monocotyledons and gymnosperms derived from protein sequences. Taxon 35: 469—475.

——. 1988. A molecular evolutionary clock for angiosperms. Taxon 37: 364—377.

Merkle, S. A. and Adams, W. T. 1987. Patterns of allozyme variation within and among Douglas-fir breeding zones in southwest Oregon. Can. J. For. Res. 17: 402—407.

Merkle, S. A., Adams, W. T. and Campbell, R. K. 1988. Multivariate analysis of allozyme variation patterns in coastal Douglas-fir from southwest Oregon. Can. J. For. Res. 18: 181—187.

Millar, C. I. 1983. A steep cline in *Pinus muricata*. Evolution 37: 311—319.

——. 1989. Allozyme variation of bishop pine associated with pygmy-forest soils in northern California. Can. J. For. Res. 19: 870—879.

Millar, C. I. and Marshall, K. A. 1990. Allozyme variation of Port-Orford-Cedar (*Chamaecyparis lawsoniana*): implications for genetic conservation. For. Sci. (in press)

Millar, C. I. and Westfall, R. D. 1992. Allozyme markers in forest genetic conservation. This issue, pp. 347—371.

Millar, C. I., Strauss, S. H., Conkle, M. T. and Westfall, R. D. 1988. Allozyme differentiation and biosystematics of the Californian closed-cone pines (*Pinus* subsect. Oocarpae). Syst. Bot. 13: 351—370.

Mirov, N. T., Zavarin, E., Snajberk, K. and Costello, K. 1966. Further studies of turpentine composition of *Pinus muricata* in relation to its taxonomy. Phytochemistry 5: 343—355.

Moore, N. J. and Moran, G. F. 1989. Microgiographical patterns of allozyme variation in *Casuarina cunninghamiana* Miq. within and between the Murrumbidgee and coastal drainage systems. Aust. J. Bot. 37: 181—192.

Moran, G. F. and Adams, W. T. 1989. Microgeographical patterns of allozyme differentiation in Douglas-fir from southwest Oregon. For. Sci. 35: 3—15.

Moran, G. F. and Bell, J. C. 1987. The origin and genetic diversity of *Pinus radiata* in Australia. Theor. Appl. Genet. 73: 616—622.

Moran, G. F., Bell, J. C. and Eldridge, K. G. 1988. The genetic structure and the conservation of the five natural populations of *Pinus radiata*. Can. J. For. Res. 18: 506—514.

Moran, G. F., Bell, J. C. and Turnbull, J. W. 1989. A cline in genetic diversity in river she-oak *Casuarina cunninghamiana*. Aust. J. Bot. 37: 169—180.

Moran, G. F. and Hopper, S. D. 1983. Genetic diversity and the insular population structure of the rare granite rock species, *Eucalyptus caesia* Benth. Aust. J. Bot. 31: 161—172.

Mueller, L. D. and Ayala, F. J. 1982. Estimation and interpretation of genetic distance in empirical studies. Genet. Res. Camb. 40: 127—137.

Neale, D. B. 1992. Potential for marker assisted selection in forest trees. This issue, pp. 391—407.

Neale, D. B. and Sederoff, R. R. 1989. Paternal inheritance of chloroplast DNA and maternal inheritance of mitochondrial DNA in loblolly pine. Theor. Appl. Genet. 77: 212—216.

Nei, M. 1972. Genetic distance between populations. Am. Nat. 196: 283—292.

——. 1978. Estimation of average heterozygosity and genetic distance from a small number of individuals. Genetics 89: 583—590.

——. 1987. Molecular Evolutionary Genetics. Columbia Univ. Press, New York. 512 pp.

Nei, M. and Li, W.-H. 1979. Mathematical model for studying genetic variation in terms of restriction endonucleases. Proc. Natl. Acad. Sci. USA. 76: 5269—5273.

Nei, M. and Tajima, F. 1983. Maximum likelihood estimation of the number of nucleotide substitutions from restriction site data. Genetics 105: 207—217.

Nei, M., Stephens, J. C. and Saitou, N. 1985. Methods for computing the standard errors of branching points in an evolutionary tree and their application to molecular data from humans and apes. Mol. Biol. Evol. 2: 66—85.

Nei, M. and Gojobori, T. 1986. Simple methods for estimating the numbers of synonymous and nonsynonymous substitutions. Mol. Biol. Evol. 3: 418—426.

Nei, M. and Miller, J. C. 1990. A simple method for estimating average number of nucleotide substitutions within and between populations from restriction data. Genetics 125: 873—879.

Niebling, C. R. and Conkle, M. T. 1990. Diversity of Washoe pine and comparisons with allozymes of ponderosa pine races. Can. J. For. Res. 20: 298—308.

Nikolic, D. and Tucic, N. 1983. Isoenzyme variation within and among populations of European black pine (*Pinus nigra* Arnold). Silvae Genet. 32: 80—89.

O'Reilly, G. J., Parker, W. H. and Cheliak, W. M. 1985. Isozyme differentiation of upland and lowland *Picea mariana* stands in northern Ontario. Silvae Genet. 34: 214—221.

Pamilo, P. 1990. Statistical tests of phenograms based on genetic distances. Evolution 44: 689—697.

Perry, D. J. and Knowles, P. 1989. Allozyme variation in sugar maple at the northern limit of its range in Ontario, Canada. Can. J. For. Res. 19: 509—514.

Plessas, M. E. and Strauss, S. H. 1986. Allozyme differentiation among populations, stands, and cohorts in Monterey pine. Can. J. For. Res. 16: 1155—1164.

Pollack, J. C. and Dancik, B. P. 1985. Monoterpene and morphological variation and hybridization of *Pinus contorta* and *P. banksiana* in Alberta. Can. J. For. Res. 63: 201—210.

Price, R. A. and Lowenstein, J. M. 1989. An immunological comparison of the Sciadopityaceae, Taxodiaceae, and Cupressaceae. Syst. Bot. 14: 141—149.

Price, R. A., Olsen-Stojkovich, J. and Lowenstein, J. M. 1987. Relationships among the genera of Pinaceae: an immunological comparison. Syst. Bot. 12: 91—97.

Rajora, O. P. 1989. Genetic structure and identification of *Populus deltoides* clones based on allozymes. Genome 32: 440—448.

——. 1990. Marker allozyme genes and alleles for differentiation of *Populus deltoides, P. nigra, P. maximowiczii,* and their interspecific hybrids. Can. J. Bot. 68: 990—998.

Rajora, O. P. and Zsuffa, L. 1989. Multilocus genetic structure, characterization, and relationships of *Populus* x *canadensis* cultivars. Genome 32: 99—108.

Rehfeldt, J. 1984. Microevolution of conifers in the northern Rocky Mountains: a view from common gardens, pp. 132—146. In: Lanner, R. M. (Ed) Proc. 8th North American Forest Biology Workshop, Dept. of Forest Resources, Utah State Univ., Logan, UT.

Ritland, K. and Clegg, M. T. 1987. Evolutionary analysis of plant DNA sequences. Am. Nat. 130: S74-S100.

Rogers, J. S. 1972. Measures of genetic similarity and genetic distance, pp. 145—153. In: Studies in Genetics VII. Univ. of Texas Publication 7213. Austin, Texas.

Rose, M. R. and Doolittle, W. F. 1983. Molecular biological mechanisms of speciation. Science 220: 157—162.

Saitou, N. and Nei, M. 1987. The neighbor-joining method: a new method for constructing phylogenetic trees. Mol. Biol. Evol. 4: 406—425.

Saitou, N. and Imanishi, T. 1989. Relative efficiencies of the Fitch-Margoliash, Maximum-Parsimony, Maximum-Likelihood, Minimum-Evolution, and Neighbor-joining methods of phylogenetic tree construction in obtaining the correct tree. Mol. Biol. Evol. 6: 514—525.

Saylor, L. C. and Smith, B. W. 1966. Meiotic irregularity in species and interspecific hybrids of *Pinus*. Am. J. Bot. 53: 453—468.

Schiller, G., Conkle, M. T. and Grunwald, C. 1986. Local differentiation among Mediterranean populations of Aleppo pine in their isoenzymes. Silvae Genet. 35: 11—19.

Schmid, C. W. and Jelinek, W. R. 1982. The Alu family of dispersed repetitive sequences. Science 216: 1065—1070.

Schnabel, A. and Hamrick, J. L. 1990. Comparative analysis of population genetic structure in *Quercus macrocarpa* and *Q. gambelii* (Fagaceae). Syst. Bot. 15: 240—251.

Selander, R. K. and Whittam, T. S. 1983. Protein polymorphism and the genetic structure of populations, pp. 89—114. In: Nei, M. and Koehn, R. K. (Eds) Evolution of Genes and Proteins. Sinauer Associates, Inc. Publishers, Sundlander, Massachusetts.

Sessions, S. K. and Larson, A. 1987. Developmental correlates of genome size in plethodontid salamanders and their implications for genome evolution. Evolution 41: 1239—1251.

Sidow, A. and Wilson, A. C. 1990. Compositional statistics: an improvement of evolutionary parsimony and its application to deep branches in the tree of life. J. Mol. Evol. 31: 51—68.

Sigurgeirsson, A. and Szmidt, A. 1988. Chloroplast DNA variation among North-American *Picea* species, and its phylogenetic implications, pp. 49—65. In: Proc. of the Frans Kempe Symposium, Umea, Sweden.

Smith, R. H. 1977. Monoterpenes of ponderosa pine xylem resin in western United States. USDA For. Serv., Pacific Southwest Forest and Range Exp. Sta., Berkeley, California. Tech. Bull. 1532.

Smouse, P. E. and Neel, J. V. 1977. Multivariate analysis of gametic disequilibrium in the Yanomama. Genetics 85: 733—752.

Sneath, P. H. A. and Sokal, R. R. 1973. Numerical Taxonomy. W.H. Freeman, San Francisco.

Sorensen, F. C., Campbell, R. K. and Franklin, J. F. 1990. Geographic variation in growth and phenology of seedlings of the *Abies procera/ A. magnifica* complex. Forest Ecol. Manage. 36: 205—232.

Sourdis, J. and Nei, M. 1988. Relative efficiencies of the Maximum Parsimony and

Distance-Matrix methods in obtaining the correct phylogenetic tree. Mol. Biol. Evol. 5: 298 311.

Steinhoff, R. J., Joyce, D. G. and Fins, L. 1983. Isozyme variation in *Pinus monticola*. Can. J. For. Res. 13: 1122—1132.

Strauss, S. H. and Doerksen, A. H. 1990. Restriction fragment analysis of pine phylogeny. Evolution 44: 1081—1096.

Strauss, S. H., Doerksen, A. H. and Byrne, J. R. 1990. Evolutionary relationships of Douglas-fir and its relatives (genus *Pseudotsuga*) from DNA restriction fragment analysis. Can. J. Bot. 68: 1052—1510.

Strauss, S. H., Neale, D. B. and Wagner, D. B. 1989. Genetics of the chloroplast in conifers. J. Forestry 87: 11—17.

Surles, S. E., Hamrick, J. L. and Bongarten, B. C. 1989. Allozyme variation in black locust (*Robinia pseudoacacia*). Can. J. For. Res. 19: 471—479.

Swofford, D. L. 1990. PAUP-Phylogenetic analysis using parsimony. Illinois Natural History Survey, Champaign, IL.

Sytsma, K. J. 1990. DNA and morphology: inference of plant phylogeny. Trends Ecol. Evol. 5: 104—110

Sytsma, K. J. and Schaal, B. A. 1985a. Genetic variation, differentiation, and evolution in a species complex of tropical shrubs based on isozymic data. Evolution 39: 582—593.

——. 1985b. Phylogenetics of the Lisianthius skinneri (Gentianaceae) species complex in Panama utilizing DNA restriction fragment analysis. Evolution 39: 594—608.

Szmidt, A. E., El-Kassaby, Y. A., Sigurgeirsson, A., Aldèn, T., Lindgren, D. and Hällgren, J.-E. 1988a. Classifying seedlots of *Picea sitchensis* and *P. glauca* in zones of introgression using restriction analysis of choloroplast DNA. Theor. Appl. Genet. 76: 841—845.

Szmidt, A. E., Sigurgeirsson, A., Wang, X.-R., Hallgren, J.-E. and Lindgren, D. 1988b. Genetic relationships among *Pinus* species based on chloroplast DNA polymorphism, pp. 33—47. In: Proc. of the Frans Kempe Symposium in Umea, Sweden.

Tajima, F. and Nei, M. 1984. Estimation of evolutionary distance between nucleotide sequences. Mol. Biol. Evol. 1: 269—285.

Thielges, B. A. 1969. A chromatographic investigation of interspecific relationships in *Pinus* (Subsection Sylvestres). Am. J. Bot. 56: 406—409.

von Rudloff, E. and Lapp, M. S. 1987. Chemosystematic studies in the genus *Pinus*. VI. General survey of the leaf oil terpene composition in lodgepole pine. Can. J. For. Res. 17: 1013—1025.

von Rudloff, E. and Rehfeldt, G. E. 1980. Chemosystematic studies in the genus *Pseudotsuga*. IV. Inheritance and geographical variation in the leaf oil terpenes of Douglas-fir from the Pacific Northwest. Can. J. Bot. 58: 546—556.

Wagner, D. B. Furnier, G. R., Saghai-Maroof, M. A., Williams, S. M., Dancik, B. P. and Allard, R. W. 1987. Chloroplast DNA polymorphisms in lodgepole and jack pines and their hybrids. Proc. Natl. Acad. Sci. USA 84: 2097—2100.

Wang, X.-R. and Szmidt, A. E. 1990. Evolutionary analysis of *Pinus densata* [Masters], a putative Tertiary hybrid. II. A study using species-specific chloroplast DNA markers. Theor. Appl. Genet. (in press).

Wang, X.-R., Szmidt, A. E., Lewandowski, A. and Wang, Z.-R. 1990. Evolutionary analysis of *Pinus densata* Masters, a putative Tertiary hybrid. I. Allozyme variation. Theor. Appl. Genet. (in press).

Wanntorp, H.-E., Brooks, D. R., Nilsson, T., Nylin, S., Ronquist, F., Stearns, S. C. and Wedell, N. 1990. Phylogenetic approaches in ecology. Oikos 57: 119—132.

Weber, J. C. and Stettler R. F. 1981. Isoenzyme variation among ten populations of *Populus trichocarpa* Torr. et Gray in the Pacific Northwest. Silvae Genet. 30: 82—87.

Wellendorf, H. and Simonsen, V. 1979. A chemotaxonomic study in *Picea* with isoenzymes in the seed endosperm, pp. 182—197. In: Proceedings of the Conference on Biochemical Genetics of Forest Trees, Umea, Sweden.

Wendel, J. F. and Parks, C. R. 1985. Genetic diversity and population structure in *Camellia japonica* L. (Theaceae). Am. J. Bot. 72: 52—65.

Westfall, R. D. 1992. Allozyme markers in breeding zone designation. This issue, pp. 279—300.

Wheeler, N. C. and Guries, R. P. 1982a. Population structure, genic diversity, and morphological variation in *Pinus contorta* Dougl. Can. J. For. Res. 12: 595—606.

——. 1982b. Biogeography of lodgepole pine. Can. J. Bot. 60: 1805—1814.

——. 1987. A quantitative measure of introgression between lodgepole and jack pines. Can. J. Bot. 65: 1876—1885.

Wheeler, N. C., Guries, R. P. and O'Malley, D. M. 1983. Biosystematics of the genus *Pinus*, Subsection Contortae. Biochem. Syst. Ecol. 11: 333—340.

Wiley, E. O. 1981. Phylogenetics: The Theory and Practice of Phylogenetic Systematics. John Wiley and Sons, New York, 305 pp.

Williams, P. L. and Fitch, W. M. 1990. Phylogeny determination using dynamically weighted parsimony method. In: Doolittle, R. F. (Ed) Molecular Evolution: Computer Analysis of Protein and Nucleic Acid Sequences. Methods Enzymol. 183: 615—626.

Wilson, M. A., Gaut, B. and Clegg, M. T. 1990. Chloroplast DNA evolves slowly in the palm family (Arecaceae). Mol. Biol. Evol. 7: 303—314.

Wu, C.-I. and Li, W.-H. 1985. Evidence for higher rates of nucleotide substitution in rodents than in man. Proc. Natl. Acad. Sci. USA. 82: 1741—1745.

Yacine, A. and Lumaret, R. 1989. Genetic diversity in holm-oak (*Quercus ilex* L.): insight from several enzyme markers. Silvae Genet. 38: 140—148.

Yeh, F. C. 1988. Isozyme variation of *Thuja plicata* (Cupressaceae) in British Columbia. Biochem. Syst. Ecol. 16: 373—377.

Yeh, F. C. and Arnott, J. T. 1986. Electrophoretic and morphological differentiation of *Picea sitchensis, Picea glauca*, and their hybrids. Can. J. For. Res. 16: 791—798.

Yeh, F. C., Cheliak, W. M., Dancik, B. P., Illingworth, K., Trust, D. C. and Pryhitka, B. A. 1985. Population differentiation in lodgepole pine, *Pinus contorta* spp. *latifolia*: a discriminant analysis of allozyme variation. Can. J. Genet. Cytol. 27: 210—218.

Yeh, F. C., Khalil, M. A. K., El-Kassaby, Y. A. and Trust, D. C. 1986. Allozyme variation in *Picea mariana* from Newfoundland: genetic diversity, population structure, and analysis of differentiation. Can. J. For. Res. 16: 713—720.

Zavarin, E. and Snajberk, K. 1973. Geographic variability of monoterpenes from cortex of *Pseudotsuga menziesii*. Pure and Appl. Chem. 34: 411—434.

Zimmer, E. A., Hamby, K. R., Arnold, M. L., Leblanc, D. A. and Theriot, E. C. 1989. Ribosomal RNA phylogenies and flowering plant evolution. In: Fernholm, B., Bremer, and Jörnvall, K. (Eds) The Hierarchy of Life. Elsevier, NY, pp. 205—214.

Zinkel, D. F. 1977. Pine resin acids as chemotaxonomic and genetic indicators, pp. 53—56. In: TAPPI Conference Papers, Forest Biology Wood Chemistry Conference, Madison, WI.

Zuckerkandl, E. 1986. Polite DNA: functional density and functional compatibility in genomes. J. Mol. Evol. 24: 12—27.

Zuckerkandl, E. and Pauling, L. 1965. Molecules as documents of evolutionary history. J. Theor. Biol. 8: 357—366.

Zurawski, G. and Clegg, M. T. 1987. Evolution of higher plant chloroplast encoded genes; implications for structure-function and phylogenetic studies. Annu. Rev. Plant Physiol. 38: 391—418.

New Forests **6**: 159—178, 1992.

Review paper

Applications of terpene analysis in forest genetics

JAMES W. HANOVER
Department of Forestry, Michigan State University, East Lansing, Michigan 48824, USA

Received 9 October 1990; accepted 3 April 1992

Key words: monoterpenes, conifers, genetic markers, tree fitness

Application. Terpenoid compounds of conifers offer some advantages over other molecular markers and should be considered for use in fingerprinting genetic materials, measuring genetic diversity, selecting for resistance to pests, and determining systematic relationships and adaptation of species and races.

Abstract. Terpenoid substances in forest trees are versatile biochemical systems for use as genetic markers and for studying genetic regulation at the biochemical level. Univariate, multivariate and correlative data analyses from numerous studies on different coniferous species are presented. These data indicate that fitness values for most terpenes examined are probably so low as to be unmeasurable except in long evolutionary time spans. Variation patterns for monoterpenes often closely parallel racial differentiation for morphological traits and reflect the migratory history of species and species-complexes. Random drift and gene flow are likely the major influences on genetic organization of terpene variability. Mutation in regulatory genes may be responsible for quantitative variation within major gene systems for terpenes. The mono-, sesqui- and diterpene systems in conifers offer good opportunities for research on the nature of gene regulation in plants and its evolutionary significance.

Introduction

The terpenoid substances in coniferous species offer unique but relatively unexplored opportunities for increasing our knowledge about eukaryotic gene regulation at all levels from DNA to whole plant responses (Hanover 1966a; Hanover 1971; Squillace 1971; Bernard-Dagan 1987). Monoterpenes (Fig. 1), in particular, have already proven to be very useful for studying a wide array of other biological problems including evolution and hybridization (von Rudloff 1975a; Adams 1977; Bongarten and Hanover 1982; Schaefer and Hanover 1986; Zavarin and Snajberk 1986), resistance to insects and diseases (Hanover 1975a; Bridgen et al. 1979b; Reed et al. 1986; Michelozzi et al. 1990), and biochemical fingerprinting (Rottink and Hanover 1972; Bridgen et al. 1979a; Hanover 1983). It is the use of terpenes as biochemical markers that is addressed here in terms

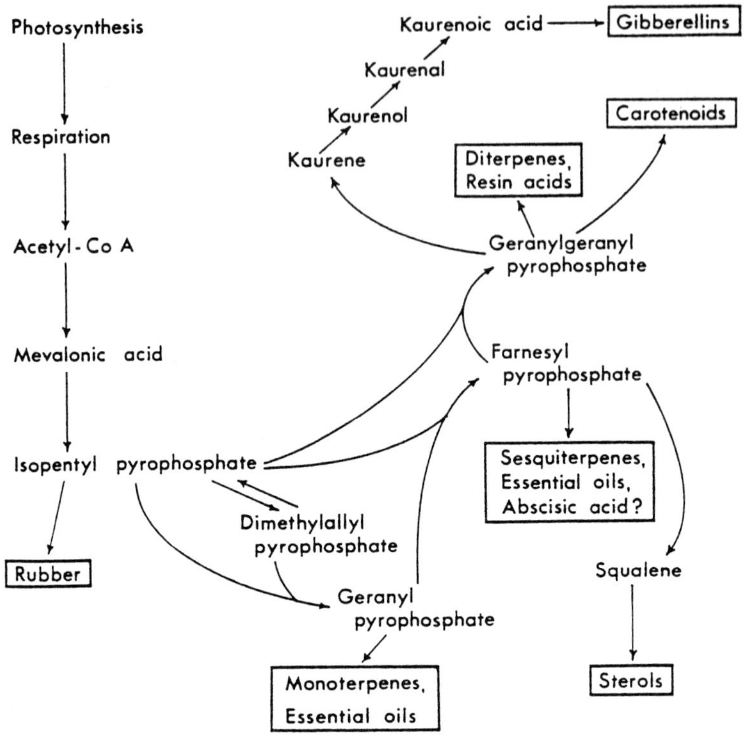

Fig. 1. General biosynthetic relationships among plant terpenes.

of their (1) general suitability as genetic markers, (2) environmental stability, (3) epigenetic variation, and (4) applications. The relationship between terpenes and tree fitness is also discussed.

Terpenes as genetic markers

The terpenoid substances of plants comprise a diverse group (Fig. 1) which include the simple C_{10} monoterpenes and their derivations; the more complex sesqui- (C_{15}) and diterpenes (C_{20}); polymeric carotenoid and chlorophyll (phytol) pigments; and the structurally more complex growth regulators, gibberellic and abscisic acids. Thus, whereas many terpenes have discrete, vital functions in the plant, others, including mono-, sesqui- and diterpenes, are classed as secondary or non-vital compounds in the metabolic or developmental sense. The physiological and genetic interrelations among the classes of terpenes are very poorly understood.

Researchers have focused almost exclusively on monoterpenes as genetic markers, although sesquiterpenes and diterpenes have been shown to be under strong genetic control in tree species (Hanover 1975b; Bridgen and Hanover 1982a; Bernard-Dagan 1987). Therefore, the discussion below will deal only with monoterpenes of coniferous species.

Suitability as markers

Monoterpenes possess characteristics somewhat intermediate between DNA, isoenzymes and morphological traits in terms of their suitability for use as genetic markers (Table 1). Their mode of inheritance varies from single genes of major effect that show dominance (e.g., 3-carene) to apparently polygenic, continuous variation (α-pinene). Regulation of concentration seems to exist at several levels of biochemical organization (Fig. 2). For example, Hanover (1975b) pointed out that regulatory control exists over all biosynthetic pathways for monoterpene versus sesqui- and diterpene pools in cortical oleoresin systems. This control puts a constraint on the quantitative or dosage effect of alleles of individual monoterpenes of a species, race, or individual tree. Even when there are major gene effects, such as for 3-carene (Hanover 1966a, b), there is evidence that the dosages are strongly regulated. Lack of recognition of this fact has lead to false interpretations of terpene inheritance data (Birks and Kanawski 1988).

In addition to their strong and diverse genetic control, monoterpenes are also very easy to measure qualitatively and quantitatively by gas-liquid (GC) chromatography, and more recently by GC coupled with mass selective detectors for positive identification. Samples are easily and quickly collected and transported to the laboratory. Collection can be done in genetic plantations or in wild stands because environmental effects are small. Very young seedlings can be sampled and minimal sample preparation is required unless foliar tissues, rather than exuded oleoresin, are analyzed. Despite the ease and simplicity of analysis of oleoresin terpenes there are important precautions that must be taken if one wishes to obtain precise, accurate measurements of terpene gene expression (Squillace 1976). Many of these precautions are elaborated in the discussion to follow.

The number of potential terpene markers in trees is large because there are literally thousands of derivatives of the basic isoprene molecule, C_5H_8. Even in a single species there are at least tens and even hundreds of qualitative possibilities, most of which also show quantitative variation. Although certainly theoretically less than the number of DNA markers, terpenes probably exceed morphological marker possibilities, and perhaps

Table 1. Some relative comparisons of tree traits for suitability as genetic markers

Marker attribute	Traits			
	DNA	Isozymes	Monoterpenes	Physiological & morphological traits
1. Inheritance	Qualitative, simple (single 'genes')	Quantitative, simple (single genes)	Qualitative/ Quantitative, simple to complex	Quantitative, complex, polygenic
2. Structural vs. regulator genes	Both	Structural	Both	Both
3. Pleiotropic effects	None	None	Some	Large
4. Epistatic effects	None	None	Small	Large
5. Environmental effects	None	Small	Small	Small to large
6. Epigenetic variation	None	Some	Some	Large
7. Early expression	Early	Early	Early	Early to late

Table 1. (Continued)

Marker attribute	Traits			
	DNA	Isozymes	Monoterpenes	Physiological & morphological traits
8. Functional value	High	High	Low	Low to high
9. No. of potential markers	Very large	Large	Large	Small
10. Ease of measurement	Difficult	Easy	Easy	Easy to difficult
11. Cost of measurement	High	Moderate	Moderate	Low to moderate
12. Correspondence with morphology/physiology	Low	Low	High	High
13. Sampling intensity required	Low to high	High	Moderate	High
14. Relation to fitness	Low to high	Moderate	Low	High

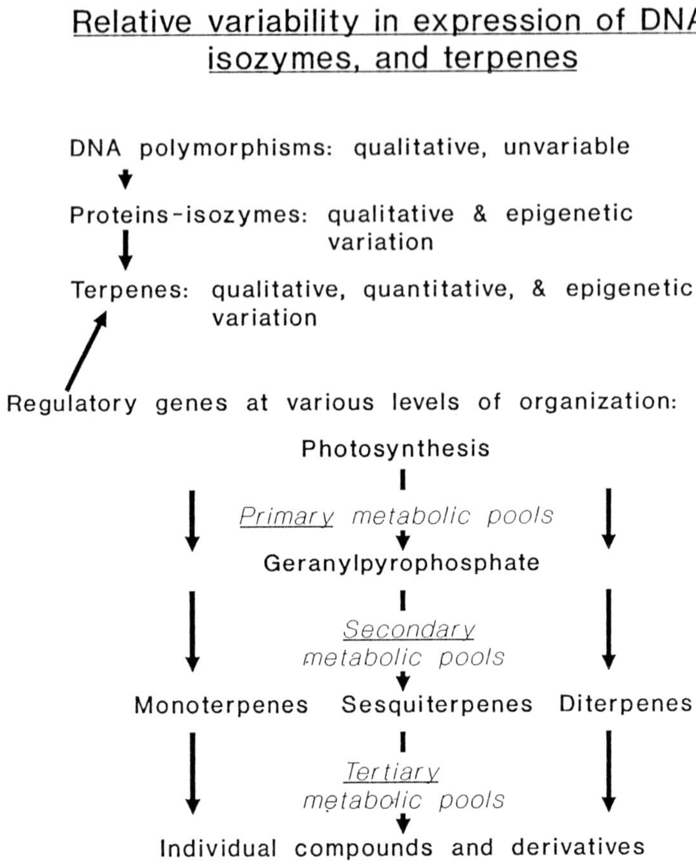

Fig. 2. Relative variability in expression of DNA, proteins, and terpenes.

isozymes, at least from the practical standpoint. However, because of the strong correlations among terpene concentrations, the number of statistically independent dimensions they provide is substantially less than the total number of chemical species observed (e.g., Strauss and Critchfield 1982) — complicating comparisons among the different genetic markers.

Environmental stability

Critical to the use of terpenes as genetic markers is their stability and lack of sensitivity in the face of environmental factors. Recognizing this fact early in my own research on terpene genetics, I established four grafted chemotype clones of western white pine (*Pinus monticola* Dougl.) at three highly contrasting sites in northern Idaho (Hanover 1966c). After several years growing in each location the major monoterpenes in each clone were analyzed (Table 2). The data show remarkably little variability between copies of the same clone after acclimation to their respective sites. Moreover, there are no detectable effects of rootstocks (randomly selected white pines) on scion terpene composition. Many other studies have since substantiated these results.

Table 2. Monoterpene levels in four clones of western white pine growing at three locations (Hanover 1966c)

Clone no.	Site no.	α-Pinene	β-Pinene	Myrcene	3-Carene	Limonene	Total terpenes[a]
		Percent of oleoresin[b]					
I	1	5.0	0.4	5.0	1.2	16.4	28.0
	2	4.6	0.2	4.8	0.8	14.6	25.0
	3	5.4	1.4	4.6	2.2	13.2	26.8
II	1	4.6	13.4	2.9	12.0	2.3	35.2
	2	4.7	14.3	2.2	11.0	1.6	33.8
	3	4.4	13.5	2.6	11.0	2.1	33.6
III	1	5.5	19.2	3.9	0.5	3.3	32.4
	2	7.2	23.5	3.8	0.3	1.7	36.5
	3	7.2	20.8	3.8	1.8	3.6	36.2
IV	1	5.2	5.4	6.6	0.4	13.4	31.0
	2	4.8	5.3	7.6	0.3	10.8	28.0
	3	6.0	4.0	6.0	0.8	10.3	27.1

[a] Does not include camphene, or unknowns 1 and 2.
[b] Each value represents the average of two trees.

To further define environmental effects on terpene composition at the family level, we sampled 45 families (12 trees per family) of blue spruce (*Picea pungens* Engelm.) from throughout its range, one-half of which were grown in a greenhouse and planted at age 9 months while the other half were grown for three years in an outdoor nursery, then outplanted at the same Michigan site (Reed and Hanover 1982; Hanover, unpublished). The trees were eight years old when sampled. Apart from some minor differences that were statistically significant, we found very little variation in the quantitative levels of eleven terpenes. This correspondence between a diverse array of half-sib families grown under radically different pre-planting conditions attests to the environmental stability and strong genetic control characterizing terpene biosynthesis in conifers. There have been many reports of environmental variation in terpene composition due to factors such as moisture stress (Langenheim et al. 1979), nutritional status (Muzika et al. 1989), fungal infection (Paine et al. 1987), and insect defoliation (Tobolski 1968). However, when appropriate sampling precautions are taken to avoid environmental extremes or to partition genetic versus environmental variances, environmental, compared to genetic variation, is virtually negligible. Thus, monoterpene concentrations in conifers can be reliable markers in trees nearly irrespective of their environmental preconditioning.

Epigenetic variation

Within-tree or developmental variation, often termed epigenetic variation, is an important attribute of monoterpenes. Epigenetic variation includes that associated with tree age, organ, and tissue conditioning. Since trees are large, perennial, long-lived organisms it is not surprising that genes determining terpene synthesis may be turned "on" or "off", or otherwise regulated, during development. Obviously, it is important to recognize epigenetic variation patterns in order to effectively use terpenes as markers.

The largest and perhaps most interesting source of epigenetic variation in terpene constituents is that which occurs between different tissues on a tree. The sources and magnitude of this variation are illustrated in Tables 3 through 5 from Moore (1980) and Moore and Hanover (1987). These data show that there are large but predictable quantitative, and some qualitative, differences between foliage, bark, xylem, root, and cone terpenes at a given time of the year.

There are also some significant changes in terpene composition in the same tissue (xylem, foliage, bark) but at different positions within a tree. This is also developmental variation attributable to age effects (Moore 1980). Other examples of within-tree tissue variation in terpenes have been reported by Cermak (1987) (within seeds of *Abies*); Roberts (1970) (roots, stems, branches of *Pinus elliotii* Engelm.); Adams and Hagerman (1976) (juvenile versus mature leaves of *Juniperus scopulorum* Sarg.); and Forrest (1980) (stem versus branch cortex in *Picea sitchensis* [Bong] Carr.). Clearly, genes regulate the concentrations of terpenes in each tissue system but the factors causing their differential expression within the tree are largely unknown.

Another manifestation of epigenetic variation in terpenes is seasonal changes in concentrations in the same tissue systems. Our work with 12 mature trees of blue spruce sampled 10 times over the growing season illustrates this type of variation in needles (Fig. 3). Similar patterns were observed in stems and cones (Moore 1980).

The seasonal patterns of variation in blue spruce terpenes represent

Table 3. Mean volatile oil composition over all positions in crown in different tissues of blue spruce (Moore 1980)

Monoterpene	Tissue		
	Xylem	Foliage	Bark
	--- Percent of total monoterpenes[a] ---		
Santene	—	0.2	0.1
Tricyclene	—	0.7	—
α-pinene	71.4	8.3	31.7
Camphene	1.0	12.6	1.2
β-pinene	20.0	0.9	6.0
Sabinene	2.4	—	1.7
Myrcene	1.5	9.8	7.1
3-carene	0.4	0.5	22.2
Limonene	3.6	30.9	19.5
β-phellandrene	—	1.0	1.4
γ-terpinene	—	—	1.0
1:8 cineole	—	0.8	—
Terpinolene	—	0.8	4.3
Camphor	—	11.2	—
Borneol	—	3.5	—
Terpinen-4-ol	—	0.9	1.0
α-terpineol	—	2.8	0.6
Bornyl acetate	—	14.0	2.6
	--- μl/g dry weight[a] ---		
Total unknown compounds	—	—	1.83
Total monoterpenes	0.62	2.38	6.89

[a] Means are of eight trees averaged over three positions in the crown.

quite consistent and significant changes, especially during the early phases of growth. Nevertheless, genetic variation among trees is clearly discernable despite large within-tree changes, and after periods of rapid development such as during bud-burst, concentrations always return to a base level characteristic of the tree and tissue. Similar conclusions were drawn in studies by Brooks et al. (1987) comparing foliar and bud monoterpenes in *Picea sitchensis*; von Rudloff (1972, 1975b) also working with *Picea pungens* and *Picea glauca* (Moench) Voss; Kossuth and Muse (1986) with *Pinus elliotii*; Bernard-Dagan (1987) with *Pinus pinaster* Ait.; and many others. The conclusion from all of the results on seasonal and tissue variation in terpenes is that one needs to define the patterns for a species and be consistent in sampling in time and tissue — preferably avoiding rapidly developing tissues — if the "basal" gene expression of a tree is desired. Of

Table 4. Mean volatile oil composition of twelve blue spruce trees in different organs at end of growing season (Moore 1980)

	Tissue		
Monoterpene	Needle	Stem	Cone
	--- Percent of total monoterpenes ---		
Santene	0.2	—	—
Tricyclene	1.1	—	—
α-pinene	8.2	23.3	31.1
Camphene	13.9	0.8	0.6
β-pinene	0.7	6.3	15.7
Sabinene	0.1	3.0	1.7
Myrcene	9.6	8.2	4.0
3-carene	0.2	27.9	1.3
Limonene	30.3	17.6	40.0
β-phellandrene	0.5	1.1	1.2
γ-terpinene	—	1.0	0.3
1:8 cineole	0.7	—	—
Terpinolene	1.0	4.9	0.8
Camphor	8.9	0.9	0.5
Borneol	2.1	—	—
Terpinen-4-ol	0.7	1.0	0.6
α-terpineol	2.2	0.9	1.0
Bornyl acetate	15.3	2.1	1.1
	----- μl/g dry weight -----		
Total unknown volatile componds	0.09	1.28	3.85
Total monoterpenes	2.70	4.48	11.81

course, epigenetic variation in terpenes provides a wide array of possible chemical profiles for interaction with insects and diseases of conifers which is probably important for the tissue- and time-specificity shown in pest development on a host. Thus, if one is to use monoterpenes in studies of insect resistance, concentrations in the preferred tissue at the normal time of attack must be studied. Another epigenetic variable in a tree is age, but variation due to age is generally much less than variation among either tissues or seasonal growth phases.

Applications of monoterpenes as genetic markers

With the above considerations and constraints in mind, the terpenes of

Table 5. Variation in the monoterpene composition in stems and roots with position in blue spruce one meter in height (Moore 1980)

Monoterpene	Position in the tree			Roots
	Top	Middle	Bottom	
	--- Percent of total monoterpenes ---			
Santene	0.0 a[e]	0.0 a	0.0 b	1.5 c
α-pinene	43.0 a	61.6 b	64.7 c	65.9 d
Camphene	1.1 a	1.4 b	1.4 b	2.4 c
β-pinene	8.0 a	12.2 b	14.7 c	17.2 d
Sabinene	2.2 a	1.3 b	2.2 a	2.1 a
Myrcene	4.0 a	2.8 b	2.6 c	2.1 d
3-carene	18.5 a	5.1 b	1.2 d	1.4 c
Limonene	14.6 a	9.5 b	6.8 c	2.2 d
β-phellandrene	1.2 a	1.0 a	1.0 a	1.1 a
γ-terpinene	0.4 a	0.0 a	0.0 a	0.0 a
Terpinolene	3.3 a	1.2 d	1.5 c	1.8 b
Bornyl acetate	1.7 a	2.1 b	1.8 c	0.2 d

[e] Means across columns followed by the same letter are not significantly different at the 5% level.

conifers offer geneticists a very useful molecular complement, and in many cases a primary tool, for various types of studies. Some major applications for terpenes include:

1. Systematic and evolutionary relationships.
2. Geographic variation and biodiversity measurements: inter- and intra-population analyses.
3. Fingerprinting: clones, varieties, seed sources, hybrids.
4. Pest resistance and indirect selection.

Systematic and evolutionary relationships

There is a long history of uses of terpenes in systematic and evolutionary studies that began before the advent of gas-liquid chromatography (GLC). Much of this work is summarized by Mirov (1967) for the genus *Pinus.* Mirov himself contributed substantially to the biochemical systematics of the pines (Mirov 1948, 1961), and others have continued to apply terpene chemistry to systematics of pines (Hanover 1975b; Zavarin and Snajberk 1986); *Juniperus* (Adams et al. 1980); *Cupressus* (Zavarin et al. 1971); spruces (von Rudloff 1967; Schaefer and Hanover 1986); *Agathis* (Smith

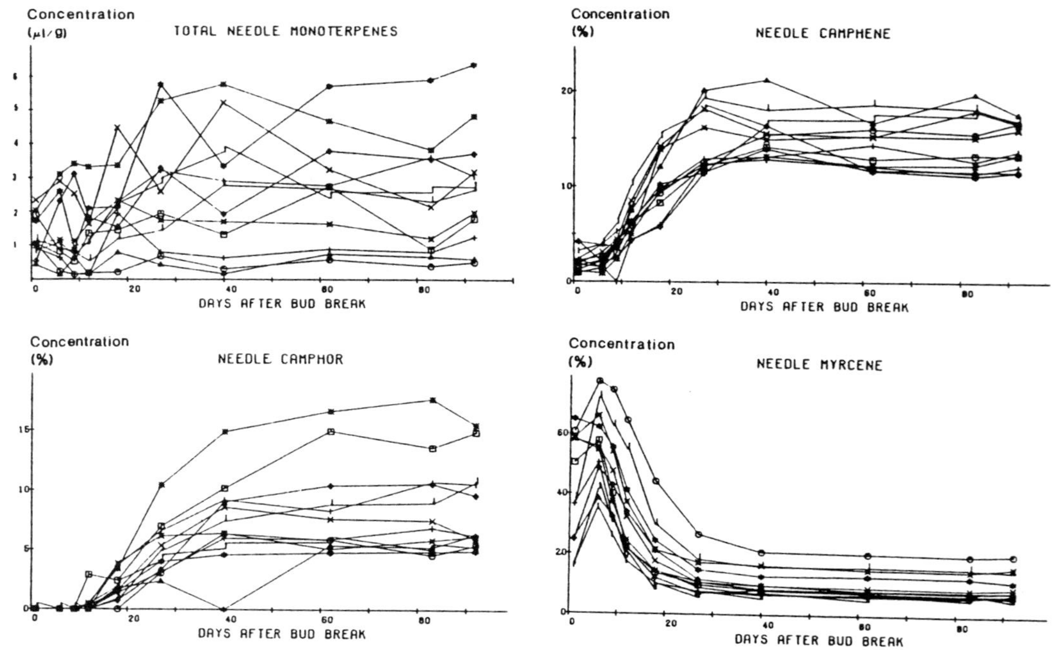

Fig. 3. Seasonal variation (bud break to dormancy) in several monoterpenes in needles of 12 mature blue spruce (Moore 1980).

et al. 1981); and *Abies* (Zavarin et al. 1975). Terpenes have served well in resolving many questions of relationships within species in the above, as well as numerous other studies. There is rarely much conflict between purported relationships based upon terpene data and morphological data (Tobolski and Hanover 1971; Adams 1977; Snajberk and Zavarin 1986), and where morphological data are ambiguous the terpenes often are more definitive.

Geographic variation and assessment of biodiversity

Closely related to studies of the evolutionary relationships between species is the problem of defining tree variation within species that is associated with geography and population dynamics. For many species clear patterns of genetic variation due to geographic location have been described [Forde and Blight 1964 (*Pinus muricata* D. Don); Tobolski and Hanover 1971 (*Pinus sylvestris* L.); Wilkinson et al. 1971 (*Picea glauca*); Coyne and Keith 1972 (*Pinus taeda* L. and *P. achinata* Mill.); Wilkinson and Hanover 1972 (*Picea rubens* Sarg.); Hanover 1974 (*Picea pungens*); Zavarin et al. 1975 (*Abies concolor* [Gord. and Glend.] Lindl.); Snajberk and Zavarin 1976 (*Pseudotsuga menziesii* Mirb. [Franco]); Squillace and Wells 1981 (*Pinus taeda*); Schiller and Grunwald 1987 (*Pinus brutia* Ten.)].

Some studies done in my laboratory show the power of terpenes for studying geographic and population variation in several species. They also shed some light on the relation of terpenes to tree fitness. For example, eastern (*Pinus strobus* L.) and western white pine are two closely related species whose ranges have been separated since miocene times. They have very similar monoterpenes, diterpenes, and other oleoresin characteristics, although significant quantitative differences between them have evolved over this long time span (Hanover 1975b). Rangewide chemical analyses of white spruce (*Picea glauca*) (Wilkinson et al. 1971), red spruce (*Picea rubens*) (Wilkinson and Hanover 1972), and black spruce (*Picea mariana* [Mill.] B.S.P.) (unpublished) indicate that these species are basically similar qualitatively, but genetic differences in their concentrations of monoterpenes clearly distinguish them, as well as geographic races within the large ranges of white and black spruce. The latter species exhibit clear eastern and western complexes from terpene data at approximately the same zone in central Canada.

A comprehensive terpene "map" of the large range of Scotch pine (*Pinus sylvestris*) reveals widely varying terpene patterns which closely correspond to the varietal classification scheme for the species (Tobolski and Hanover 1971; Ruby and Wright 1976). There is a large, clear north-south cline in frequency of the 3-carene gene in Scotch pine, but within Sweden, Yazdani et al. (1985) have identified discrete differences among populations that don't always parallel the overall species pattern. Apparently, genetic variation in terpenes is rather closely correlated with the evolutionary history of the species and its extensive variation in adaptive traits.

Multivariate cluster analyses performed on rangewide terpene data of a spruce (*Picea pungens*) and a pine (*Pinus flexilis-strobiformis*) indicate a clear separation of growth races at the same geographic area (Colorado-New Mexico) in southwestern United States (Hanover 1974). Steinhoff (1964) had provided taxonomic evidence for separation of the *flexilis-strobiformis* species complex in this area, and the terpene markers have strongly corroborated his data. Studies of geographic variation of terpenes in various species have shown strong differentiation that is often correlated with that of other traits (Smith et al. 1969; Squillace and Wells 1981; Forrest 1987).

A number of terpene investigators have attempted to measure genetic variation at the local level in order to assess biodiversity and detect population differentiation over relatively small distances. Townsend and Hanover (1972) found considerable variability in monoterpenes, photosynthesis, and growth of western white pine in Idaho, but the variation was not associated with environmental gradients (altitude or latitude), but rather with populations and individual trees. Hall and Langenheim (1987) found considerable within and among population variation in monoterpenes of *Sequoia sempervirens* (D. Don) Endl. throughout its range, and emphasized the utility of monoterpenes in assessing biodiversity in forest trees. Kinloch et al. (1986) used both terpenes and isozymes to define the genetic structure of Scotch pine in northern Scotland. Average heterozygosity values were 0.33 for monoterpenes and 0.30 for isozymes, with more than 95% of the variation being within populations. Although populations differed, and one distinctive population was identified, there was no pattern with geography.

Schaefer and Hanover (1986) intensively sampled two *Picea* species, blue and Engelmann (*Picea engelmannii* Parry) spruces, growing in a single southwest Colorado drainage to detect species differences in morphology and monoterpenes. Uni- and multivariate analyses showed much variation within each species, but they, nonetheless, were clearly separated by multivariate analysis of monoterpenes (Fig. 4) as well as by morphological data.

Fingerprinting clones, varieties, seed sources and hybrids

The usefulness of monoterpenes as markers in trees extends to clones,

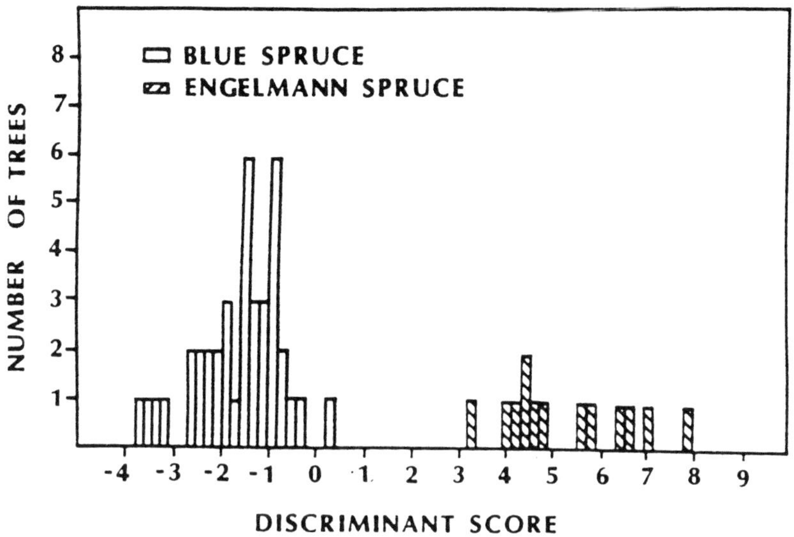

Fig. 4. Discriminate analysis of mature blue and Engelmann spruce based on the concentrations of 13 monoterpenes in cortical tissue (Schaefer and Hanover 1986).

varieties, seed sources and hybrids. Because of their strong genetic control, high variability, and large numbers, they can often be useful for marking individual tree genotypes. There are good examples of chemical marking for varietal seed source identification in Scotch pine (Bridgen et al. 1979a); clonal cultivar identification in blue spruce (Rottink and Hanover 1972); artificial hybrid verification in white, red and blue spruces (Bongarten and Hanover 1982); and putative natural hybridization in blue and Engelmann spruces (Schaefer and Hanover 1986, 1990). Because of the relatively low cost and ease of application, biochemical fingerprinting with terpenes should become much more widely used in the future. There have been no studies to date, however, of their effectiveness for fingerprinting of genotypes with reference to large base populations (e.g., natural stands).

Pest resistance and indirect selection

Terpenes have often been studied for their potential role in resistance or susceptibility to attack by diseases, insects, and animals (Whitney and Denyer 1969; Rockwood 1973; Hanover 1975a; Bridgen et al. 1979b; Harris et al. 1983; Raffa et al. 1985; Wilkinson 1985; Reed et al. 1986; Brooks et al. 1987; and Michelozzi et al. 1990). In several of these cases rather clear associations have been found between terpenes and pest attack, and the use of terpenes for indirect selection to enhance resistance has been suggested (Rehfeldt and von Rudloff 1978; Bridgen and Hanover 1982b; Hanover 1983). However, due to the complexity of resistance mechanisms and sometimes conflicting evidence for and against certain host chemistry-insect relationships (Bridgen et al. 1979a), the requirements for a good terpene marker to incorporate into a breeding program have not yet been met. There is ample evidence for the involvement of tree terpenes in regulating insect behavior (Hanover 1975a; Byers 1983; Raffa and Berryman 1987; Reed et al. 1986). It would seem that indirect selection methods will eventually be developed for tree resistance to both insects and diseases, and that some of these methods will involve terpenes. But since resistance may not be sufficiently correlated with single terpenes but rather with several factors, some form of indirect index selection based upon multivariate and genetic correlation analyses will be required (Zavarin and Cobb 1970).

Are trees which possess oleoresin systems and specific terpene profiles better off (i.e., more fit) than those which do not? To answer this question we must, of course, be able to show that within a single species terpene differences between individuals in variable populations convey different fitness values. In other words, individual trees must possess a certain terpene or combination of terpenes which render those trees better able to survive and leave more offspring. There is little evidence that terpenes contribute to tree fitness in terms of growth and development traits, although in my own studies I frequently observe that genetic differences in growth rate are negatively correlated with concentrations of terpenes (Hanover 1971). Even if this is a true biological relationship between terpenes and growth rate, it would be very difficult to prove that there is also a differential fitness relationship.

In the case of survival traits such as resistance to insects or diseases, there is evidence of individual tree variation within species being correlated with resistance as described earlier. However, here again the ultimate proof of a relationship to differential fitness is lacking. It seems likely that we may be close to achieving that proof, for example, with the gene for 3-carene and its possible role in resistance to attack by bark beetles (Reed et al. 1986). Even in this system, though, the impact of selection may have to be measured in very long evolutionary time spans. But knowledge of any biological relationship between terpenoids and survival or economic traits can be extremely useful to the tree breeder for indirect selection. Nonetheless, the effective genetic neutrality of terpenes in most circumstances, and their environmental stability, makes them excellent markers for a variety of purposes.

References

Adams, R. P. 1977. Chemosystematics — analyses of populational differentiation and variability of ancestral and recent populations of *Juniperous ashei*. Ann. Missouri Bot. Gard. 64: 184—209.

Adams, R. P. and Hagerman, A. 1976. A comparison of the volatile oils of mature versus young leaves of *Juniperus scopulorum*: chemosystematic significance. Biochem. Systematics and Ecology 4: 75—79.

Adams, R. P., von Rudloff, E., Zanoni, T. A. and Hogge L. 1980. The terpenoids of an ancestral/advanced species pair of *Juniperus*. Biochem. Systematics and Ecology 8: 35—37.

Bernard-Dagan, C. 1987. Biosynthesis of lower terpenoids: genetic and physiological controls in woody plants, pp. 329—351. In: Hanover, J. W. and Keathley, D. (Eds) Genetic Manipulation of Woody Plants. Plenum Press, New York.

Birks, J. S. and Kanowski, P. J. 1988. Interpretation of the composition of coniferous resin. Silvae Genetica 37: 29—39.

Bongarten, B. C. and Hanover, J. W. 1982. Hybridization among white, blue, and white × blue spruces. For. Sci. 28: 129—134.

Bridgen, M. R. and Hanover, J. W. 1982a. Genetic variation in oleoresin physiology of Scotch pine. For. Sci. 28: 582—589.

Bridgen, M. R. and Hanover, J. W. 1982b. Indirect selection for pest resistance using terpenoid compounds, pp. 161—168. In: Heybroek, H. M., Stephan, B. R. and von Weissenberg, K. (Eds) Resistance to Diseases and Pests in Forest Trees. Centre for Agricultural Publishing and Documentation (Pudoc), Wageningen, Netherlands.

Bridgen, M. R., Hanover, J. W. and Noecker, N. 1979a. Varietal identification of eight-year old Scotch pine by monoterpene analysis. Mich. Christmas Tree J. Feb. 1979: 33—34.

Bridgen, M. R., Hanover, J. W. and Wilkinson, R. C. 1979b. Oleoresin characteristics of eastern white pine seed sources and relationship to weevil resistance. For. Sci. 25: 175—183.

Brooks, J. E., Borden, J. H., Pierce, Jr., H. D. and Lister, G. R. 1987. Foliar and cortical monoterpenes in Sitka spruce: potential indicators of resistance to the white pine weevil, *Pissodes strobi* Peck (Coleoptera: Curculionidae). Can. J. For. Res. 17: 740—745.

Byers, J. A. 1983. Bark beetle conversion of a plant compound to a sex-specific inhibitor of pheromone attraction. Science 220: 624—626.

Cermak, J. 1987. Monoterpene hydrocarbon contents of the resin from seeds of silver fir (*Abies alba* Mill.). Trees 1: 94—101.

Coyne, J. F. and Keith, G. C. 1972. Geographic survey of monoterpenes in loblolly and shortleaf pines. USDA For. Serv. Res. Paper SO-79. 12 pp.

Forde, M. B. and Blight, M. M. 1964. Geographical variation in the turpentine of Bishop pine. New Zealand J. Bot. 2: 44—52.

Forrest, G. I. 1980. Seasonal and spatial variation in cortical monoterpene composition of Sitka spruce oleoresin. Can. J. For. Res. 10: 452—457.

Forrest, G. I. 1987. A rangewide comparison of outlying and central lodgepole pine populations based on oleoresin monoterpene analysis. Biochemical Systematics and Ecology 15: 19—30.

Hall, G. D. and Langenheim, J. H. 1987. Geographic variation in leaf monoterpenes of *Sequoia sempervirens*. Biochemical Systematics and Ecology 15: 31—43.

Hanover, J. W. 1966a. Genetics of terpenes. I. Gene control of monoterpene levels in *Pinus monticola* Doug. Heredity 21: 73—84.

Hanover, J. W. 1966b. Inheritance of 3-carene concentration in *Pinus monticola*. For. Sci. 12: 447—450.

Hanover, J. W. 1966c. Environmental variation in the monoterpenes of *Pinus monticola* Doug. Phytochem. 5: 713—717.

Hanover, J. W. 1971. Genetics of terpenes. II. Genetic variances and interrelationships of monoterpene concentrations in *Pinus monticola*. Heredity 27: 237—245.

Hanover, J. W. 1974. Biochemical analysis of tree speciation. Proc. Third North Amer. For. Biol. Workshop, Fort Collins, CO. pp. 106—131.

Hanover, J. W. 1975a. Physiology of tree resistance to insects. Ann. Rev. Entomol. 20: 75—95.

Hanover, J. W. 1975b. Comparative physiology of eastern and western white pines: oleoresin composition and viscosity. For. Sci. 22: 214—221.

Hanover, J. W. 1983. Biochemical methods for accelerating progress in tree improvement. Proc. 17th Southern Forest Tree Improvement Conf. pp. 312—324.

Harris, L. H., Borden, J. H., Pierce, Jr., H. D. and Oehlschlager, A. C. 1983. Cortical resin monoterpenes in Sitka spruce and resistance to the white pine weevil, *Pissodes strobi* (Coleoptera: Curculionidae). Can. J. For. Res. 13: 350—352.

Kinloch, B. B., Westfall, R. D. and Forrest, G. I. 1986. Caldonian Scots pine: origins and genetic structure. New Phytol. 104: 703—729.

Kossuth, S. V. and Muse, H. D. 1986. Cortical monoterpene variation among slash pine ramets by season, aspect, crown position, and bud vigor. For. Sci. 32: 605—613.

Langenheim, J. H., Stubblebine, W. H. and Foster, C. E. 1979. Effect of moisture stress on composition and yield in leaf resin of *Hymenaea courbaril*. Biochem. Systematics and Ecology 7: 21—28.

Michelozzi, M., Squillace, A. E. and White, T. L. 1990. Monoterpene composition and fusiform rust resistance in slash pine. For. Sci. 36: 470—475.

Mirov, N. T. 1948. The terpenes in relation to the biology of the genus *Pinus*. Ann. Rev. Biochem. 17: 521—540.

Mirov, N. T. 1961. Composition of gum turpentines of pines. USDA Tech. Bull. No. 1239. 158 pp.

Mirov, N. T. 1967. The genus *Pinus*. Ronald Press, NY. 602 pp.

Moore, P. P. 1980. Developmental variation in volatile oil of blue spruce. Ph.D. Thesis, Michigan State University, East Lansing, Michigan. 138 pp.

Moore, P. P. and Hanover, J. W. 1987. Variation in yield of blue spruce monoterpenes associated with crown position and frequency of resin canals. Forest Science 33: 1081—1088.

Muzika, R. M., Pregitzer, K. S. and Hanover, J. W. 1989. Changes in terpene production following nitrogen fertilization of grand fir (*Abies grandis* (Dougl.) Lindl.) seedlings. Oecologia 80: 485—489.

Paine, T. D., Blance, C. A., Nebeker, T. E. and Stephen, F. M. 1987. Composition of loblolly pine resin defenses: comparison of monoterpenes from induced lesion and sapwood resin. Can. J. For. Res. 17: 1202—1206.

Raffa, K. F. and Berryman, A. A. 1987. Interacting selective pressures in conifer-bark beetle systems: a basis for reciprocal adaptations? Amer. Nat. 129: 234—262.

Raffa, K. F., Berryman, A. A., Simasko, J., Teal, W. and Wong, B. L. 1985. Effects of grand fir monoterpenes on the fir engraver, *Scolytus ventralis* (Coleoptera: Scolytidae) and its symbiotic fungus. Environ. Entomol. 14: 552—556.

Reed, A. N. and Hanover, J. W. 1982. Genetic variation in cortical monoterpene composition of *Picea pungens* Engelm. Proc. Seventh North Amer. For. Biol. Workshop, Lexington, KY. 332 pp. (Abstract).

Reed, A. N., Hanover, J. W. and Furniss, M. M. 1986. Douglas-fir and western larch: chemical and physical properties in relation to Douglas-fir bark beetle attack. Tree Physiology 1: 277—287.

Rehfeldt, G. E. and von Rudloff, E. 1978. Terpenes for indirect selection of growth potential in Rocky Mountain Douglas-fir. USDA For. Res. Note INT-244. 8 pp.

Roberts, D. R. 1970. Within-tree variation of monoterpene hydrocarbon composition of slash pine oleoresin. Phytochemistry 9: 809—815.

Rockwood, D. L. 1973. Monoterpene-fusiform rust relationships in loblolly pine. Phytopathology 63: 551—553.

Rottink, B. A. and Hanover, J. W. 1972. Identification of blue spruce cultivars by analysis of cortical oleoresin monoterpenes. Phytochemistry 11: 3255—3257.

Ruby, J. L. and Wright, J. W. 1976. A revised classification of geographic varieties in Scots pine. Silvae Genetica 25: 169—175.

Schaefer, P. R. and Hanover, J. W. 1986. Taxonomic implications of monoterpene compounds of blue and Engelmann spruces. For. Sci. 32: 725—734.

Schaefer, P. R. and Hanover, J. W. 1990. An investigation of sympatric populations of blue and Engelmann spruces in the Scotch Creek drainage, Colorado. Silvae Genetica 39: 72—81.

Schiller, G. and Grunwald, C. 1987. Cortex resin monoterpene composition in *Pinus brutia* provenances grown in Israel. Biochemical Systematics and Ecology 15: 389—394.

Smith, R. H., Peloquin, R. C. and Passof, P. C. 1969. Local and regional variation in the monoterpenes of Ponderosa pine wood oleoresin. USDA For. Serv. Res. Paper PSW-56. 10 pp.

Smith, R. M., Marty, R. A. and Peters, C. F. 1981. The diterpene acids in the bled resins of three Pacific Kauri, *Agathis vitiensis, A. lanceolata* and *A. macrophylla*. Phytochemistry 20: 2205—2207.

Snajberk, K. and Zavarin, E. 1976. Mono- and sesqui-terpenoid differentiation of *Pseudotsuga* of the United States and Canada. Biochem. Systematics and Ecology 4: 159—163.

Snajberk, K. and Zavarin, E. 1986. Monoterpene differentiation in relation to the morphology of *Pinus remota*. Biochem. Systematics and Ecology 14: 155—163.

Squillace, A. E. 1971. Inheritance of monoterpene composition in cortical oleoresin of slash pine. For. Sci. 17: 381—387.

Squillace, A. E. 1976. Analyses of monoterpenes by gas-liquid chromatography, pp. 120—157. In: Miksche, J. P. (Ed) Modern Methods in Forest Genetics. Springer-Verlag, Berlin.

Squillace, A. E. and Wells, O. O. 1981. Geographic variation of monoterpenes in cortical oleoresin of loblolly pine. Silvae Genetica 30: 127—135.

Steinhoff, R. J. 1964. Taxonomy, nomenclature, and variation within the *Pinus flexilis* complex. Ph.D. Thesis, Michigan State Univ., East Lansing, MI. 81 pp.

Strauss, S. H. and Critchfield, W. B. 1982. Inheritance of β-pinene in xylem oleoresin of knobcone × Monterey pine hybrids. Forest Sci. 28: 687—696.

Tobolski, J. J. 1968. Variations in monoterpenes in Scotch pine. Ph.D. Thesis, Michigan State University, East Lansing, Michigan. 125 pp.

Tobolski, J. J. and Hanover, J. W. 1971. Genetic variation in the monoterpenes of Scotch pine. For. Sci. 17: 293—299.

Townsend, A. M. and Hanover, J. W. 1972. Altitudinal variation in photosynthesis, growth and monoterpene composition of western white pine (*Pinus monticola* Dougl.) seedlings. Silvae Genetica 21: 133—139.

von Rudloff, E. 1967. Chemosystematic studies in the genus *Picea* (Pinaceae) II. The leaf oil of *Picea glauca* and *P. mariana*. Can. J. Bot. 45: 1703—1714.

von Rudloff, E. 1972. Seasonal variation in the composition of the volatile oil of the leaves, buds, and twigs of white spruce (*Picea glauca*). Can. J. Bot. 50: 1595—1603.

von Rudloff, E. 1975a. Volatile leaf oil analysis in chemosystematic studies of North American conifers. Biochem. System. Ecol. 2: 131—167.

von Rudloff, E. 1975b. Seasonal variation of the terpenes of the leaves, buds, and twigs of blue spruce (*Picea pungens*). Can. J. Bot. 53: 2978—2982.

Whitney, R. D. and Denyer, W. B. G. 1969. Resin as a barrier to infection of white spruce by heartrotting fungi. For. Sci. 15: 266—267.

Wilkinson, R. C. 1985. Comparative white-pine weevil attack susceptibility and cortical monoterpene composition of western and eastern white pines. For. Sci. 31: 39—42.

Wilkinson, R. C. and Hanover, J. W. 1972. Geographical variation in the monoterpene composition of red spruce. Phytochem. 11: 2007—2010.

Wilkinson, R. C., Hanover, J. W., Wright, J. W. and Flake, R. H. 1971. Genetic variation in the monoterpene composition of white spruce. For. Sci. 17: 83—90.

Yazdani, R., Nilsson, J. E. and Ericsson, T. 1985. Geographical variation in the relative proportion of monoterpenes in cortical oleoresin of *Pinus sylvestris* in Sweden. Silvae Genetica 34: 201—208.

Zavarin, E. and Cobb, Jr., F. W. 1970. Oleoresin variability in *Pinus ponderosa*. Phytochemistry 9: 2509—2515.

Zavarin, E., Lawrence, L. and Thomas, M. C. 1971. Compositional variations of leaf monoterpenes in *Cupressus macrocorpa, C. Pygmaea, C. Goveniana, C. Abramsiana,* and *C. Sargentii*. Phytochem. 10: 379—393.

Zavarin, E. and Snajberk, K. 1986. Monoterpenoid differentiation in relation to the morphology of *Pinus discolor* and *Pinus johannis*. Biochem. Syst. and Ecol. 14: 1—11.

Zavarin, E., Snajberk, K. and Fisher, J. 1975. Geographic variability of monoterpenes from cortex of *Abies concolor*. Biochem. Syst. and Ecol. 3: 191—203.

New Forests **6**: 179–196, 1992.

Review paper

Evidence for the adaptive significance of allozymes in forest trees

ROBIN M. BUSH[1] and PETER E. SMOUSE[2]

[1] *Department of Ecology and Evolutionary Biology, University of California, Irvine, CA 92717, USA;* [2] *Center for Theoretical and Applied Genetics, Cook College, Rutgers University New Brunswick, NJ 08903—0231, USA*

Received 9 October 1990; accepted 30 May 1991

Key words: adaptive distance, heterosis, inbreeding depression, overdominance, rare alleles

Application. Rare allozyme alleles may be associated with decreased fitness in pines. We should consider designing orchards consisting only of clones carrying highly polymorphic alleles, with replication designed to attain allele frequencies mimicking frequencies of natural stands.

Abstract. Positive correlations between allozyme heterozygosity and fitness measures, primarily growth and fecundity, have been reported for a number of forest tree species. Because the amount of variation in growth explained by allozyme genotype is usually on the order of only a few percent, there has been little effort made towards using electrophoretic screening of allozymes as a tool in early selection on seedlings in production nurseries. We review the progress made in studies of heterozygosity in forest trees, focusing on how recent studies have utilized careful experimental design to allow testing of hypotheses as to the causative nature of the heterozygosity-fitness phenomena. We discuss evidence suggesting a deleterious nature for rare allozyme alleles, and present a case of apparent balancing selection across life history stages acting to maintain rare alleles in *Pinus taeda*. We also review the apparently common trend in natural stands toward increasing heterozygosity over time, and suggest how gains might be made through artifical selection based on allozyme survey data.

Introduction

Studies of allozymes have shown that forest trees, particularly conifers, are among the most genetically variable organisms known (Hamrick and Godt 1989). There has been a great deal of interest, from both the theoretical and applied points of view, in whether allozyme variability reflects the work of natural selection. Evolutionary ecologists are interested in the adaptive nature of allozyme variability, with an eye toward understanding the mechanisms reponsible for maintaining genetic diversity in natural populations. A greater understanding of the adaptive nature of allozyme loci would be of benefit to breeders if it would allow some relief from time and labor intensive field studies through selection on simply assayed allozyme loci.

In spite of considerable work in the area, we still know little about the phenotypic effects associated with particular allozyme alleles in forest trees. Moreover, we have had limited success in finding adaptationist explanations for the large amounts of genetic diversity observed in most populations. This is hardly surprising, given that selective coefficients might not be large enough to yield detectable differences in fitness among individuals differing genotypically at a single locus. There is, however, indirect evidence from population survey data to suggest that allozymes may not be selectively neutual markers, much of this evidence coming from studies of correlations between fitness components and multiple-locus heterozygosity. We begin by reviewing the basic patterns of allozyme variability in forest trees, relating the ways in which this variability can be viewed as a reflection of natural selection. We focus on conifers, for which most data are available. We then turn to studies of the relationship between allozyme heterozygosity and components of fitness in forest trees. We close with a discussion of the sorts of practical information that can be gleaned from such studies and some suggestions for future work.

Patterns of allozyme variability

Population structure

Numerous forest tree species have been surveyed for allozyme variation, and it is now abundantly clear, from studies using techniques such as Nei's (1972) genetic distance and Wright's (1969) F-statistics, that most allozyme variation is found within, rather than among populations, even when samples are taken from across the geographic range of a species (Lundkvist 1979; Yeh and O'Malley 1980; Steinhoff et al. 1983; Ross and Hawkins 1986; Schiller et al. 1986; Dickinson et al. 1988; see review by Loveless and Hamrick 1984). Low levels of interpopulational variation indicate that many of the same alleles are seen at similar frequencies in the different populations sampled. The cause of this homogeneity is an open question; some explanations include a common history, widespread gene flow, or similar types of natural selection acting on alleles across the sampled populations. What are apparently the same allozyme alleles are often shared by closely related species (Dancik and Yeh 1983; Yeh and

Arnott 1986; Conkle et al. 1988; Niebling and Conkle 1990), arguing for considerable antiquity of the alleles in question.

Correlations between allozymes and environmental variables

Homogeneity of frequencies across geographic space is not the universal condition. Genetic differentiation both within and among stands of several species (Grant and Mitton 1977; Mitton et al. 1977; Linhart et al. 1981; Hamrick et al. 1989) could be due either to spatial clustering of related individuals, microgeographic adaptation or drift. For example, Furnier and Adams (1986b) found that variation in allele frequencies of *Pinus jefferyi* Grev. and Balf. seemed a reflection of adaptation to ultramafic soils, although soil type was largely confounded with geography. On a macrogeographic scale, one way to test the hypothesis that allozymic variation results from natural selection is to look for correlations between allele frequencies and environmental variables. As an example, Guries and Ledig (1981) found significant correlations of allele frequencies and climatic variables such as winter temperature and snowfall that parallel clines in cone serotiny (Ledig and Fryer 1972), wood specific gravity (Ledig et al. 1975), and height growth (Ledig et al. 1976), all measured in a provenance trial of *Pinus rigida* Mill.

Various authors have proposed adaptation to local environments as an explanation for changes in allele frequencies across large geographic distances, altitude, or changes in slope (Grant and Mitton 1977; Mitton et al. 1977, 1980; Bergmann 1978; Lundkvist 1979), but others have failed to find correlations between allele frequencies and environmental clines (El-Kassaby and Sziklai 1982; Knowles and Grant 1985; Neale and Adams 1985; Alden and Loopstra 1987). There is frequently a confounding of geographic continuity and environmental continuity; simple divergence and gene flow could account for some of these reported patterns. Studies of acid phosphatase in *Picea abies* (L.) Karst., however, involving two independent clines with climatic variation argue strongly for selection (Bergmann 1978). Although a few other studies (Grant and Mitton 1977; Lagercrantz and Ryman 1990) have found significant congruence between geographic variability in morphological characters and allozyme loci, there does not appear to be any clear or general relationship between allozymes and environmental variability. Conifers are far more apt to vary morphologically or physiologically over a short distance than they are to vary at allozyme loci (Wheeler and Guries 1982; Millar 1983; Neale and Adams 1985; Linhart et al. 1989).

Temporal changes in heterozygosity

Temporal studies of heterozygosity in forest trees have provided additional insight into the possible adaptive significance of allozyme loci. Several workers have compared embryo and parental genotypes of wind-pollinated, predominantly outcrossing tree species having inferred levels of selfing ranging from 5% to 20% (D. DeHayes (pers. comm.); Moran and Brown 1980; Furnier and Adams 1986a; Shea 1987; and references therein). In general, there is excess homozygosity within seed lots, an excess that often disappears with increasing age of the stand, representing severe selection against selfs. Although there are a few studies showing excess homozygosity in mature stands as well (Dancik and Yeh 1983; Dickinson et al. 1988), excess heterozygosity (although not always statistically significant) is much more common (O'Malley et al. 1979; Yeh and Layton 1979; Linhart et al. 1981; Cheliak et al. 1984; Yeh et al. 1986).

Evidence for temporally increasing heterozygosity comes from comparisons of seed lots with parents (Fins and Libby 1982; Shaw and Allard 1982; El-Kassaby et al. 1987; Neale and Adams 1985; Boyle and Morgenstern 1986), comparisons of seeds and seedlings (Farris and Mitton 1984), comparisons of different age classes within a stand (Plessas and Strauss 1986), and from surveys of seed of differing ages retained on the tree over time (Cheliak et al. 1985; Moran and Brown 1980; Snyder et al. 1985; Perry and Dancik 1986). Thus, it seems that selection against inbred individuals over time is a common phenomenon in many species. Interestingly, inbreeding coefficients vary somewhat among loci (Yeh and Morgan 1987, and references therein), sometimes being positive and sometimes negative, suggesting preferential selection against particular homozygous genotypes, in addition to across-the-board elimination of inbred individuals *per se*.

Heterozygosity and morphological variability

Lerner (1954) proposed that higher levels of heterozygosity should be associated with phenotypic intermediacy and thus with increased developmental homeostasis, the latter often considered to promote fitness in variable environments. Support for this hypothesis comes from negative correlations between heterozygosity and growth variability in *Pinus contorta* Dougl. ex Loud. (Mitton et al. 1981; Knowles and Mitton 1980) and reproductive variability in *Pinus ponderosa* Dougl. ex Laws. (Linhart and Mitton 1985). Other studies have found positive correlations between heterozygosity and growth variability for *Populus tremuloides* Michx. (Mitton and Grant 1980), *Pinus ponderosa* (Knowles and Grant 1981;

Mitton et al. 1981; Linhart and Mitton 1985) and *Pinus attenuata* Lemm. (Strauss 1987). Ledig et al. (1983) found no correlation between heterozygosity and growth variation in *Pinus rigida*. Overall, these studies provide an ambiguous picture of the relationship between genetic and growth variation. Mitton et al. (1981) suggested that part of the problem may be that variability in growth is often considered independently of variability in reproductive output. It is not at all clear how year to year variability in growth should be related to fitness in forest trees. Much of the thinking on this subject comes from the idea that growth variability is maladaptive, based on studies of bilateral asymmetry in animals. One might easily make the argument that the ability to grow differentially, in the face of year to year environmental variation, would be adaptive in forest trees, as discussed by Strauss (1987). Neither the evidence nor the predictions are clear cut.

Heterozygosity and fitness

Heterozygosity and growth

Increases in population heterozygosity over time imply superior survival of heterozygotes; for forest trees, a large component of ultimate survival is the capacity for early growth (Spurr and Barnes 1980). Many authors have reported positive correlations between allozyme heterozygosity and growth, although the amount of variance in individual growth rates explained is small for *Populus tremuloides* (Mitton and Grant 1980; Mitton et al. 1981), *Pinus rigida* (Ledig et al. 1983; Bush et al. 1987), *Pinus attenuata* (Strauss 1986), *Pinus radiata* D. Don (Strauss and Libby 1987), *Picea engelmannii* (Parry) Engelm. and *Abies lasiocarpa* (Hook.) Nutt. (Shea 1989). Moran et al. (1989) found a positive correlation between heterozygosity and height at the population level in *Casuarina cunninghamiana* Miq. Not all workers report positive correlations between heterozygosity and growth; Mitton et al. (1981) found no correlation for *Pinus ponderosa* or *Pinus contorta*.

Experimental control

There remains much to be learned through studies of the heterozygosity-fitness phenomenon; much of the past work has been hampered by the limitations inherent in studies of natural populations, particularly variable levels of inbreeding, and the lack of control over microhabitat differences, density, and age; confounding factors that generate large-scale, non-genetic variation in fitness measures. An example of this is found in our earlier work on *Pinus rigida*, for which variation in growth rate due to age differences among trees and variation in competitive environments could be only partially removed (Bush et al. 1987). The same criticism could be applied to most other studies, some recent work by Strauss (1986, 1987), and Strauss and Libby (1987) being conspicuous exceptions. An illustration of the genetic resolution that can be gained from careful experimental control is available from our recent plantation study of selfed and outcrossed *Pinus taeda* L. progeny (Bush and Smouse 1991). The amount of variation in growth we could explain by cross type (selfed vs outcrossed) increased from 4% to 30% upon removal of the growth variation due to differences among blocks and due to variation in the number of surviving neighbors.

Most studies could also be improved by ensuring that the measure under consideration has actually undergone selection and that fitness is measured during an appropriate life-history stage. In this vein, Ledig et al. (1983) showed that the correlation between heterozygosity and growth increased with stand age, density, and environmental unpredictability in *Pinus rigida*, observations suggesting that heterozyotes were better able to withstand greater environmental variation over the years. Competition during crown closure would have accentuated growth differences among the surviving trees.

Hypothesis testing

Our primary interest in this line of research stems from a desire to determine the root causes of the heterozygosity-fitness phenomenon, documented for many sorts of organisms, notably bivalves and forest trees (see Mitton and Grant 1984 for a review). The phenomenon could be due either to homozygosity for deleterious recessive alleles (dominance), or to differing degrees of heterozygote advantage for the allozyme loci themselves (overdominance), or to some combination of the two. The analytical techniques used in almost all such studies, usually regression of some fitness-related measure for an individual on its allozyme heterozygosity, fail to use the available information very effectively for hypothesis testing. There have only been a few attempts to test the predictions of the competing hypotheses (dominance and overdominance), but some of the best of these have been conducted with forest trees.

The dominance hypothesis postulates that the heterozygosity of a sample of allozyme loci is representative of genomic heterozygosity, indirectly reflecting the degree of inbreeding (Ledig 1986). This hypothesis assumes that effects on fitness are caused by deleterious recessive alleles throughout the genome, not by the allozyme loci themselves, which are viewed as mere neutral genetic indicators of genomic inbreeding. Whether allozyme variation is a credible indicator of genomic variation is an issue that needs some serious attention. Mitton and Pierce (1980) and Chakraborty (1981) have shown that heterozygosity calculated from a small number of randomly selected, independently segregating loci (the usual allozyme sample) is poorly correlated with genomic heterozygosity in panmictic populations. Smouse (1986) has argued that measured heterozygosity of such a set is correlated with that of the whole genome only as the part is correlated with the whole and that the correlation between the markers and the rest of the genome is zero. With inbreeding or relatedness within the sample, allozyme variation *is* a partial indicator of genomic heterozygosity, but only a few studies of trees provide information on the relationships among individuals (Strauss 1986, 1987; Bush and Smouse 1991), and there are more often excesses of heterozygosity than decrements. The net value of extrapolation to the whole genome remains unclear.

The overdominance hypothesis postulates that the allozyme loci themselves directly influence fitness, arguing that heterozygosity for the assayed loci allows individuals to code for slightly different polypeptides, resulting in improved fitness. Studies by Koehn's group on LAP in *Mytilus edulis*, by Watts' group on PGI in *Colias* butterflies, and by Powers' group on LDH in *Fundulus heteroclitus* provide some physiological evidence for the adaptive significance of specific allozyme alleles in the homozygous state (see reviews by Koehn 1987; Powers 1987), however, little is known about the physiology of heterozygotes. There is no information of this type available for forest trees.

The overdominance hypothesis also predicts differential contributions to fitness by different allozyme loci and by different genotypes at single loci. Evidence for differential contributions of different loci violates the assumption of the dominance hypothesis that the allozyme loci function only as neutral markers. Although linkage of allozyme markers with loci carrying deleterious alleles of varying effects on fitness could cause differential contributions of allozyme loci to the relationship with fitness, this would only occur if there was no variation among individuals in the coupling-repulsion association of the two loci.

The relevant data to detect differential contributions of various alleles and loci are readily available, but these data are obscured by the usual regression of mean heterozygosity on fitness. Studies that have looked for specific-locus effects in forest trees have found them (Bush et al. 1987; Shea 1989; Bush and Smouse 1991). Strauss (1986) found that correlations between allozyme heterozygosity and canonical scores for growth variable ranged from −0.13 to +0.20 among inbred *Pinus attenuata* individuals, evidence that is suggestive of differential single-locus effects although these differences were not tested for statistical significance. Some of the most convincing evidence for the selective relevance of allozyme loci is that of Koehn et al. (1988), who found that the loci contributing most to the correlation of heterozygosity with growth rate were those involved in protein catabolism and glycolysis, rather that loci involved in other metabolic pathways for the bivalve *Mulinia lateralis.*

Strauss and Libby (1987), in one of the first studies to control for both inter-tree spacing and age, used a plantation of *Pinus radiata* to test two predictions that should help distinguish between the dominance and overdominance hypotheses: First, in a predominantly outcrossing species, if dominance were the correct hypothesis, heterozygote advantage should be strongest in populations with the greatest excesses of homozygotes, as they will be the most inbred and will thus have more easily detectable inbreeding depression. Also, the heterozygosity-fitness correlation should be largely confined to classes of below-average heterozygosity, if the phenomenon is caused by differences in inbreeding levels, since the higher heterozygosity classes are composed almost exclusively of outcrosses.

Strauss and Libby (1987) make three subsidiary predictions about the higher heterozygosity classes.

— No correlation is expected if the allozymes are neutral.
— A positive correlation is expected under overdominance.
— A negative correlation is expected if rare alleles are deleterious.

They examined mean early growth of 9 ramets cloned from an average of 8 trees per stand from 10 stands. The clones were grown to either 9 or 14 years of age in an evenly-spaced plantation. They found that populations with higher fixation indices had lower average growth, and found a positive correlation between clonal heterozygosity and mean growth at the lower heterozygosity levels, but a negative correlation at high heterozygosity levels. There was no correlation between the level of polymorphism of a locus and its contribution to the heterozygosity-growth correlation. All of these factors were interpreted as evidence against overdominance.

Control of breeding structure

Differentiating between the dominance and overdominance hypotheses requires knowledge of the breeding structure of the population. A correlation between heterozygosity and fitness can result from a mixed mating system (e.g., outcrossing mixed with selfing) even in the neutral case; the inbred offspring will be more homozygous at the marker loci than the outcrossed progeny from the same parent. Unfortunately, most studies have been conducted in total ignorance of the breeding history of the individuals examined. Outcrossed full-sib families should not show internal correlations between heterozygosity and fitness if variation among individuals in inbreeding coefficients is the main cause of the correlation; sibs all have virtually the same F-value, averaged over the genome. If there is a correlation, then we have to entertain the possibility that the marker loci are either themselves involved or that they are very closely linked to the loci that are responsible. Pierce and Mitton (1982) and Leary et al. (1987) report correlations among full-sibs for heterozygosity and measures of fitness for salamanders and trout, but a number of studies of sib groups in bivalves have failed to show such correlations (Beaumont et al. 1983; Gaffney and Scott 1984; Foltz and Chatry 1986).

Strauss (1986) found positive correlations between heterozygosity and growth within both selfed and outcrossed progeny groups of *Pinus attenuata*. Strauss reported a negative correlation between heterozygosity and cone production in the outcrosses but a positive correlation in the selfs. Allozyme heterozygosity explained an average of 2.3% of the individual variation in a number of fitness estimates for the selfed progeny. Somewhat counter to theory, the correlation was slightly larger in the outcrosses, but Strauss noted that all relationships between heterozygosity and both growth and cone production in the outcrosses were caused by three unusually slow-growing, highly fecund trees which made up the class of completely homozygous trees. Strauss (1987) found that the selfed progeny were less variable in growth than were the outcrosses. Among the outcrossed progeny, but not the selfs, the most heterozygous individuals exhibited the most year to year variability. He concluded that in light of the small correlations obtained, heterozygote superiority due to overdominance would be very difficult to detect in a natural population of a predominantly outcrossing species.

Adaptive distance

Formal evidence for the overdominance hypothesis can be obtained with an adaptive distance analysis (Smouse 1986), which tests the hypothesis that the fitnesses of the various homozygotes are proportional to their frequencies in the population, with the more common homozygotes being more fit. We first used the adaptive distance analysis to provide some evidence for overdominance of allozyme loci in our work on growth rate in *Pinus rigida* (Bush et al. 1987). Adaptive distance analysis consistently explained more of the variation (on average, 29%) than a simple count of heterozygous loci (2.5%), with some loci contributing more to the correlations than others, but with different loci playing a prominent role in different populations. Shea (1989) has conducted similar analyses for growth and cone production in two populations each of *Picea engelmanii* and *Abies lasiocarpa*, showing that about 25% of the variation could be accounted for under the assumptions of the adaptive distance model; a simple count of heterozygosity accounted for only 2% of the variation. A pair of particular loci in each species (DIA and GDH in *Picea engelmannii* and MPI and IDH in *Abies lasiocarpa*), made particularly notable contributions (Shea pers. comm.). In both cases, the improvement in prediction was in excess of that expected from merely fitting more parameters. Neither of these field studies controlled for breeding system, of course, but together they suggest that we should be looking at specific loci and genotypes.

Our most recent effort utilizes the adaptive distance model to test for overdominance using an experimental plot similar to that used by Strauss (1986, 1987). We used a plantation population of *Pinus taeda* consisting of the progeny of controlled self and controlled outcrosses from 45 mother trees, out-planted in a standard, blocked, field design in 1967 as part of a study of genetic load (Franklin 1968, 1969, 1972). We measured the survivors and assayed them for eight allozyme loci at age 19. With this controlled study, we were able to separate the effects of inbreeding *per se* from the effects of the marker loci themselves. As anticipated, we found profound inbreeding depression; coupled with Franklin's (1968) tally of losses before outplanting, our results show that by age 19, the survival of selfed progeny is only 8% that of outcrossed progeny, and further (and differential) losses were still occurring at the time of measurement (Bush and Smouse 1991).

The most interesting feature of the study, however, was that survival, growth, and fecundity differentials were decidedly non-random with respect to allozyme genotype for both selfed and outcrossed progeny. There was only one allele among the eight assayed that was an obvious candidate for a balanced polymorphism: PGI2-0010 (seen in the parental trees at a frequency of 0.18), when in heterozygous form with the most common allele, PGI2-0100, was mildly overdominant with respect to height growth and, probably, fecundity, but suffered a survival disadvantage relative to the common homozygote. The rarer alleles at this and other loci suffered serious survival disadvantages in homozygous form and suffered less severe survival deficits even in heterozygotes, but exhibited subtle growth and fecundity advantages. The net effect would seem to be overdominance over the whole life cycle. Projection of the genetic composition of this stand through computer simulation across life history stages results in a balance of forces that maintains the alleles at their (observably) low population frequencies.

While no one phase of the life cycle yields evidence for balancing selection, the net effect over the life cycle is just that. Generalized inbreeding depression is undeniable for forest trees, but that cannot be the whole explanation for the allozyme-specific effects we see here. There is no credible evidence for inbreeding either in the mother trees or resulting from the cross pollinations; Franklin (1971) estimated the inbreeding coefficient among the maternal trees to be only $F = 0.0003$, too small to account for the sizes of the effects we see if the allozymes were neutral markers of the genomic condition.

Deleterious rare alleles and segregation distortion

A common and intriguing observation emerging from recent studies is the general fitness deficit experienced by individuals carrying alleles that are relatively rare in the population. Strauss and Libby (1987) found a positive correlation between growth and heterozygosity for low levels of heterozygosity (mostly involving highly polymorphic alleles); at high heterozygosity levels (involving additional loci with rarer alleles), the correlation between growth and heterozygosity was negative. They suggested that heterozygosity for low-frequency alleles caused the decline, although they did not find differences between growth of individuals carrying different alleles in single-locus tests. Our own work on *Pinus taeda* (Bush and Smouse 1991) showed severe survival selection against low-frequency alleles, both in heterozygous and homozygous form.

These are not the only reports of deleterious effects associated with rare allozyme alleles in conifers. Strauss and Conkle (1986) found deficits of rare alleles in 10 of 11 cases of segregation distortion at the gametophyte level in *Pinus attenuata*. Deficits of rare alleles have also been reported in gametophytes of *Pinus sylvestris* L. (Rudin 1977), *Pinus muricata* D. Don (Millar 1985), and *Calocedrus decurrens* (Torr.) Florin (Harry 1986). Reports of segregation distortion at the gametophyte level are common in conifers; unfortunately, most reports do not include population allele frequencies, so we do not know whether the rare alleles are those being lost. One suspects that the information is available, though often not published. If this rare-allele pattern should turn out to be general, however, it will mean that while highly polymorphic alleles are obvious candidates for balancing selection, rare alleles are not.

Associative overdominance

Our work on *Pinus taeda* supports both dominant and overdominant explanations to differing degrees (Bush and Smouse 1991). We found that homozygosity for rare alleles is correlated with decreased fitness, as is expected under dominance, however, given the lack of credible correlation between a sample of neutral markers and genomic heterozygosity (for those unlinked parts of the genome thought to be reponsible for fitness) in the absence of population structure (Mitton and Pierce 1980; Chakraborty 1981; Smouse 1986), we must continue to entertain the hypothesis that some part of the correlations between heterozygosity and fitness are due either to the marker loci themselves or to unseen but linked genetic loci that are responsible for the fitness effects. Those linked loci might themselves be fitness-overdominant, in which case we have traded an overdominant locus we can see for a phantom locus we cannot observe. Alternatively, we can envisage a set of loci, closely linked to a neutral allozyme marker, each of which is subject to dominance selection against one or more rare deleterious alleles. Although each locus individually behaves in dominant fashion, recombination is so limited that the block of genes behaves as an overdominant unit, leading to associative overdominance (Ohta 1971). Is it possible that we are really observing the effects of small blocks of genes around our marker loci?

Koehn et al. (1988) report two tightly linked loci having very different correlations with growth rate in the bivalve *Mulinia lateralis*, evidence against the associative overdominance hypothesis, which requires substantial linkage disequilibrium (Zouros et al. 1980). Comparable data on tightly linked loci are not available in forest trees, and standard estimates of N_e would seem to require small recombination rates for the linkage argument to have much force. Any locus used as a marker will have other loci closely linked, of course, loci that will probably be in linkage disequilibrium with the marker locus. Rare alleles, in particular, might be expected to show disequilibrium with their flanking loci; neutral theory suggests that they are younger than the more common alleles, which might lead to the rare-allele effects we have described. The problem with this explanation is that measured inbreeding coefficients among reproducing adults are so small (even negative in many cases) that there is no reason to expect disequilibria except at very short recombinational distances. Moreover, the low-frequency allozyme alleles under discussion are widespread throughout a given species, sometimes even putatively occurring in related species.

There has probably been more than enough evolutionary time for rare alleles to come to linkage equilibrium with the flanking loci; the claim to linkage disequilibrium and associative overdominance has been facile, but may not stand up to close examination. The truth is that we need some hard data on this point. As Strauss (pers. comm.) has pointed out, gene transfer studies could help in resolving this issue. As a practical matter, we cannot distinguish between overdominance and associative overdominance at the survey level, but at that same level, the distinction between whether selection is imposed by the biological features of the allozymes themselves or by the corresponding features of their closely linked neighbors is unimportant.

Practical implications

While there has been much speculation, most of it negative (Bongarten et al. 1985; Mitton and Jeffers 1989), on the efficacy of using allozyme genotype as a guide to artificial selection, as yet no one has reported an attempt to do so. Experiments contrasting growth of individuals selected for superior early growth with run-of-the-mill nursery seedlings are only an indirect approach to the problem (Mitton and Jeffers 1989). How much of the variation in fitness traits can be explained by allozyme heterozygosity will remain an open question in the absence of studies utilizing progeny of controlled uniparental crosses, grown under homogeneous conditions, that have been assayed from the embryo stage on. It is quite possible that by selection based on allozyme genotypes, we could increase potential for future growth, but in planning future studies it should be remembered that non-random genetic losses are cumulative and that a stand becomes progressively skewed toward more heterozygous individuals with increasing age, so that studies of young trees, or studies done under artificial or benign conditions will probably not tell us anything useful.

In view of the devastating effects of air pollution on forests, especially those of eastern Europe, it does not seem to be too soon to increase the amount of effort being expended in determining how genetic markers can be used to identify and breed pollution tolerant trees. Recent work on trees differentially sensitive to air pollution has not produced strong evidence for the role of heterozygosity or other aspects of genotype in response to environmental toxins (Meinartowicz 1983; Müller-Stark 1985; Geburek et al. 1987), however, work of this nature could have applicability for breeding and for reforestation in heavily polluted environments.

It is pointless to argue about how much of the variation explained by allozymes is actually due to closely linked loci until we have a much better knowledge of the structure of the conifer genome. For instance, we currently know little about the amount of linkage disequilibrium existing among allozyme loci in conifers. Lacking a set of closely linked allozyme markers, this question will probably need to be addressed using RFLP markers.

The use of RFLP markers could facilitate the study of the relationship between heterozygosity and fitness in a number of ways. The question of whether heterozygosity of a genetic marker itself is responsible for increased fitness might be addressed by comparing restriction site heterozygosity in coding and non-coding regions of DNA, assuming that the non-coding regions are not involved in functions such as regulation. The use of RFLP's would also reduce the problem of identifying similarly migrating allozyme bands as allelic, however, there is no assurance that DNA sequences that appear to be allelic on the basis of similar identical restriction sites do not differ in some other manner.

If it becomes evident that rare alleles are in general deleterious, application for selective breeding should include eliminating trees carrying extremely low frequency alleles from seed orchards. We might wish to consider creating orchards composed only of clones carrying highly polymorphic alleles, with clonal replication designed to result in allele frequencies similar to those found in natural stands. Such a strategy runs quite counter to usual notions of gene conservation, and measures would have to be taken to preserve rare alleles outside of seed orchards while more is discovered about their functions.

Acknowledgments

The authors would like to thank Tom Ledig, Jeff Mitton, Steve Strauss and Nick Wheeler for critical review of the manuscript; and Kathy Shea and Don Dettayes for sharing their unpublished data. This work was supported by NJAES/USDA-32102.

References

Alden, J. and Loopstra, C. 1987. Genetic diversity and population structure of *Picea glauca* on an altitudinal gradient in interior Alaska. Can. J. For. Res. 17: 1519—1526.

Beaumont, A. R., Beveridge, C. M. and Budd, M. D. 1983. Selection and heterozygosity within single families of the mussel *Mytilus edulis* (L.). Mar. Biol. Lett. 4: 151—161.

Bergmann, F. 1978. The allelic distribution at an acid phosphatase locus in Norway spruce (*Picea abies*) along similar climatic gradients. Theor. Appl. Genet. 52: 57—64.

Bongarten, B. C., Wheeler, N. C. and Jech, K. S. 1985. Isozyme heterozygosity as a selection criterion for yield improvement in Douglas-fir, pp. 121—128. In: Caron, F., Corriveau, A. G. and Boyle, T. J. B. (Eds) Proc. 20th meeting of the Canadian Tree Improvement Assoc., Quebec City, Quebec, Canada, August 19—22, 1985. 2. New ways in forest genetics. Canadian Forestry Service, Ottawa, Canada.

Boyle, T. J. B. and Morgenstern, E. K. 1986. Estimates of outcrossing rates in six populations of black spruce in central New Brunswick. Silvae Genet. 35: 102—106.

Bush, R. M., Smouse, P. E. and Ledig, F. T. 1987. The fitness consequences of multiple-locus heterozygosity: The relationship between heterozygosity and growth rate in pitch pine (*Pinus rigida* Mill.). Evolution 41: 787—798.

Bush, R. M. and Smouse, P. E. 1991. The impact of electrophoretic genotype on life history traits in *Pinus taeda*. Evolution 45: 481—498.

Chakraborty, R. 1981. The distribution of the number of heterozygous loci in an individual in natural populations. Genetics 98: 461—466.

Cheliak, W. M., Pitel, J. A. and Murray, G. 1984. Population structure and the mating system of white spruce. Can. J. For. Res. 15: 301—308.

Cheliak, W. M., Dancik, B. P., Morgan, K., Yeh, F. C. H. and Strobeck, C. 1985. Temporal variation of the mating system in a natural population of jack pine. Genetics 109: 569—584.

Conkle, M. T., Schiller, G. and Grunwald, C. 1988. Electrophoretic analysis of diversity and phylogeny of *Pinus brutia* and closely related taxa. Syst. Bot. 13: 411—424.

Dancik, B. P. and Yeh, F. C. 1983. Allozyme variability and evolution of lodgepole pine (*Pinus contorta* var. *latifolia*) and jack pine (*P. banksiana*) in Alberta. Can. J. Genet. Cytol. 25: 57—64.

Dickinson, T., Knowles, P. and Parker, W. H. 1988. Data set congruence in northern Ontario tamarack (*Larix laricina*, Pinaceae). Syst. Bot. 13: 442—455.

El-Kassaby, Y. A. and Sziklai, O. 1982. Genetic variation of allozyme and quantitative traits in a selected Douglas-fir [*Pseudotsuga menziesii* var. *menziesii* (Mirb.) Franco] population. For. Ecol. Manag. 4: 115—126.

El-Kassaby, Y. A., Meagher, M. D., Parkinson, J. and Portlock, F. T. 1987. Allozyme inheritance, heterozygosity and outcrossing rate among *Pinus monticola* near Ladysmith, British Columbia. Heredity 58: 173—181.

Farris, M. A. and Mitton, J. B. 1984. Population density, outcrossing rate, and heterozygote superiority in ponderosa pine. Evolution 38: 1151—1154.

Fins, L. and Libby, W. J. 1982. Population variation in *Sequoiadendron*: Seed and seedling studies, vegetative propagation, and isozyme variation. Silvae Genet. 31: 102—110.

Foltz, D. W. and Chatry, M. 1986. Genetic heterozygosity and growth rate in Louisiana oysters (*Crassostrea virginica*). Aquaculture 57: 261—269.

Franklin, E. C. 1968. Artificial self-pollination and natural inbreeding in *Pinus taeda* L. Ph.D. thesis. North Carolina State Univ., Raleigh.

——. 1969. Mutant forms found by self-pollination in loblolly pine. J. Hered. 60: 315—320.

——. 1971. Estimates of frequency of natural selfing and of inbreeding coefficients in loblolly pine. Silvae Genet. 20: 194—195.

——. 1972. Genetic load in loblolly pine. Am. Nat. 106: 262—265.

Furnier, G. R. and Adams, W. T. 1986a. Mating system in natural populations of Jeffrey pine. Am. J. Bot. 73: 1002—1008.

Furnier, G. R. and Adams, W. T. 1986b. Geographic patterns of allozyme variation in Jeffrey pine. Am. J. Bot. 73: 1009—1015.

Gaffney, P. M. and Scott, T. M. 1984. Genetic heterozygosity and production traits in natural and hatchery populations of bivalves. Aquaculture 42: 289—302.

Geburek, T., Scholz, F., Knabe, W. and Vornweg, A. 1987. Genetic studies by isozyme gene loci on tolerance and sensitivity in an air polluted *Pinus sylvestris* field trial. Silvae Genet. 36: 49—53.

Grant, M. C. and Mitton, J. B. 1977. Genetic differentiation among growth forms of Engelmann spruce and subalpine fir at tree line. Arct. Alp. Res. 9: 259—263.

Guries, R. P. and Ledig, F. T. 1981. Genetic structure of populations and differentiation in forest trees, pp. 42—47. In: Conkle, M. T. (tech. coord.) Proceedings of the symposium on isozymes of North American forest trees and forest insects. USDA For. Serv. Gen. Tech. Rep. PSW-48.

Hamrick, J. L., Blanton, H. M. and Hamrick, K. J. 1989. Genetic structure of geographically marginal populations of ponderosa pine. Am. J. Bot. 76: 1559—1568.

Hamrick, J. L. and Godt, M. J. W. 1990. Allozyme diversity in plant species, pp. 43—63. In: Brown, A. H. D., Clegg, M. T., Kahler, A. L. and Weir, B. S. (Eds) Plant Population Genetics, Breeding, and Genetic Resources. Sinauer Associates, Sunderland, MA.

Harry, D. E. 1986. Inheritance and linkage of isozyme variants in incense-cedar. J. Hered. 77: 261—266.

Knowles, P. and Grant, M. C. 1981. Genetic patterns associated with growth variability in ponderosa pine. Am. J. Bot. 68: 942—946.

Knowles, P. and Grant, M. C. 1985. Genetic variation of lodgepole pine over time and microgeographical space. Can. J. Bot. 63: 722—727.

Knowles, R. and Mitton, J. B. 1980. Genetic heterozygosity and radial growth variability in *Pinus contorta*. Silvae Genet. 29: 114—118.

Koehn, R. K. 1987. The importance of genetics to physiological ecology, pp. 170—185. In: Feder, M. E., Bennett, A. F., Burggren, W. W. and Huey, R. B. (Eds) New Directions in Ecological Physiology. Cambridge University Press, Cambridge.

Koehn, R. K., Diehl, W. J. and Scott, T. M. 1988. The differential contribution by individual enzymes of glycolysis and protein catabolism to the relationship between heterozygosity and growth rate in the coot clam, *Mulinia lateralis*. Genetics 118: 121—130.

Lagercrantz, U. and Ryman, N. 1990. Genetic structure of Norway spruce (*Picea abies*): Concordance of morphological and allozymic variation. Evolution 44: 38—53.

Leary, R. F., Allendorf, F. W. and Knudsen, K. L. 1987. Differences in inbreeding coefficients do not explain the association between heterozygosity at allozyme loci and developmental stability in rainbow trout. Evolution 41: 1413—1415.

Ledig, F. T. and Fryer, J. H. 1972. A pocket of variability in *Pinus rigida*. Evolution 26: 259—266.

Ledig, F. T., Zobel B. J. and Matthias, M. F. 1975. Geoclimatic patterns in specific gravity and tracheid length in wood of pitch pine. Can J. For. Res. 5: 318—329.

Ledig, F. T., Lambeth C. C. and Linzer, D. I. H. 1976. Nursery evaluation of a pitch pine provenance trial, pp. 93—108. In: Proc. 23rd Northeast. forest tree improv. conf., New Brunswick, NJ.

Ledig, F. T., Gruies, R. P. and Bonefeld, B. A. 1983. The relation of growth to heterozygosity in pitch pine. Evolution 37: 1227—1238.

Lerner, I. M. 1954. Genetic Homeostasis. Oliver and Boyd, Edinburgh, 134 pp.

Linhart, Y. B., Grant, M. C. and Montazer, P. 1989. Experimental studies in ponderosa pine. I. Relationship between variation in proteins and morphology. Am. J. Bot. 76: 1024—1032.

Linhart, Y. B. and Mitton, J. B. 1985. Relationships among reproduction, growth rates, and protein heterozygosity in ponderosa pine. Am. J. Bot. 72: 181—184.

Linhart, Y. B., Mitton, J. B., Sturgeon, K. B. and Davis, M. L. 1981. Genetic variation in space and time in a population of ponderosa pine. Heredity 46: 407—426.

Loveless, M. D. and Hamrick, J. L. 1984. Ecological determinants of genetic structure in plant populations. Ann. Rev. Ecol. Syst. 15: 65—95.

Lundkvist, K. 1979. Allozyme frequency distributions in four Swedish populations of Norway spruce (*Picea abies* K.). I. Estimations of genetic variation within and among populations, genetic linkage and a mating system parameter. Hereditas 90: 127—143.

Mejnartowicz, L. E. 1983. Changes in genetic structure of Scots pine (*Pinus silvestris* L.)

population affected by industrial emission of fluoride and sulfur dioxide. Genet. Pol. 24: 41–50.

Millar, C. I. 1983. A steep cline in *Pinus muricata*. Evolution 37: 311–319.

Millar, C. I. 1985. Inheritance of allozyme variants in bishop pine (*Pinus muricata* D. Don). Biochem. Genet. 23: 933–946.

Mitton, J. B. and Grant, M. C. 1980. Observations on the ecology and evolution of quaking aspen, *Populus tremuloides*, in the Colorado Front Range. Am. J. Bot. 67: 202–209.

Mitton, J. B. and Grant, M. C. 1984. Associations among protein heterozygosity, growth rate, and developmental homeostasis. Ann. Rev. Ecol. Syst. 15: 479–499.

Mitton, J. B. and Jeffers, R. M. 1989. The genetic consequences of mass selection for growth rate in Engelmann spruce. Silvae Genet. 38: 6–12.

Mitton, J. B., Knowles, P. Sturgeon, K. B., Linhart, Y. B. and Davis, M. 1981. Associations between heterozygosity and growth rate variables in three western forest trees, pp. 27–34. In: Conkle, M. T. (tech. coord.) Proceedings of the Symposium on isozymes of North American forest trees and forest insects. USDA For. Serv. Gen. Tech. Rep. PSW-48.

Mitton, J. B., Linhart, Y. B., Hamrick, J. L. and Beckman, J. S. 1977. Observation on the genetic structure and mating system of ponderosa pine in the Colorado Front Range. Theor. Appl. Genet. 51: 5–13.

Mitton, J. B. and Pierce, B. A. 1980. The distribution of individual heterozygosity in natural populations. Genetics 95: 1043–1054.

Mitton, J. B., Sturgeon, K. B. and Davis, M. L. 1980. Genetic differentiation in ponderosa pine along a steep elevational transect. Silvae Genet. 29: 100–103.

Moran, G. F., Bell, J. C. and Turnbull, J. W. 1989. A cline in genetic diversity in river she-oak *Casuarina cunninghamiana*. Aust. J. Bot. 37: 169–180.

Moran, G. F. and Brown, A. H. D. 1980. Temporal heterogeneity of outcrossing rates in alpine ash (*Eucalyptus delegatensis* R. T. Bak.). Theor. Appl. Genet. 57: 101–105.

Müller-Starck, G. 1985. Genetic differences between “tolerant” and “sensitive” beeches (*Fagus sylvatica* L.) in an environmentally stressed adult forest stand. Silvae Genet. 34: 241–247.

Neale, D. B. and Adams, W. T. 1985. Allozyme and mating-system variation in balsam fir (*Abies balsamea*) across a continuous elevational transect. Can. J. Bot. 63: 2448–2453.

Nei, M. 1972. Genetic distance between populations. Am. Nat. 106: 283–292.

Niebling, C. R. and Conkle, M. T. 1990. Diversity of Washoe pine and comparisons with allozymes of ponderosa pine races. Can. J. For. Res. 20: 298–308.

O'Malley, D. M., Allendorf, F. W. and Blake, G. M. 1979. Inheritance of isozyme variation and heterozygosity in *Pinus ponderosa*. Biochem. Genet. 17: 233–250.

Ohta, T. 1971. Associative overdominance caused by linked detrimental mutations. Genet. Res. Camb. 18: 277–286.

Perry, D. J. and Dancik, B. P. 1986. Mating system dynamics of lodgepole pine in Alberta, Canada. Silvae Genet. 35: 190–195.

Pierce, B. A. and Mitton, J. B. 1982. Allozyme heterozygosity and growth in the tiger salamander, *Ambystoma tigrinum*. J. Hered. 73: 150–153.

Plessas, M. E. and Strauss, S. H. 1986. Allozyme differentiation among populations, stands, and cohorts in Monterey pine. Can. J. For. Res. 16: 1155–1164.

Powers, D. A. 1987. A multidisciplinary approach to the study of genetic variation within species, pp. 102–130. In: Feder, M. E., Bennett, A. F., Burggren, W. W. and Huey, R. B. (Eds) New Directions in Ecological Physiology. Cambridge University Press, Cambridge.

Ross, H. A. and Hawkins, J. L. 1986. Genetic variation among local populations of jack pine (*Pinus banksiana*). Can. J. Genet. Cytol. 28: 453–458.

Rudin, D. 1977. Leucine-amino-peptidases (LAP) from needless and macrogametophytes of *Pinus sylvestris* L. Inheritance of allozymes. Hereditas 85: 219–226.

Schiller, G., Conkle, M. T. and Grunwald, C. 1986. Local differentiation among Mediterranean populations of Aleppo pine in their isoenzymes. Silvae Genet. 35: 11—19.

Shaw, D. V. and Allard, R. W. 1982. Estimation of outcrossing rates in Douglas-fir using isozymes markers. Theor. Appl. Genet. 62: 113—120.

Shea, K. L. 1987. Effects of population structure and cone production on outcrossing rates in Engelmann spruce and subalpine fir. Evolution 41: 124—136.

Shea, K. L. 1989. The relationship between heterozygosity and fitness in Engelmann spruce and subalpine fir. Am. J. Bot. 76 (suppl.): 153—154 (Abstr.)

Smouse, P. E. 1986. The fitness consequences of multiple-locus heterozygosity under the multiplicative overdominance and inbreeding depression models. Evolution 40: 946—957.

Snyder, T. P., Steward, D. A. and Stricker, A. F. 1985. Temporal analysis of breeding structure in jack pine *Pinus banksiana* (Lamb.). Can. J. For. Res. 15: 1159—1166.

Spurr, S. H. and Barnes, B. V. 1980. Forest Ecology. 3rd edn, Wiley, New York.

Steinhoff, R. J., Joyce, D. G. and Fins, L. 1983. Isozyme variation in *Pinus monticola*. Can. J. For. Res. 13: 1122—1132.

Strauss, S. H. 1986. Heterosis at allozyme loci under inbreeding and crossbreeding in *Pinus attenuata*. Genetics 113: 115—134.

Strauss, S. H. 1987. Heterozygosity and developmental stability under inbreeding and crossbreeding in *Pinus attenuata*. Evolution 41: 331—339.

Strauss, S. H. and Conkle, M. T. 1986. Segregation, linkage, and diversity of allozymes in knobcone pine. Theor. Appl. Genet. 72: 483—493.

Strauss, S. H. and Libby, W. J. 1987. Allozyme heterosis in radiata pine is poorly explained by overdominance. Am. Nat. 130: 879—890.

Wheeler, N. C. and Guries, R. P. 1982. Population structure, genic diversity, and morphological variation in *Pinus contorta* Dougl. Can. J. For. Res. 12: 595—606.

Wright, S. 1969. Evolution and the Genetics of Populations. Volume 2. The Theory of Gene Frequencies. Univ. of Chicago Press, Chicago.

Yeh, F. C. and Layton, C. 1979. The organization of genetic variability in central and marginal populations of lodgepole pine *Pinus contorta* spp. *latifolia*. Can. J. Genet. Cytol. 21: 487—503.

Yeh, F. Ch.-H. and O'Malley, D. 1980. Enzyme variations in natural populations of Douglas-fir, *Pseudotsuga menziesii* (Mirb.) Franco, from British Columbia. 1. Genetic variation patterns in coastal populations. Silvae Genet. 29: 83—92.

Yeh. F. C. and Arnott, J. T. 1986. Electrophoretic and morphological differentiation of *Picea sitchensis, Picea glauca*, and their hybrids. Can. J. For. Res. 16: 791—798.

Yeh, F. C., Khalil, M. A. K., El-Kassaby, Y. A. and Trust, D. C. 1986. Allozyme variation in *Picea mariana* from Newfoundland: genetic diversity, population structure, and analysis of differentiation. Can. J. For. Res. 16: 713—720.

Yeh, F. C. and Morgan, K. 1987. Mating system and multilocus associations in a natural population of *Pseudotsuga menziesii* (Mirb.) Franco. Theor. Appl. Genet. 73: 799—808.

Zouros, E., Singh, S. M. and Miles, H. W. 1980. Growth rate in oysters: an overdominant phenotype and its possible explanations. Evolution 34: 856—867.

New Forests **6**: 197—216, 1992.

Review paper

The dynamic mating systems of conifers

J. B. MITTON
Department of Environmental, Population and Organismic Biology, Campus Box 334, The University of Colorado, Boulder, Colorado 80309, USA

Received 8 October 1990; accepted 26 December 1991

Key words: breeding systems, allozyme variation, natural selection

Application. Increasingly sophisticated analyses of the mating systems of forest trees have reported observations incompatible with the assumptions employed by the estimation procedures. These observations include:

1) heterogeneity of male and female mating success
2) selection acting upon genotypes in seeds
3) heterogeneity in the pollen pool over time and among individuals
4) heterogeneity in rates of outcrossing among individuals

Further studies of mating systems are needed to understand the dynamics within natural stands, to most efficiently manage seed orchards, and to prescribe more effective gene conservation practices.

Abstract. Conifer mating systems vary among species and within species; both ecological variables and genetic variation cause mating systems to be dynamic. Within species, estimates of rates of outcrossing vary among populations, among loci, and among individuals within stands. The level of outcrossing varies with stand density, age, and the abundance of local and foreign pollen. Variation in the mating success of both males and females violates the assumption that populations are randomly mating and at equilibrium, and justifies more quantitative analyses of mating systems. Allelic frequencies in the pollen pool and in the pool of receptive female surfaces may vary through a season and among seasons, producing positively assortative mating in time. Seed and seedling viability selects against homozygotes for lethal alleles and favors heterozygous genotypes, biasing estimates of outcrossing. Particularly fruitful topics of research include the genetic consequences of polyembryony, differential male and female mating success, and associations between specific genotypes and floral phenology.

Introduction

The importance of mating systems has long been apparent to theoreticians, field biologists, and plant breeders (Wright 1931, 1969; Allard et al. 1968; Lande and Schemske 1985; Brown 1990). Mating systems determine the distribution of genotypes within populations, and influence the degree of differentiation among populations. While outcrossing promotes gene flow, and brings genotypic proportions to Hardy—Weinberg equilibrium, selfing reduces gene flow, and brings genotypic distributions to an equilibrium described by Wright's equilibrium law (Wright 1931, 1969). Selfing reduces heterozygosity by half each generation, reducing the effective rate of recombination, and thereby promoting genetic organization within populations (Allard et al. 1968). Lower levels of gene flow permit higher levels of differentiation among populations. The importance of the mating system is reflected in the ongoing interest in its evolution (Lande and Schemske 1985; Schemske and Lande 1985), in the advent of paternity analysis (Ellstrand 1984; Ellstrand and Marshall 1985; Hamrick and Schnabel 1985; Meagher 1986; Meagher and Thompson 1986; Ellstrand 1992), and in numerous studies characterizing population structure (Hamrick 1983; Loveless and Hamrick 1984), components of the mating system (Clegg 1980; Brown 1990), and gene flow (Levin and Kerster 1974; Adams and Birkes 1990a, b).

For a first approximation, mating systems can be grouped into 5 categories: predominantly selfing, predominantly outcrossing, mixed selfing and outcrossing, apomictic, and haploid selfing (Brown 1990). The mating systems of conifers fall into the mixed mating category. That is, some proportion of seed is produced by selfing (s), and the complementary proportion ($t = 1 - s$) is produced by outcrossing. There is dramatic variation among species, with most species almost completely outcrossed (Muller 1977a, b; Mitton et al. 1977, 1981; Adams and Joly 1980; Moran et al. 1980; Shaw and Allard 1981, 1982; Shen et al. 1981; El-Kassaby et al. 1981, 1986, 1987, 1988, 1992; Epperson and Allard 1984; Farris and Mitton 1984; King et al. 1984; Cheliak et al. 1985; Snyder et al. 1985; Friedman and Adams, 1985a; Neale and Adams, 1985a, b: Furnier and Adams, 1986; Shea, 1987; Denti and Schoen, 1988; Erickson and Adams, 1989, 1990; Gibson and Hamrick, 1991), and others, such as tamarack, *Larix laricina*, and eastern white cedar, *Thuja occidentalis*, having population outcrossing rates as low as $t = 0.53$ (Knowles et al. 1987) and $t = 0.51$ (Perry and Knowles 1990), respectively.

This paper focuses upon the variation in the mating system within a species. It presents a brief overview of methods, notes the assumptions employed in the estimation procedures, and discusses interesting violations of the assumptions. Readers seeking more general or more comprehensive treatments of mating systems are referred to the reviews by Allard et al. (1977), Ritland (1983), Brown et al. (1985, 1989), and Brown (1990).

Many conifers suffer marked inbreeding depression (Sorensen 1969, 1971, 1982; Franklin 1970) and the fitness differentials between selfed and outcrossed progeny probably influence the evolution of plant mating systems (Lande and Schemske 1985). Selfed genotypes exhibit lower germinability, slower growth rates, and higher mortality. While the fitness consequences of selfing are profound, the genetic consequences are not really apparent from the perspective of one or a few loci. Consider a locus with 2 alleles, A and a, with allelic frequencies equal to 0.5. A population in outcrossing equilibrium would exhibit the AA, Aa, and aa genotypes in the proportions 0.25, 0.50, and 0.25, respectively. A single round of selfing, which reduces heterozygosity by 50%, alters the genotypic proportions to 0.375, 0.125, and 0.375, respectively. From this single locus perspective, selfing has produced a subtle shift in the frequencies of genotypes, but no new genotypes have appeared. A multilocus model with 2 or a few loci gives similar results — selfing only alters the frequencies of the genotypes already present. When we approach biological reality by considering hundreds or thousands of polymorphic loci, however, the genotypic distributions of selfed and outcrossed progeny are quite distinct (Mitton and Jeffers 1989). In highly variable forest trees, selfing produces many different multilocus homozygotes that are not expected in outcrossing populations. This loss in heterozygosity, in combination with an enhanced expression of rare deleterious recessive alleles, produces inbreeding depression.

Quantitative studies of the mating system

The earliest measurements of mating systems necessarily relied upon morphological markers, but current studies rely primarily upon protein polymorphisms. Future studies will no doubt also employ protein polymorphisms, but will probably include as well the hypervariable sites detected with DNA fingerprinting (Jeffreys et al. 1985a, b).

F and H statistics

Wright (1931, 1969) defined the inbreeding coefficient F, which measures the probability that a genotype is homozygous identical by descent, a direct measure of the increase in homozygosity due to inbreeding. To quantify hierarchical population structure within a species, he elaborated several measures of F, F_{it}, F_{is}, and F_{st}, and described the relationship among them:

$$(1 - F_{it}) = (1 - F_{is})(1 - F_{st}) \tag{1}$$

F_{is} measures the degree of inbreeding within a subpopulation, F_{st} measures the variance in allelic frequencies among subpopulations, and F_{it} measures the apparent value of F, attributable to both F_{is} and F_{st}, when the subpopulations are lumped into a single population. These statistics are calculated directly for a locus segregating 2 alleles, but the treatment of loci with many alleles is cumbersome. The H statistics of Nei (1973) can also be used to quantify the apportionment of genetic variation and they are easier to calculate when loci segregate many alleles. F and H statistics estimate the degree of inbreeding within subpopulations, the amount of variation within the entire species, within subpopulations, and among subpopulations (Hamrick 1983, 1987, 1989; Loveless and Hamrick 1984).

F statistics also give an estimate of degree of selfing, for, at equilibrium the relationship between F and s are:

$$F = s/(2 - s) \tag{2}$$

F and H statistics have the advantage that they employ genetic data sampled at a single time in the life cycle, say, from genetic data obtained from needle tissue from adults. However, the inference of selfing may be confounded by selection (Allard et al. 1977). This is especially a problem when data are gathered from needle tissue of adult conifers, for viability selection selects against selfed genotypes and favors heterozygous genotypes in conifers (Shaw and Allard 1982b; Farris and Mitton 1984; Plessas and Strauss 1986; Knowles et al. 1987; Muona et al. 1987; Govindaraju and Dancik 1987; Mitton and Jeffers 1989; Mopper et al. 1991).

Estimates of the mating system from progeny arrays

Single locus data gathered from progeny arrays can be used to estimate frequencies in the effective pollen pool and to estimate the proportion of seed set by selfing (Jones 1916; Fyfe and Bailey 1951; Brown and Allard 1970). This estimate uses an iterative maximum likelihood estimation procedure that infers the maternal genotype, and estimates allelic frequencies in the effective pollen pool and the rate of outcrossing. When the maternal genotype is known, the Expectation—Maximation algorithm (Cheliak et al. 1983) can be used to obtain maximum likelihood estimates of parameters of the mating system. This algorithm is more convenient when there are 3 or more alleles, and it also avoids the occasional errors inherent in the inference of the maternal genotype from the distribution of genotypes in a progeny array (Mitton et al. 1981). In addition to providing estimates of the proportion of seed set by outcrossing, these procedures provide estimates of the frequencies in the successful pollen pool. Tests of homogeneity of frequencies in pollen and in mature individuals gives greater insight into the mating system, and allows some of the assumptions employed by these procedures (below) to be tested.

When data from several loci are available, single locus estimates of selfing can be averaged across loci. Alternatively, information from all of the loci can be used to obtain multilocus estimates of selfing (Green et al. 1980; Shaw et al. 1981; Ritland and Jain 1981; Yeh and Morgan 1987). Multilocus methods are more efficient at identifying outcrosses, and usually return estimates of outcrossing that exceed single locus estimates by 5—10% (Brown 1990). This discrepancy in the estimates of outcrossing may be attributable to outcrosses between related individuals. The effective selfing model of Ritland (1984, 1986) attempts to partition the inbreeding into components of selfing and biparental inbreeding.

A modification of the maximum likelihood estimation of Green et al. (1980) estimates the mating system from a single progeny array (Neale and Adams 1985b; Ritland and El-Kassaby 1985). This procedure can be used to test for homogeneity in the mating systems of trees within a stand.

Gene flow

Gene flow influences neighborhood size within populations and the magnitude of differentiation among populations. Most wind dispersed conifer seeds fall within a few hundred feet (Wright 1976), but seeds harvested and cached by birds will have the potential for moving kilometers rather than meters. Wind borne pollen has the potential to move dozens or hundreds of kilometers, but it is not presently known how successful long distance disperal is. The important question is not how far can pollen travel, but what is the distribution of distances of the effective pollen? Forest biologists have used chemical dyes and radioactively labeled pollen to estimate gene flow (reviewed in Levin and Kerster 1974), and have utilized trees bearing rare alleles to study effective pollen flow (Muller 1977a). Because trees bearing rare alleles are, by definition, rare, this technique is not generally useful.

Indirect measures of gene flow

The distribution of genetic variation among populations can be used to estimate amounts of gene flow among populations (Slatkin 1985, 1987). The number of individuals migrating between populations within a generation, *Nm*, can be estimated from the variance of allelic frequencies among populations, F_{st} with the following relationship:

$$\mathrm{Nm} = (1/F_{st} - 1)/4 \tag{3}$$

When this measure is applied to genetic data from wind pollinated forest trees, *Nm* is usually in the range of 5—20 (Hamrick 1987; Schuster et al. 1989). Indirect measures of gene flow reflect historical rates of gene flow upon the present distribution of genetic variation, and may not represent the pattern of gene flow in any single year or any single generation. In addition, the estimate of gene flow can be biased upwards by stabilizing selection upon genetic variation or be biased downwards by diversifying selection. Finally, indirect measures infer gene flow among populations, but not within populations. To learn about the year to year variation in gene flow, and to measure gene flow within populations and within managed stands, direct measures of gene flow are needed.

Direct measures of gene flow

Paternity analyses allow direct measures of the mean and variance of the distance of pollen flow within populations, and allow identification of the proportion of pollinations originating outside of the population. A seed collected from any plant contains equal contributions from female and male parents, and sufficient genetic variation allows unambiguous identification of the male parent (Ellstrand 1984). This technique is rich in promise and fraught with difficulties. Paternity studies have returned a low proportion, usually around 5%, of seed for which a single male parent is identified. Other seeds, bearing more common genotypes, have several to dozens of potential male parents. Clearly, the most direct solution to this problem is to utilize methods that identify sufficient genetic variation to uniquely identify paternal parents.

When unambiguous identification of male parents is not possible, paternity can be assigned to the genotype most likely to have segregated the pollen haplotype under investigation (Meagher 1986). This procedure may be biased toward homozygous genotypes (Devlin et al. 1988); but the bias may be mitigated by adding distance to the algorithm that assigns paternity (Adams, Griffin, and Moran, in prep.). The distance to unambiguous paternal parents can be used to describe a likely distribution of pollination distances, and then both distance and likelihood of segregating the specific pollen haplotype can be used to assign paternity.

Paternity analyses have been conducted in natural populations (Hamrick and Schnabel 1985; Schuster and Mitton, in prep.), but have been used to best advantage in the analyses of seed orchards (Friedman and Adams 1985b; Adams and Birkes 1990a, b). These studies have identified high levels of pollen contamination (> 50%) in seed orchards and have revealed variation in male mating success (Schoen and Stewart 1986).

Coefficient of relationship

The potential for selfing in forest trees and their limited dispersal of seeds make it likely that related individuals cluster together (Govindaraju 1988). For example, disagreement counts, or the number of enzyme bands not shared between individuals, increased with the distance between individuals in stands of *Thujopsis dolabrata* (Sakai and Miyazaki 1972). The relationship coefficient, calculated from the proportions of filled seeds in different types of crosses, revealed family structure in stands of tamarack, *Larix laricina* (Park and Fowler 1982) and white spruce, *Picea glauca* (Coles and Fowler 1976; Park et al. 1984). Phenological data and allozyme frequencies indicated that clustered ponderosa pine, *Pinus ponderosa*, are related (Linhart et al. 1981).

Measures of relatedness that utilize genetic data (Pamilo 1989) are ideal for studies of population structure of forests, but they are rarely applied to natural plant populations. The number of allozyme alleles in common was used to estimate the degree of relationship in the open-pollinated progeny of two species of leguminous trees, *Robinia pseudo-acacia* and *Gleditsia triacanthos* (Surles et al. 1990). This analysis suggested that the degree of relationship in open-pollinated progenies was intermediate between half-sib and full-sib. The method of Queller and Goodnight (1989) utilizes allozyme data to estimate the coefficient of relationship, r, and it has been applied to natural populations of limber pine, *Pinus flexilis* (Schuster and Mitton 1991) to show that individuals that fuse together are more closely related ($r = 0.35 +/- 0.099$) than clustered individuals that do not fuse ($r = 0.05 +/- 0.114$). This measure will allow direct examination of the degree of relatedness within stands of forest trees.

Ecological factors affecting mating systems

Mating systems, like other traits, are not constants, but are influenced by both environmental and genetic variation. Although there is still much variation that we do not yet understand, there are some patterns that we can predict. For example, the rate of outcrossing is expected to increase with stand density. Truly isolated trees, separated by many kilometers from conspecifics, can receive only their own pollen and set only selfed seed (Fowler 1965). Similarly, in stands with low density, there will be low levels of outcrossing, for each tree stands in a mist of its own pollen, only partly diluted by the pollen of nearest neighbors. For this reason, lower levels of outcrossing are expected in years of low pollen production. In stands with normal or high density, the abundance of pollen sources necessarily dilutes selfing pollen, and outcrossing climbs to nearly 100%. An example of the association between stand density and outcrossing was found in ponderosa pine at the forest-grassland ecotone in eastern Colorado (Farris and Mitton 1984). In a stand of normal density within the continuous forest, the average of estimates of outcrossing from allozymes and from albino seedlings was 0.96, and none of the estimates differed from 1.00 (Mitton et al. 1981). Just a few kilometers to the east, a stand with nearest neighbor distances of 25—200 m had a level of outcrossing of only 0.80 (Farris and Mitton 1984). However, the expected relationship between stand density and rate of outcrossing is not always observed (Neale and Adams 1985b).

Because pollen is carried by the wind, the wind conditions at the time of pollen release will influence distances of gene flow and levels of outcrossing. Total lack of wind would cause each tree to be enveloped in selfing pollen, and to receive little outcrossing pollen. At the other extreme, very turbulent air would carry pollen exceptional distances, promoting outcrossing.

Variation among trees in floral phenology will also contribute to variation in rates of outcrossing. For example, ponderosa pine in eastern Colorado exhibit a gentle switch in reproductive effort associated with age and size (Linhart and Mitton 1992). Small young trees act primarily as males, producing pollen but few cones. Large, old trees produce relatively little pollen, but large cone crops. From this switch in reproductive effort we predict that the rates of outcrossing in old trees will exceed those in young trees.

The timing of pollen release and female receptivity will also affect variation in rates of outcrossing. In a study of Douglas-fir growing in a seed orchard, Erickson and Adams (1989, 1990) found considerable clone to clone variation in the timing of pollen release and female receptivity. Outcrossing was highest in clones whose pollen release followed female receptivity, and was lowest in clones in which male and female activities were coincident.

Assumptions of the mating system estimation procedures

The methods for estimating parameters of the mating system make a series of assumptions about the systems and data that they treat. These assumptions are discussed in Shaw et al. (1981) and Brown et al. (1985). The assumptions are:

1) The alleles at a locus segregate in a Mendelian fashion.
2) Selection does not intervene in the time between pollination and the time that seeds or seedlings are sampled.
3) Frequencies in the pollen pool are homogeneous over space and over time.
4) The probability of outcrossing is independent of the maternal genotype.

Violations of the assumptions

With time, assumptions grow into assertions, which then might be accepted as fact. The empirical observations and inferences below remind us that these are merely assumptions, and although they are convenient, they may not always be correct.

There have been occasional reports of segregation distortion for allozyme loci. Thus, this assumption is still worth testing in studies of mating systems, for segregation distortion could easily bias estimates of allelic frequencies in the pollen pool or rates of outcrossing.

The time elapsed between pollination and the time that seeds or seedlings are examined for electrophoresis is on the order of a year, but if seed are stored or taken from serotinous cones the time may be several years. Several observations are not consistent with the assumption of no selection in this time. The first point, admittedly speculative, comes from a tabulation rates of selfing in conifers (Charlesworth and Charlesworth 1987). They estimated s, the rate of selfing, as

$$s = (p_x - p_w)/(p_x - p_s) \quad (4)$$

where p_x, p_s, and p_w are the proportions of filled seeds in outcrossed cones, self pollinated cones, and open pollinated cones. Their estimates for *Pinus attenuata*, *Picea glauca*, *Larix decidua* and *Sequoia sempervirens* were 50%, 60%, 71%, and 59%, respectively. In comparison to estimates based upon allozymes in progeny arrays, in which rates of selfing in conifers average 10% (see the compilation in Surles et al. 1990), these rates of selfing seem suspiciously high. For example, Denti and Schoen (1988) estimated the mean rate of selfing in white spruce to be 7%, with a range of 0% to 22%. Allozyme studies sample genotypes at a later stage in the life cycle than tabulations of filled seed, and they might be missing some selection acting upon selfed genotypes.

A common pattern in rates of outcrossing through time suggests that the viability of seeds is associated with their heterozygosity. In studies of the mating system of lodgepole pine (Hamrick and Schnabel 1985; Perry and Dancik 1986) and jack pine (Cheliak et al. 1985) seeds from serotinous cones provided estimates of the mating system 1, 2, 3, 4, etc. years earlier. In each of these studies, the rates of outcrossing increase with the age of the seeds. A similar pattern is seen in *Eucalyptus delagatensis* (Moran and Brown 1980) and *Eucalyptus stellulata* (Brown et al. 1985). Although the decrease in outcrossing in recent years could be a fortuitous concidence among studies, it is more parsimonious to hypothesize that seed viability is associated with heterozygosity. This association could be generated by selection against selfed genotypes, or selection against homozygous outcrossed genotypes, or a combination of these modes of selection.

The assumption of homogeneity of the pollen cloud through space and over time is challenged by several sorts of observations. The pollen pool of black locust, *Robinia pseudoacacia*, is heterogeneous across its natural range (Surles et al. 1990). This heterogeneity is not unexpected in a broadly distributed species. But variation within stands has been detected in *Pseudotsuga menziesii* (Shaw and Allard 1982), *Eucalyptus citriodora* (Yeh et al. 1983), *Picea glauca* (Cheliak et al. 1985), *Picea mariana* (Boyle and Morgenstern 1986), *Pinus taeda* (Friedmann and Adams 1985a), and *Liriodendron tulipifera* (Brotschol et al. 1986). In natural stands, it is difficult to determine whether pollen heterogeneity over space is associated primarily with family structure or microgeographic variation. Heterogeneity in the pollen pool over geographic space is expected to bias rates of outcrossing downwards (Ennos and Clegg 1982; Schoen and Clegg 1984).

A study of allelic frequencies in the pollen pool of *Eucalyptus regnans* demonstrated that allelic frequencies can vary in time. Fripp et al. (1987) studied 19 trees in a remnant of a natural stand in Victoria, Australia. The trees were known to differ in flowering time, with no overlap in flowering between some of the trees in the stand. Fecundity varied from 500 to 1.4 million flowers per tree. Studies of 3 allozyme polymorphisms revealed substantial variation over time at MDH, but little or no variation over time for ADH and AAT. In a period of 45 days, frequencies of the MDH 2—1 allele climbed abruptly from 0.50 to 0.91, and then dropped back to 0.50.

A study of the mating system of maize (Bijlsma et al. 1986) also suggests that pollen pools may vary over time. Genotypes were determined for 100 seeds for each of 5 plants in an experimental garden. These plants were detasselled, so all of their seeds were outcrossed. They were surrounded by and intermixed with tasselled plants. The mating system was estimated with 6 polymorphic allozymic loci, and 4 of the loci returned estimates of t not significantly lower than 1.0. Estimates of *t* from the remaining 2 loci, Prx and Est, were both significantly lower than 1.00 (0.81 and 0.80, respectively). The allelic frequencies in the effective pollen pools of the 5 detasseled plants were very highly heterogeneous (Table 1). The distances between plants was 1 meter for plants 1 and 2 and for plants 2 and 3, and the greatest distance between these 5 plants was 8 meters. Prx pollen frequencies differed by as much as 0.80 between plants 1 meter apart, and Est pollen frequencies differed by as much as 0.74. Further analyses of the seeds revealed a vast excess of homogametic matings. That is, homozygotes for a particular allele were most likely to receive pollen bearing that same allele. After considering several possible causes for the dramatic heterogeneity in the pollen pool and the aberrant estimates of outcrossing, Bijlsma et al. (1986) favored the following hypothesis. Consider a locus with 2 alleles, with genotypes associated with flowering time, so that the flowering of one homozygote is early, the other late. In a perfectly outcrossed population, this association will produce positive assortative mating in time, and the progeny arrays will appear to be produced by a mixture of selfing and outcrossing. This hypothesis provides an intuitive explanation for the aberrant estimates of outcrossing obtained from PRX and EST. Associations between marker loci and flowering time are common in cultivated species (Allard 1990).

A careful analysis of the mating system of Table Mountain pine, *Pinus pungens*, revealed dramatic heterogeneity in the pollen sampled by eight trees in a population (Gibson and Hamrick 1991). All of the six polymorphic loci used to test for heterogeneity in the pollen pool revealed significant heterogeneity, but just two polymorphisms were used here to illustrate the heterogeneity (Table 1). Trees 1 and 2 were just 4 meters apart, and the most distant trees, 1 and 8, were separated by approximately 50 meters (P. Gibson, pers. comm.). Allelic frequencies differ by 0.62 and 0.50 for alleles 1 and 3 at the 6Pgd polymorphism, and by 0.76 for the alleles at the Mnr polymorphism. Additional examples of heterogeneous pollen pools are presented in Hamrick (1989).

Heterogeneity among single locus estimates of outcrossing is common in studies of plant mating systems, including studies of forest trees (Allard et al. 1977; Mitton et al. 1981; Brown et al. 1985; Shaw and Allard 1982; Neale and Adams 1985a; Muona et al. 1987). There are several reasons for this heterogeneity (Brown and Allard 1970; Shaw et al. 1981; Brown et al. 1985), including random variation and violation of the underlying assumptions. Although a better estimate of outcrossing may be obtained with a multilocus estimation procedure, the heterogeneity of single locus outcrossing rates is interesting in itself, and bears further examination.

Table 1. Heterogeneous pollen pools

Zea mays

Enzyme	Allele	Plant number 1	2	3	4	5	Probability
Prx	1	0.65	0.05	0.85	0.31	0.59	
	2	0.29	0.53	0.06	0.35	0.39	***
	3	0.06	0.42	0.09	0.34	0.02	
Est	1	0.70	0.17	0.68	0.38	0.91	
	2	0.08	0.61	0.04	0.62	0.00	***
	3	0.22	0.22	0.28	0.00	0.09	

Pinus pungens

Enzyme	Allele	Plant number 1	2	3	4	5	6	7	8	Probability
6Pgd2	1	0.26	0.00	0.02	0.62	0.09	0.07	0.06	0.12	
	2	0.74	0.50	0.98	0.38	0.91	0.91	0.94	0.81	***
	3	0.00	0.50	0.00	0.00	0.00	0.02	0.00	0.07	
Mnr	1	0.42	0.16	0.14	0.08	0.09	0.85	0.31	0.34	***
	2	0.58	0.84	0.86	0.92	0.91	0.15	0.69	0.66	

*** = $P < 0.001$. Data on *Zea mays* from Bijlsma et al. 1986. Data on *Pinus pungens* from Gibson and Hamrick, 1991.

From the example in Bijlsma et al. (1986), it seems worthwhile to examine the loci that return exceptionally low estimates of outcrossing. Are genotypes at these loci associated with flowering time? Is there evidence of a Wahlund effect or assortative mating in time?

Studies of the mating system of ponderosa pine (Mitton et al. 1981) revealed significant heterogeneity of estimates of outcrossing from 6 allozyme polymorphisms. Estimates ranged from $t = 1.15$ for PGM to $t = 0.60$ for SDH. Estimates of outcrossing were obtained for homozygous genotypes (Table 2) by the estimation procedure of Imam and Allard (1965):

$$t = H/(1 - p) \tag{5}$$

where t is the rate of outcrossing, H is the proportion of heterozygotes in the progeny of mothers homozygous for a particular allele, and p is the

Table 2. Estimates of outcrossing in ponderosa pine from shikimate dehydrogenase

Maternal genotype	Offspring genotype			t	SE
	22	23	33		
22	50	12	—	0.68	0.040
33	—	18	10	0.89	0.016

frequency of the allele. The outcrossing rates estimated from alternate homozygous genotypes at the same locus are significantly heterogeneous. The causes for this heterogeneity are not known, but the pattern of variation at this locus appears similar to the patterns exhibited by the PRX and EST loci in maize (Bijlsma et al. 1986, Table 1). If genotypes at a specific locus tend to have different flowering rates, then the pollen cloud will be heterogeneous over time and genotypes may differ in their probability of outcrossing.

The assumption of equal probability of outcrossing among genotypes can be tested directly by estimating rates of outcrossing for individual trees. Four studies of outcrossing of individual Douglas-fir (Sorensen 1971; Shaw and Allard 1982a; El-Kassaby et al. 1986; Erickson and Adams 1990) reported variation in rates of outcrossing from 29% to more than 100%. The rate of outcrossing in nine eastern white cedar varied from 25% to approximately 100% (Perry and Knowles 1990). The rate of outcrossing is lowest in trees in which male and female flowering are coincident (Erickson and Adams 1990). We do not know whether the rates of outcrossing are reliably heterogeneous among specific single locus or multilocus genotypes.

Questions to be addressed in future research

To better understand the processes occurring in natural populations, and to better manage resources in seed orchards, we need to understand the forces and mechanisms that make mating systems so dynamic. Heterogeneity appears at virtually every level, and many of the assumptions employed to estimate mating system parameters are violated in natural populations and in seed orchards. Differential male and female mating success is apparent in seed orchards (Müller-Starck and Ziehe 1984; Schoen and Stewart 1986; Schoen and Cheliak 1987; Apsit et al. 1989), and it is probably common in natural populations (Linhart and Mitton

1992). Pollen pools are heterogeneous in space, and perhaps also in time. Estimates of outcrossing are heterogeneous among stands, among loci, and among individuals. Some of this variation is associated with floral phenology. What are the ecological and genetic determinants of all this heterogeneity?

It is most convenient to assume that the genetic markers used to estimate mating systems are adaptively neutral, but there are observations that suggest the allozyme variation influences both physiology (Mitton and Grant 1984; Mitton 1983, 1989) and demography (Mitton and Jeffers 1989; Bush and Smouse, this volume, pp. 179—196). Some allozyme polymorphisms yield suspiciously low estimates of outcrossing. Are genotypes at those loci associated with flowering time? The rates of outcrossing estimated from seeds increase with the time that the seeds are stored. This casts suspicion on the assumption that selection will not intervene between the time of fertilization and the time that genotypes are assayed. What are the genetic consequences of the mortality associated with polyembryony? Which variables contribute to differential male reproductive success?

Ecological variation certainly influences mating systems. But mating systems are polygenic characters that evolve in response to both natural and artificial selection. To come to a full understanding of the variability in mating systems, studies of mating systems must incorporate ecological, genetic, physiological and demographic perspectives.

References

Adams, W. T. 1982. Clonal variation in pollen-related characteristics of Douglas fir. Can. J. For. Res. 12: 403—408.

Adams, W. T. and Birkes, D. S. 1990a. Mating patterns in seed orchards, pp. 75—86. In: Proceedings of 20th Southern Forest Tree Improvement Conference, June 26—30, 1989, Charleston, South Carolina.

Adams, W. T. and Birkes, D. S. 1990b. Estimating mating patterns in forest tree populations. Proceedings, International Workshop on Plant Biology, Biochemical Markers in Population Genetics of Forest Trees. Institute for Agroforestry of the National Research Council of Italy. Porano—Orvieto, Italy, October 11—13, 1988.

Adams, W. T. and Joly, R. J. 1980. Allozyme studies in loblolly pine seed orchards: clonal variation and frequency of progeny due to self-fertilization. Silvae Genet. 29: 1—4.

Allard, R. W. 1990. Future directions in plant population genetics, evolution, and breeding, pp. 1—19. In: Brown, A. H. D., Clegg, M. T., Kahler, A. L. and Weir, B. S. (Eds) Plant Population Genetics, Breeding, and Genetic Resources. Sinauer Associates, Inc., Sunderland, Ma.

Allard, R. W., Jain, S. K. and Workman, P. L. 1968. The genetics of inbreeding populations. Advances in Genetics 14: 55—131.

Allard, R. W., Kahler, A. L. and Clegg, M. T. 1977. Estimation of mating cycle components of selection in plants, pp. 1—19. In: Christiansen, F. B. and Fenchel, T. M. (Eds) Measuring Selection in Natural Populations. Springer-Verlag, Berlin.

Apsit, V. J. and Nakamura, R. R. and Wheeler, N. C. 1989. Differential male reproductive success in Douglas fir. Theor. Appl. Genet. 77: 681—684.

Bijlsma, R., Allard, R. W. and Kahler, A. L. 1986. Nonrandom mating in an open-pollinated maize population. Genetics 112: 669—680.

Brotschol, J. V., Roberds, J. H. and Namkoong, G. 1986. Allozyme variation among North Carolina populations of *Liriodendron tulipifera* L. Silvae Genetica 35: 131—138.

Brown, A. H. D. 1979. Enzyme polymorphism in plant populations. Theor. Popul. Biol. 15: 1—42.

Brown, A. H. D. 1990. Genetic characterization of plant mating systems, pp. 145—162. In: Brown, A. H. D., Clegg, M. T., Kahler, A. L. and Weir B. S. (Eds) Plant Population Genetics, Breeding, and Genetic Resources. Sinauer Associates, Inc., Sunderland, Ma.

Brown, A. H. D. and Allard, R. W. 1970. Estimation of the mating system in open-pollinated maize populations using isozyme polymorphisms. Genetics 66: 133—145.

Brown, A. H. D., Barrett, S. C. H. and Moran, G. F. 1985. Mating system estimation in forest trees: Models, methods, and meanings, pp. 32—49. In: Gregorius, H. R. (Ed) Population Genetics in Forestry. Springer-Verlag, Berlin.

Brown, A. H. D., Burdon, J. J. and Jarosz, A. M. 1989. Isozyme analysis of plant mating systems, pp. 73—86. In: Soltis, D. E. and Soltis, P. S. (Eds) Isozymes in Plant Biology. Dioscorides Press, Portland, OR.

Boyle, J. B. and Morgenstern, E. K. 1986. Estimates of outcrossing rates in six populations of black spruce in central New Brunswick. Silvae Genetica 35: 102—106.

Bush, R. M. and Smouse, P. E. 1992. Evidence for the adaptive significance of allozymes in forest trees. This issue (pp. 179—196).

Charlesworth, D. and Charlesworth, B. 1987. Inbreeding depression and its evolutionary consequences. Ann. Rev. Ecol. Syst. 18: 237—268.

Charlesworth, D., Schemske, D. W. and Sork, V. L. 1987. The evolution of plant reproductive characters: sexual versus natural selection, pp. 317—335. In: Stearns, S. C. (Ed) The Evolution of Sex and its Consequences. Birkhauser, Basel.

Cheliak, W. M., Morgan, K., Strobeck, C., Yeh, F. C. H. and Dancik, B. P. 1983. Estimation of mating system parameters in plant populations using the EM algorithm. Theor. Appl. Genet. 65: 157—161.

Cheliak, W. M., Dancik, B. P., Morgan, K., Yeh, F. C. H. and Strobeck, C. 1985. Temporal variation of the mating system in a natural population of jack pine. Genetics 109: 569—584.

Clegg, M. T. 1980. Measuring plant mating systems. BioScience 30: 814—818.

Clegg, M. T. and Epperson, B. K. 1985. Recent developments in population genetics. Adv. Genet. 23: 235—269.

Coles, J. F. and Fowler, D. P. 1976. Inbreeding in neighboring trees in two white spruce populations. Silvae Genet. 25: 29—34.

Denti, D. and Schoen, D. J. 1988. Self-fertilization rates in white spruce: effect of pollen and seed production. J. Hered. 79: 284—288.

Devlin, B., Roeder K. and Ellstrand, N. C. 1988. Fractional paternity assignment: Theoretical development and comparison to other methods. Theor. Appl. Genet. 76: 369—380.

El-Kassaby, Y. A., Meagher, M. D. and Davidson, R. 1992. Temporal variation in the outcrossing rate in a natural stand of western white pine. submitted to Silvae Genetica.

El-Kassaby, Y. A., Meagher, M. D., Parkinson, J. and Portlock, F. T. 1987. Allozyme inheritance, heterozygosity and outcrossing rate among *Pinus monticola* near Ladysmith, British Columbia. Heredity 58: 173—181.

El-Kassaby, Y. A., Parkinson, J. and Devitt, W. J. B. 1986. The effect of crown segment on the mating system in a Douglas-fir (*Pseudotsuga menziesii* (Mirb.) Franco) seed orchard. Silvae Genet. 35: 149—155.

El-Kassaby, Y. A., Ritland, K. Fashler, A. M. K. and Devitt, W. J. B. 1988. The role of reproductive phenology upon the mating system of a Douglas-fir seed orchard. Silvae Genet. 37: 76–82.

El-Kassaby, Y. A., Yeh, F. C. and Sziklai, O. 1981. Estimation of the outcrossing rate of Douglas-fir (*Pseudotsuga menziesii* (Mirb.) Franco) using allozyme polymorphisms. Silvae Genet. 30: 182–184.

Ellstrand, N. C. 1984. Multiple paternity within the fruits of the wild radish *Raphanus sativus*. Am. Nat. 123: 819–828.

Ellstrand, N. C. 1992. Gene flow among seed plant populations. This issue (pp. 241–256).

Ellstrand, N. C. and Marshall, D. L. 1985. Interpopulation gene flow by pollen in wild radish. Am. Nat. 126: 596–605.

Ennos, R. A. and Clegg, M. T. 1982. Effect of population substructuring on estimates of outcrossing rate in plant populations. Heredity 48: 283–292.

Epperson, B. K. and Allard, R. W. 1984. Allozyme analysis of the mating system in lodgepole pine populations. J. Hered. 75: 212–214.

Erickson, V. J. and Adams, W. T. 1989. Mating success in a coastal Douglas-fir seed orchard as affected by distance and floral phenology. Can. J. For. Res. 19: 1248–1255.

Erickson, V. J. and Adams, W. T. 1990. Mating system variation among individual ramets in a Douglas-fir seed orchard. Can. J. For. Res. 20: 1672–1675.

Farris, M. A. and Mitton, J. B. 1984. Population density, outcrossing rate, and heterozygote superiority in ponderosa pine. Evolution 38: 1151–1154.

Fowler, D. P. 1965. Effects of inbreeding in red pine, *Pinus resinosa* Ait. III. Factors affecting natural selfing. Silvae Genet. 14: 36–46.

Franklin, E. C. 1970. Survey of mutant forms and inbreeding depression in species of the family Pinaceae. USDA For. Serv. Res. Pap. SE–61.

Friedman, S. T. and Adams, W. T. 1985a. Levels of outcrossing in two loblolly pine seed orchards. Silvae Genetica 34: 157–162.

Friedman, S. T. and Adams, W. T. 1985b. Estimation of gene flow into two seed orchards of loblolly pine (*Pinus taeda* L.). Theor. Appl. Genet. 69: 609–615.

Fripp, Y. J., Griffin, A. R. and Moran G. F. 1987. Variation in allelic frequencies in the outcross pollen pool of *Eucalyptus regnans* F. Muell. throughout a flowering season. Heredity 59: 161–172.

Furnier, G. R. and Adams, W. T. 1986. Mating system in natural populations of Jeffrey pine. Amer. J. Bot. 73: 1009–1015.

Fyfe, J. L. and Bailey, N. T. J. 1951. Plant breeding studies in leguminous forage crops. I. Natural crossbreeding in winter beans. J. Agric. Sci. 41: 371–378.

Gibson, J. P. and Hamrick, J. L. 1991. Heterogeneity in pollen allele frequencies among cones, whorls, and trees of table mountain pine (*Pinus pungens*). Am. J. Bot. 78: 1244–1251.

Govindaraju, D. R. 1988. Mating systems and the opportunity for group selection in plants. Evolutionary Trends in Plants 2: 99–106.

Govindaraju, D. R. and Dancik, B. P. 1987. Allozyme heterozygosity and homeostasis in germination seeds of jack pine. Heredity 59: 279–283.

Green, A. G., Brown, A. H. D. and Oram, R. N. 1980. Determination of outcrossing in a breeding population of *Lupinus alpus* L. Z. Pflanzenzucht. 84: 181–191.

Hamrick, J. L. 1983. The distribution of genetic variation within and among natural plant populations, pp. 335–348. In: Schonewald-Cox, C. Chambers, S., MacBryde, B. and Thomas, W. (Eds), Genetics and Conservation. Benjamin/Cummings, Menlo Park, Ca.

Hamrick, J. L. 1987. Gene flow and distribution of genetic variation in plant populations, pp. 53–67. In: Urbanska, K. M. (Ed) Differentiation Patterns in Higher Plants. Academic Press, New York.

Hamrick, J. L. 1989. Isozymes and the analysis of genetic structure in plant populations, pp. 87—105. In Soltis, D. E. and Soltis, P. S. (Eds) Isozymes in Plant Biology. Dioscorides Press, Portland, Oregon.

Hamrick, J. L. and Schnabel, A. 1985. Understanding the genetic structure of plant populations: some old problems and a new approach, pp. 50—70. In: Gregorius, H. R. (Ed), Lecture Notes in Biomathematics 60: Population Genetics in Forestry. Springer Verlag, Berlin.

Imam, A. G. and Allard, R. W. 1965. Population studies in predominantly self-pollinated species. VI. Genetic variability between and within natural populations of wild oats from differing habitats in California. Genetics 51: 49—67.

Jeffreys, A. J., Wilson, V. and Thein, S. L. 1985a. Hypervariable "minisatellite" regions in human DNA. Nature (London) 314: 67—73.

Jeffreys, A. J., Wilson, V. and Thein, S. L. 1985b. Individual-specific "fingerprints" of human DNA. Nature (London) 316: 76—79.

Jones, D. F. 1916. Natural cross-pollination in the tomato. Science 43: 509—510.

King, J. N., Dancik, B. P. and Dhir, N. K. 1984. Genetic structure and mating system of white spruce (*Picea glauca*) in a seed production area. Can. J. For. Res. 14: 639—643.

Knowles, P., Furnier, G. R., Aleksiuk, M. A. and Perry, D. J. 1987. Significant levels of self-fertilization in natural populations of tamarack. Can. J. Bot. 65: 1087—1091.

Lande, R. and Schemske, D. W. 1985. The evolution of self-fertilization and inbreeding depression in plants. I. Genetic models. Evolution 39: 24—40.

Levin, D. A. and Kerster, H. W. 1974. Gene flow in seed plants. Evolutionary Biology 7: 139—220.

Lindgren, D. 1975. The relationship between self-fertilization, empty seeds and seeds originating from selfing as a consequence of polyembryony. Stud. For. Suec. 126: 1—24.

Linhart, Y. B., Mitton, J. B., Sturgeon, K. B. and Davis, M. L. 1981. Genetic variation in space and time in a population of ponderosa pine. Heredity 46: 407—426.

Linhart, Y. B. and Mitton, J. B. 1985. Relationships among reproduction, growth rate, and protein heterozygosity in ponderosa pine. Amer. J. Bot. 72: 181—184.

Linhart, Y. B. and Mitton, J. B. 1992. Reproductive tradeoffs and their consequences in ponderosa pine. (submitted to Ecology)

Loveless, M. D. and Hamrick, J. L. 1984. Ecological determinants of genetic structure in plant populations. Ann. Rev. Ecol. Syst. 15: 65—95.

Meagher, T. R. 1986. Analysis of paternity within a natural population of *Chamaelirium leuteum*. I. Identification of most-likely male parents. Am. Nat. 128: 129—215.

Meagher, T. R. and Thompson, E. 1986. The relationship between single parent and parent pair genetic likelihoods in genealogy reconstruction. Theor. Pop. Biol. 29: 87—106.

Mitton, J. B. 1983. Conifers, pp. 443—472. In: Tanksley S. and Orton, T. (Eds), Isozymes in Plant Genetics and Breeding, Part B. Elsevier, Amsterdam.

Mitton, J. B. 1989. Physiological and demographic variation associated with allozyme variation, pp. 127—145. In: Soltis, D. E. and Soltis, P. S. (Eds) Isozymes in Plant Biology. Dioscorides Press, Portland, Oregon.

Mitton, J. B. and Grant, M. C. 1984. Relationships among protein heterozygosity, growth rate, and developmental stability. Ann. Rev. Ecol. Syst. 15: 479—499.

Mitton, J. B. and Jeffers, R. M. 1989. The genetic consequences of mass selection for growth rate in Engelmann spruce. Silvae Genetica 38: 6—12.

Mitton, J. B., Linhart, Y. B., Davis, M. L. and Sturgeon, K. B. 1981. Estimation of outcrossing in ponderosa pine, *Pinus ponderosa* Laws., from patterns of segregation of protein polymorphisms and from frequencies of albino seedlings. Silvae Genetica 30: 117—121.

Mitton, J. B., Linhart, Y. B., Hamrick, J. L. and Beckman, J. S. 1977. Observations on the genetic structure and mating system of ponderosa pine in the Colorado Front Range. Theor. Appl. Genet. 57: 5—13.

Mopper, S., Mitton, J. B., Whitham, T. G., Cobb, N. S. and Christensen, K. M. 1991. Genetic differentiation and heterozygosity in pinyon pine associated with resistance to herbivory and environmental stress. Evolution 45: 989—999.

Moran, G. F., Bell, J. C. and Matheson, A. C. 1980. The genetic structure and levels of inbreeding in a *Pinus radiata* D. Don seed orchard. Silvae Genet. 29: 190—193.

Moran, G. F., Bell, J. C. and Turnbull, J. W. 1989. A cline in genetic diversity in river she-oak *Casuarina cunningharmiana*. Aust. J. Bot. 37: 169—180.

Moran, G. F. and Brown, A. H. D. 1980. Temporal heterogeneity of outcrossing rates in alpine ash (*Eucalyptus delegatensis*). Theor. Appl. Genet. 57: 101—105.

Moran, G. F. and Griffin, A. R. 1985. Non-random contribution of pollen in polycrosses of *pinus radiata* D. Don. Silvae Genet. 34: 117—121.

Mueller-Starck, G. and Ziehe, M. 1984. Reproductive systems in conifer seed orchards. 3. Female and male fitnesses of individual clones realized in seeds of *Piunus sylvestris* L. Theor. Appl. Genet. 69: 173—177.

Muller, G. 1977a. Cross fertilization in a conifer stand inferred from gene-markers in seeds. Silvae Genetica 26: 223—226.

Muller, G. 1977b. Investigations of the degree of natural self-fertilization in stands of Norway spruce (*Picea abies* (L.) Darst.) and Scots pine (*Pinus sylvestris* L.). Silvae Genet. 26: 207—217.

Muona, O., Yazdani, R. and Rudin, D. 1987. Genetic change between life stages in *Pinus sylvestris*: allozyme variation in seeds and planted seedlings. Silvae Genet. 36: 39—42.

Neale, D. B. and Adams, W. T. 1985a. Allozyme and mating system variation in balsam fir (*Abies balsamea*) across a continuous elevational transect. Can. J. Bot. 63: 2448—2453.

Neale, D. B. and Adams, W. T. 1985b. The mating system in natural and shelterwood stands of Douglas-fir. Theor. Appl. Genet. 71: 201—207.

Nei, M. 1973. Analysis of gene diversity in subdivided populations. Proc. Natl. Acad. Sci. USA 70: 3321—3323.

Pamilo, P. 1989. Estimating relatedness in social groups. Trends Ecol. Evol. 4: 353—355.

Park, Y. S., and Fowler, D. P. 1982. Effects of inbreeding and genetic variances in a natural population of tamarack (*Larix laricina* (Du Roi) K. Koch) in eastern Canada. Silvae Genet 31: 21—26.

Park, Y. S., Fowler, D. P. and Coles, J. F. 1984. Population studies of white spruce. II. Natural inbreeding and relatedness among neighboring trees. Can. J. For. Res. 14: 909—913.

Perry, D. J. and Dancik, B. P. 1986. Mating system dynamics of lodgepole pine in Alberta, Canada. Silvae Genet. 35: 190—195.

Perry, D. J. and Knowles, P. 1990. Evidence of high self-fertilization in natural populations of eastern white cedar (*Thuja occidentalis*). Can. J. Bot. 68: 663—668.

Plessas, M. E. and Strauss, S. H. 1986. Allozyme differentiation among populations, stands and cohorts in Monterey pine. Can. J. For. Res. 16: 1155—1164.

Queller, D. C. and Goodknight, K. F. 1989. Estimating relatedness using genetic markers. Evolution 43: 258—275.

Ritland, K. 1983. Estimation of mating systems, pp. 289—302. In: Tanksley, S. D. and Orton, T. J. (Eds) Isozymes in Plant Genetics and Breeding. Part A. Elsevier, Amsterdam. 516 pp.

Ritland, K. 1984. The effective proportion of self-fertilizantion with consanguineous matings in inbred populations. Genetics 106: 139—152.

Ritland, K. 1985. The genetic-mating structure of subdivided populations. I. Open-mating model. Theor. Pop. Biol. 27: 51—74.

Ritland, K. 1986. Joint maximum likelihood estimation of genetic and mating structure using open-pollinated progenies. Biometrics: 42: 25—43.

Ritland, K. and El-Kassaby, Y. A. 1985. The nature of inbreeding in a seed orchard of Douglas fir as shown by an efficient multi-locus model. Theoret. Appl. Genet. 71: 375—384.

Ritland, K. R. and Jain, S. 1981. A model for the estimation of outcrossing rate and gene frequencies using n independent loci. Heredity 47: 35—52.

Sakai, K. I. and Miyazaki, Y. 1972. Genetic studies in natural populations of forest trees. II. Family analysis: a new method for quantitative genetic studies. Silvae Genet. 21: 149—154.

Schemske, D. W. and Lande, R. 1985. The evolution of self-fertilization and inbreeding depression in plants. II. Empirical observations. Evolution 39: 41—52.

Schoen, D. J. and Cheliak, W. M. 1987. Genetics of the polycross. 2. Male fertility variation in Norway spruce, *Picea abies* (L.) Karst. Theor. Appl. Genet. 74: 554—559.

Schoen, D. J. and Clegg, M. T. 1984. Estimation of mating system parameters when outcrossing events are correlated. Proc. Natl. Acad. Sci. USA 81: 5258—5262.

Schoen, D. J. and Stewart, S. C. 1986. Variation in male reproductive investment and male reproductive success in white spruce. Evolution 40: 1109—1120.

Schuster, W. S., Alles, D. L. and Mitton, J. B. 1989. Gene flow in limber pine: Evidence from pollination phenology and genetic differentiation along an elevational transect. Amer. J. Bot. 76: 1395—1403.

Schuster, W. S. F. and Mitton, J. B. 1991. Relatedness within clusters of a bird-dispersed pine and the potential for kin interactions. Heredity 67: 41—48.

Shaw, D. V. and Allard, R. W. 1981. Analysis of mating system parameters and population structure in Douglas-fir using single-locus and multilocus methods, pp 18—22. In: Conkle, M. T. (Ed) Isozymes of North American Forest Trees and Forest Insects. USDA For. Serv. Gen. Tech. Rep. PSW—48.

Shaw, D. V. and Allard, R. W. 1982a. Estimation of outcrossing rates in Douglas-fir using isozyme markers. Theor. Appl. Genet. 62: 113—120.

Shaw, D. V. and Allard, R. W. 1982b. Estimation of outcrossing rates in Douglas-fir using allozyme markers. Theor. Appl. Genet. 62: 113—120.

Shaw, D. V., Kahler, L. L. and Allard, R. W. 1981. A multilocus estimator of mating system parameters in plant populations. Proc. Natl. Acad. Sci. USA 78: 1298—1302.

Shaw, D. V. and Allard, R. W. 1982. Isozyme heterozygosity in adult and open-pollinated embryo samples of Douglas-fir. Silvae. Fenn. 16: 115—121.

Shea, K. L. 1987. Effects of population structure and cone production on outcrossing rates in Engelmann spruce and subalpine fir. Evolution 41: 124—136.

Shen, H.-H., Rudin, D. and Lindgren, D. 1981. Study of pollination pattern in a scots pine seed orchard by means of isozyme analysis. Silvae Genet: 30: 7—15.

Slatkin, M. 1985. Gene flow in natural populations. Ann. Rev. Ecol. Syst. 16: 393—430.

Slatkin, M. 1987. Gene flow and the geographic structure of natural populations. Science 236: 787—792.

Snyder, T. P., Steward, D. A. and Strickler, A. F. 1985. Temporal analysis of breeding structure in jack pine (*Pinus banksiana* Lamb.). Canad. J. For. Res. 15: 1159—1166.

Sorensen, F. C. 1969. Embryonic genetic load in coastal Douglas fir, *Pseudotsuga menziessii* var. *menziessii*. Am. Nat. 103: 389—398.

Sorensen, F. C. 1971. Estimate of self-fertility in coastal Douglas-fir from inbreeding studies. Silvae. Genet. 20: 115—120.

Sorensen, F. C. 1982. The role of polyembryonal vitality in the genetic system of conifers. Evolution 36: 725—733.

Surles, S. E., Hamrick, J. L. and Bongarten, B. C. 1990. Mating systems in open-pollinated families of black locust (*Robinia pseudoacacia*). Silvae Genetica 39: 35—40.

Wright, J. W. 1976. Introduction to forest genetics. Academic Press, New York.

Wright, S. 1931. Evolution in Mendelian populations. Genetics 16: 97—159.

Wright, S. 1969. Evolution and the Genetics of Populations. Volume 2. The Theory of Gene Frequencies. University of Chicago Press, Chicago. 511 pp.

Yeh, F. C., Brune, A., Cheliak, W. M. and Chipman, D. C. 1983. Mating system of *Eucalyptus citriodora* in a seed production area. Can. J. For. Res. 13: 1051—1055.

Yeh, F. C. and Morgan, K. 1987. Mating system and multilocus associations in a natural population of *Pseudotsuga menziessii* (Mirb.) Franco. Theor. Appl. Genet. 73: 799—808.

New Forests **6**: 217—240, 1992.

Review paper

Gene dispersal within forest tree populations

W. T. ADAMS

Department of Forest Science, Oregon State University, Corvallis, OR 97331—5705, USA

Received 30 January 1991; accepted 8 January 1992

Key words: gene dispersal, mating patterns, paternity analysis, genetic structure

Application. Knowledge of factors influencing gene dispersal is important for evaluating impacts of forest management on the genetics of populations. The magnitude of gene dispersal by pollen, for example, influences mating patterns and the composition of seed crops. Thus, understanding relationships between effective pollen dispersal and tree spacing is necessary for assessing the genetic implications of alternative natural regeneration systems and for designing seed orchards.

Abstract. Patterns of gene dispersal by seeds and pollen greatly influence the genetic structure of plant populations and their effective size. This paper reviews methods of measuring gene dispersal and current information on patterns of dispersal within local populations of forest trees. Recently, a number of statistical procedures for investigating gene movement based on the use of large numbers of isozyme loci have been described. These procedures include various forms of parentage analysis and the fitting of mating models to genotypic arrays of offspring from individual maternal plants. With the levels of genetic discrimination currently possible in forest trees, the model approach appears to be the most reliable means of estimating gene dispersal parameters. Too little data are available to draw general conclusions about patterns of gene movement within natural populations. Nevertheless, reports to date indicate that dispersal by both pollen and seed can be considerable. For any one mother tree, the bulk of effective pollen in conifers may come from distant males in the same stand or from surrounding stands (gene flow). In insect-pollinated angiosperms, gene flow may also be substantial, but cross—fertilization within stands may primarily be between nearest flowering trees.

Introduction

Patterns of gene movement via pollen and seed dispersal have important implications for the genetic structure of forest tree populations. The extent of gene movement not only influences levels of genetic diversity within populations and effective population size, but also the degree to which populations may become genetically subdivided as a result of selection or drift (Levin and Kerster 1974; Hamrick and Schnabel 1985). Patterns of gene movement are also of practical significance in forestry. The degree to which seed dispersal is limited and results in family sub-structuring of forest stands influences the validity of alternative plus-tree selection strategies (Ledig 1974) and the minimum acceptable distances between seed trees used for reforestation. The magnitude of pollen dispersal affects the genetic composition of open-pollinated families used in progeny testing. If open-pollinated families are the result of extensive inbreeding or of few male parents, the validity of such testing is questionable (Namkoong 1966; Squillace 1974; Gregorius 1975; Sorensen and White 1988). Knowledge of pollen dispersal patterns is also important for developing seed orchard designs and management regimes that will maximize intermating among selected genotypes and limit pollen contamination from unimproved sources (Adams and Birkes 1989; Wheeler et al. 1992).

Despite the importance of gene movement in both evolutionary biology and tree breeding, knowledge of gene dispersal patterns in forest trees is very limited (Hamrick and Schnabel 1985; Adams and Birkes 1991). To date, most estimates of gene dispersal have been obtained indirectly by measuring pollen and seed dispersal. Pollen trapping studies in wind-pollinated species indicate the concentration of pollen dispersed from individual trees drops off rapidly with distance, with the highest densities within 50 to 100 m of the source, but small amounts found up to several hundred meters away (Wang et al. 1960; Silen 1962; Levin and Kerster 1974; Bramlett 1981). These results suggest that most intermating within local populations occurs among neighboring individuals. Nevertheless, the accumulated long-distance dispersal of small amounts of pollen from many trees can result in considerable pollen distributed over long distances when whole stands are considered as pollen sources (Lanner 1966; Libby et al. 1969). Thus, the potential for gene flow between nearby populations is great.

Pollen dispersal in animal-pollinated plants is often tracked by monitoring movement of pollinators, marker dyes or marked pollen (Levin and Kerster 1974; Handel 1983). These methods are impractical when large numbers of different pollinators are involved or when host plants are large, and thus, have generally not proven useful in forest trees. In plants where such studies have been possible, the amount of pollen dispersed from individuals seems to drop off rapidly with distance, much in the same manner as observed with wind pollination.

Most of the information on seed dispersal in forest trees is for wind-dispersed species, primarily conifers (Levin and Kerster 1974). Like wind-dispersed pollen, seed dispersal also declines rapidly with distance. For example, in a Douglas-fir [*Pseudotsuga menziesii* (Mirb.) Franco] study, most seed was dispersed within 60 m of the source, with comparatively little or no distribution beyond 300 m (Isaac 1930).

Although measurements of seed and pollen dispersal provide much useful information on the potential for gene movement, actual patterns of gene dispersal may be quite different for a variety of reasons (Levin 1981, 1983; Handel 1983; Hamrick and Schnabel 1985). Estimating pollen gene dispersal on the basis of pollinator movements, for example, requires the assumption that most or all pollen picked up at one flower is deposited on the next flower visited. If pollen is carried further than one flower this method underestimates average gene dispersal distances (Levin 1981, 1983). The presumption from pollen dispersal patterns in wind-pollinated species that most mating occurs among neighboring plants does not take into account the influence of competing pollen sources (Levin and Kerster 1974). If the amount of pollen produced by near neighbors is small relative to more distant pollen sources, the advantage of proximity may be eliminated or compromised. Patterns of mating estimated from pollen dispersal alone also do not account for differential pollen fertility or floral phenology of potential male parents or their crossability with specific females (Levin 1981, 1983; Handel 1983). Gene movement by seed, even in predominantly wind-dispersed species, can be influenced by the foraging behavior of predators. Intensity of predation may increase with seed density, thus, average gene dispersal may be greater than indicated by seed dispersal distances alone (Levin 1981).

Because of the problems with indirect measures, gene dispersal is best assessed directly using genetic markers, the feasibility of which has been enhanced in recent years by the availability of isozymes (Brown 1989; Adams and Birkes 1991). To date, most mating studies using isozymes have emphasized estimating proportions of progeny due to selfing versus outcrossing. These studies reveal that selfing normally accounts for only a small proportion of viable offspring in forest trees; often less than 10% in conifers (Muona 1989; Adams and Birkes 1991; Mitton 1992). Given that outcrossing is responsible for the bulk of progeny, patterns of outcrossing must be understood if the evolutionary significance of mating is to be fully appreciated (Brown 1989). Certainly, pollen gene flow between populations can account for significant proportions of offspring (Adams and Birkes 1991; Ellstrand 1992, this issue, pp. 241—256; Wheeler et al. 1992), but the majority of outcrossing is expected to occur among individuals within local populations (Brown 1989). A number of questions regarding outcross mating patterns are of interest. For example, to what extent do individuals within populations differ in male mating success, and to what degree is male mating success influenced by their distance to females, and by differential pollen fertility, floral phenology and crossability? In addition, how are patterns of outcrossing affected by stand structure, including tree density, size, and age, and internal plant arrangement (e.g., clustered versus evenly spaced individuals) (Handel 1983)?

In this paper, I briefly describe and compare methods of estimating gene dispersal patterns within local populations using genetic markers. I then review recent literature on the application of these methods to forest trees.

Methods of estimating gene dispersal patterns with genetic markers

Three basic approaches have been used in forest trees to measure gene dispersal with genetic markers: 1) Track effective pollen or seed dispersal from individual parent trees which carry rare marker alleles (rare marker approach) (Muller 1977; Yazdani et al. 1989), 2) Use multilocus markers to infer parentage of offspring (parentage analysis) (Neale 1984; Hamrick and Schnabel 1985; Schnabel 1988), 3) Model the probability structure of entire samples of offspring genotypes and estimate the most likely dispersal parameters that would have generated the sample (model approach) (Schoen and Stewart 1986; Adams and Birkes 1989, 1991).

Rare marker approach

This is a very effective means of measuring gene dispersal patterns, particularly relationships between mating success and distance between mates. Patterns of effective pollen dispersal can be examined by the relative proportions of offspring with a marker allele detected in samples of seed from mother trees at varying distances and directions away from a male marker tree. Likewise, effective seed dispersal can be evaluated by assaying seeds or seedlings at varying distances from mother trees carrying rare markers. This approach is limited in application because of the need for individuals with unique markers. If more than one tree with the marker allele is present in the population, interpretation of results are confounded (Müller—Starck 1982; Yazdani et al. 1989). In addition, since only one, or a few, individuals in a population are likely to carry rare markers, generalizations on gene dispersal patterns may not be possible.

Parentage analysis

Parentage is inferred by comparing multilocus genotypes in offspring to those of potential parents. Thus, the genotypes of all potential parents in the population must be known and the population must be isolated and small for this method to be efficient. When seeds are sampled from known mother trees, this method can be used to assess effective pollen dispersal by inferring paternity of offspring. Parentage analysis can also be used to assign maternal parents to seeds or seedlings sampled on the ground, or to biparental assignment of established seedlings (Meagher and Thompson 1986, 1987; Schnabel 1988). Conifers are particularly favorable for parentage analysis of seeds because of the ability to identify the haploid isozyme genotypes of the maternal and paternal gametes forming viable embryos (Adams and Birkes 1991). For the remainder of this section on parentage analysis, I will refer specifically to paternity assignment which has been the primary focus of this methodology. The concepts, however, are applicable to all types of parentage assignment.

The effectiveness of paternity analysis depends on the probability of being able to exclude potential males on the basis of genetic incompatibility alone (i.e., the exclusion probability) (Chakraborty et al. 1988; Devlin et al. 1988). The exclusion probability (EP) can be determined empirically by calculating the proportion of males which are excluded for each offspring sampled and averaging over all offspring, or can be estimated on a population basis using formulas given in Chakraborty et al. (1988). EP increases with the number of alleles and evenness of allele frequencies at each locus, and the number of polymorphic loci used in the analysis. There are diminishing returns, however, with increasing number of loci. The relationship between average values of EP in a random mating population and the number of polymorphic loci is illustrated in Fig. 1. All loci are assumed to be independent with two codominant alleles in frequencies of 0.5. In the case of where paternity is inferred on the basis of diploid genotypes of offspring, it would take 12 of these highly polymorphic loci to achieve an average EP greater than 90% (Fig. 1a). Nearly twice as many loci (23) would be needed to reach 99% exclusion. If paternity is determined on the basis of pollen (haploid) gametes, as is possible with conifer seeds, the numbers of loci required for the same levels of exclusion are much less: only 9 for 90% and 17 for 99%. Unfortunately, such highly polymorphic loci are infrequent in isozyme studies of forest trees. More often, one allele dominates, with one or more additional alleles occurring at low frequency. A more realistic average polymorphic locus might have three alleles in frequencies of 0.80, 0.15 and 0.05, respectively. If all loci were of this makeup, 13 would be needed to achieve an average EP greater than 90% in the diploid case, and 12 would be needed in the haploid case. Thus, the curves in Fig. 1 most likely underestimate the number of isozyme loci needed to achieve different EP's in real situations, but perhaps, not greatly so.

Paternity analysis is ideal if complete genetic exclusion is possible, that is, if all potential fathers of an offspring can be excluded except one, so that unambiguous assignment of paternity can be made. Complete genetic exclusion, however, is very difficult to achieve. The probability of unam-

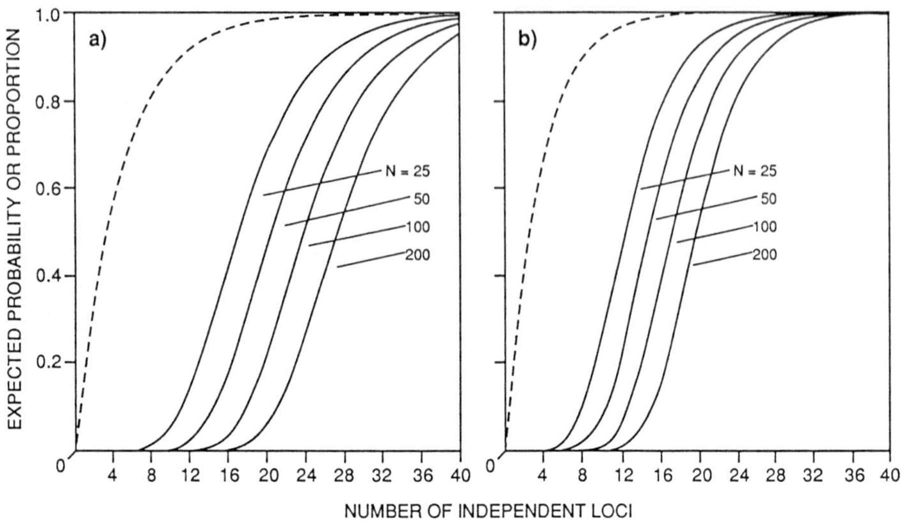

Fig. 1. Expected exclusion probabilities when increasing numbers of independent loci (each with two codominant alleles in equal frequency) are employed in paternity analysis (dashed curves), and expected proportions of offspring with unambiguous paternity (solid curves) for varying numbers of possible parents (N = 25—200). Figure 1a applies when diploid genotypes of offspring are analyzed, and Fig. 1b when haploid pollen gametes of offspring can be identified.

biguous assignment depends on both EP and the number of potential males in the population, and even when the number of potential males is small, the number of highly polymorphic loci needed to increase the frequency of unambiguous assignment to a high level is beyond the capability of isozyme studies in most forest trees. For example, to achieve a 90% probability of unambiguous assignment when all loci have two alleles in equal frequency, and assuming only 25 potential males, at least 26 loci would be required for diploid offspring and 19 for pollen gametes (Fig. 1). Employing DNA markers either alone or in combination with isozyme loci could greatly increase the genetic resolving power of parentage analysis (Smouse and Chakraborty 1986; Chakraborty et al. 1988; Brown 1989), especially if hypervariable molecular markers (Jeffreys et al. 1985; Nybom and Schaal 1990) and inexpensive methods for their use in large samples of offspring and parents were available.

When a relatively high proportion of sampled offspring can be assigned paternity unambiguously, reasonable estimates of mating pattern parameters might be obtained by discarding offspring with ambiguous paternity and basing the analysis on the remaining fraction. Offspring of fathers with rare alleles, however, will be favored in this procedure so that bias based on genotype may be introduced. For example, estimated pollen fertility of males with rare alleles will be inflated. Partial genetic exclusion, where all offspring sampled are utilized but not necessarily assigned specific parents, however, has a number of important applications in mating analysis. These include estimating proportions of selfed progeny (Shaw and Allard 1982; Neale and Adams 1985), matings due to gene flow (Smith and Adams 1983; Friedman and Adams 1985), and the minimum number of male sources contributing to individual fruits (Ellstrand and Marshall 1986).

Because unambiguous assignment is usually not possible, a number of likelihood methods for assigning paternity among non-excluded males have been proposed. One method is to simply assign paternity to the individual with the highest conditional likelihood of producing the multilocus offspring genotype (called the *transition probability* by Devlin et al. 1988). If there is no most-likely father (i.e., more than one potential male parent has the highest likelihood), the observation must be thrown out. This is referred to as the *most-likely method.* This procedure also favors males with rare markers, but not to the extent above. A more significant problem with the most-likely method is that paternity assignment is biased towards homozygotes, since males with the highest number of loci homozygous for alleles compatible with the offspring genotype will always have the highest transition probabilities (Devlin et al. 1988; Brown 1989). In addition, since there are no constraints on the most-likely male with regards to its location relative to the mother tree, or to other factors influencing paternity (e.g., pollen fecundity, floral phenology), estimates of gene dispersal parameters based on the most-likely method can be severely biased. For example, when male mating success decreases with distance from the mother tree, estimates of mean effective pollen dispersal distance will be biased upwards (Brown 1989; Adams et al. 1992b).

A second likelihood approach, called the *fractional paternity method*, does not assign paternity in ambiguous cases to single individuals, but assigns it fractionally to all non-excluded males in proportion to their transition probabilities (Devlin et al. 1988). In contrast to the most-likely method, paternity is assigned to all progeny, although some progeny will not be assigned a single male parent. The efficacy of fractional paternity assignment relative to the most-likely procedure for estimating male fertilities was tested by Devlin et al. (1988) using computer simulations. Under the assumption of equal fertility of all males, the fractional method was found to remove biases due to over-assignment of homozygotes. Biased estimates, however, were obtained when equal fertilities were not assumed; although, bias was minimized with higher exclusion probabilities and larger samples of offspring.

A third approach to assigning paternity among non-excluded males is to weight transition probabilities with probabilities of mating and choose the most-likely parent based on the combined measure (*modified most-likely method*) (Ritland 1983; Neale 1984; Adams et al. 1992b). Information on pollen fecundity, phenological overlap in flowering, and distance between mates, for example, could be used to approximate probabilities of mating. A significant criticism of this approach, is that since a major goal of paternity analysis is to estimate parameters that influence mating probabilities, the use of approximate mating probabilities in calculating likelihoods of paternity is circular in reasoning (Devlin et al. 1988; Brown 1989). Estimates of gene dispersal patterns, however, will not be seriously biased unless the approximate mating probability function used to aid in assigning paternity is somehow significantly biased. For example, while the parameter(s) specifying the function relating male mating success to distance between mates may be unknown, the general form of the function can often be reasonably assumed (e.g., mating success decreases exponentially with increasing distance from the mother plant, but at an unknown rate). One approach to utilizing the assumed distance relationship in assigning paternity is to first estimate the parameter(s) for the function from a subsample of offspring where genetic exclusion has removed all but a few possible male parents (Adams et al. 1992b). This approximate distance function can then be used by itself or with other information on mating success to estimate mating probabilities. Application of such a procedure in computer simulated populations, where mating success was assumed to be negatively related to distance between parent trees, resulted in estimates of mean distances between mates that were considerably less biased than those obtained using the most-likely method (i.e., based on transition probabilities alone) (Adams et al. 1992b).

Model approach

In the model approach, gene dispersal parameters are estimated, using maximum likelihood or other estimation procedures, from probability models designed to account for observed frequencies of multilocus genotypes in offspring samples, rather than from inferred parent-offspring pairs. Thus, estimation bias expected with parentage analysis procedures might be eliminated. Schoen and Stewart (1986) used the model approach to investigate male fertilities of clones in a white spruce [*Picea glauca* (Moench) Voss] seed orchard. Their model applies to male gametes, and thus, is specific to conifers. A more general model that can be applied to estimate fertilities of either males or male and female pairs has been described by Roeder et al. (1989).

Both of the above models assume that the population of interest is isolated from sources of genes outside the population (i.e., gene flow). An expanded mating model which can be used in situations where pollen gene flow is expected is described by Adams and Birkes (1989, 1991). The model is specific for male gametes but could readily be expanded for application to diploid offspring. In this model, called the *Neighborhood Model*, a specified area around a mother tree is called a neighborhood. The model states that pollen successful in fertilizing viable offspring of the mother tree comes from three sources: self-fertilization (with probability s), migrant pollen from outside the neighborhood (probability w), and cross-fertilization with males within the neighborhood (probability $1 - w - s$), with relative mating success (i.e., fertility) of the jth outcross male in the neighborhood being ϕ_j. The probability of observing multilocus genotype g_i in the pollen gametes of offspring from the mother tree is

$$P(g_i) = s\,P(g_i|M) + (1 - w - s)\,\Sigma^r \phi_j\,P(g_i|F_i) + w\,P(g_i|B) \tag{1}$$

where $P(g_i|M)$ is the probability that the mother tree produces gametes with genotype g_i (transition probability for the mother tree), $P(g_i|F_j)$ is the transition probability for the jth outcross male in the neighborhood (there are r outcross males in total), and $P(g_i|B)$ is the probability that pollen gametes from migrant (background) pollen sources have genotype g_i. Instead of attempting to estimate a relative mating success of each outcross male in the neighborhood, ϕ_j is expressed as a function relating mating success to factors expected to influence mating success, such as distance to the mother tree, degree of floral synchrony, and pollen fecundity. For example, an exponential function can be used to relate ϕ_j to distance from the mother tree (d) (Fig. 2). If the distance function is substituted into the above probability model, the three parameters to be estimated are s, w, and β. Similarly, additional models can be developed to describe other factors (or combinations of factors) influencing mating success. In practice, the neighborhood might be a small disjunct population or some arbitrarily defined area within a larger stand. Size of the neighborhood doesn't need to be large, since the relative mating success of near neighbors is probably of greatest interest. Furthermore, the smaller the number of potential males within the neighborhood, the greater the ability to discriminate genetically among potential males and between multilocus pollen gametes from within and outside the neighborhood.

Maximum likelihood estimation of parameters in the neighborhood model applied to computer simulated data sets indicate that the model approach gives reasonably efficient estimates, especially when compared to estimates derived from paternity analysis procedures (Adams and

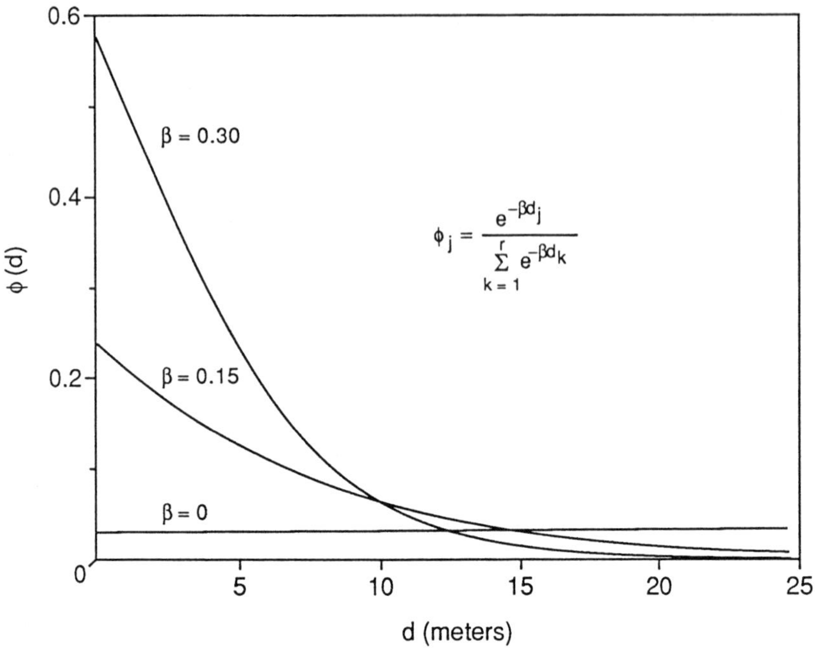

$$\phi_j = \frac{e^{-\beta d_j}}{\sum_{k=1}^{r} e^{-\beta d_k}}$$

Fig. 2. Exponential function which can be used to relate mating success of a male (ϕ) to distance from the mother tree (d), illustrated for various values of the distance parameter (β). Also given is the formula for computing ϕ for the jth male (ϕ_j), where d_j is the distance the jth male is to the mother tree and r is the number of males in the neighborhood.

Birkes 1991; Adams et al. 1992a; Adams and Birkes, in prep.). For example, offspring data samples ($N = 200$ ten-locus pollen gametes) for each of three mother trees in a Douglas-fir seed orchard were simulated 500 times using each of three sets of model parameters (s and w fixed at 0.05 and 0.10, respectively, while β was 0, 0.15, or 0.30). The neighborhoods around each mother tree were 25 m in radius and included 25 to 26 outcross males. For each sample of offspring, the mean distance between the mother tree and its outcross mates within the neighborhood (μ_d) was estimated using three methods: the most-likely and fractional paternity procedures, and by estimating β using the model approach and calculating $\hat{\mu}_d$ from the relationship between ϕ_j and d (Fig. 2). Pollen gametes identified by genetic exclusion as immigrants (i.e., from males outside the neighborhoods) were deleted from the data before calculating $\hat{\mu}_d$ in the most-likely and fractional paternity analyses. Unidentified immigrants remaining in the data sets contribute bias in estimating μ_d, but at the low level of w assumed in the simulations, the bias is expected to be negligible (Adams et al. 1992a). Mean values of $\hat{\mu}_d$ and its standard deviation over the 500 simulations were averaged over the three mother trees (Table 1). When there was no relation between mating success of outcross males in the neighborhood and distance to the mother tree (i.e., $\beta = 0$), the fractional-likelihood procedure and the model approach both provided nearly unbiased estimates of μ_d. As mating success of near neighbors increased (i.e., as β increased), however, the fractional procedure resulted in increasingly overestimated values of μ_d, while the model approach continued to provide estimates with negligible bias. The most-likely procedure gave poor estimates of μ_d at all values of β, and as expected, resulted in overestimates when $\beta > 0$.

Table 1. Comparisons of means and standard deviations (in parentheses) of estimates of the mean distance between mother trees and their outcross mates ($\hat{\mu}_d$) obtained by applying three estimation procedures to computer simulated samples of ten-locus pollen gametes[1]

Parameter values		$\hat{\mu}_d$		
β	$\hat{\mu}_d$	Most likely	Fractional likelihood	Model approach
0	16.67	14.66 (0.58)	16.63 (0.18)	16.59 (1.11)
0.15	12.07	13.43 (0.62)	15.67 (0.20)	12.06 (0.90)
0.30	8.07	13.13 (0.56)	14.75 (0.23)	7.92 (0.64)

[1] Five hundred pollen gamete samples ($N = 200$ each) were generated using the neighborhood model for each of three mother trees and three sets of parameter conditions (see text for details). The results are averages for the three mother trees which had mean exclusion probabilities for pollen gametes ranging from 0.57 to 0.74.

Comparison of estimation procedures

While the rare marker approach has the advantage of simplicity, it can be applied only in special circumstances because of the need for trees with unique markers. The parentage analysis and model approaches, on the other hand, are at least potentially applicable in a broader range of situations, as long as sufficient polymorphic marker loci are available.

The advantage of parentage analysis is that with the exception of the modified most-likely procedure, *a priori* knowledge of gene dispersal patterns is not required. Once parentage is assigned to a large sample of offspring or gametes, dispersal patterns can be inferred. These inferences, however, may be a poor reflection of true dispersal patterns if exclusion probabilities are not high. For example, when the fractional likelihood procedure was applied to simulated samples of offspring in a population composed of only 48 parents, EP's greater than 0.90 were necessary to obtain reasonable estimates of male fertilities when actual male fertilities were not constant (Devlin et al. 1988). When reliable information on factors influencing mating success is available, accuracy of parentage assignment can be increased and bias in mating parameter estimation reduced by using the modified most-likely procedure (Adams et al. 1992b). This procedure should not be employed when information on mating probabilities (e.g., the relationship between mating success and distance between mates) is questionable, since bias associated with the assumed relationship will be introduced. In addition, like other parentage procedures, high EP is essential for accurate parameter estimation with the modified most-likely method. In computer simulations testing this method, Adams et al. (1992b) found that the efficiency of effective mean pollen dispersal estimates was greatly improved when sampled offspring with more than 15—30 non-excluded males (i.e., more than 5—10% of the potential males in the population) were deleted from data sets prior to analysis.

The advantage of the mating model approach is that estimation bias may be eliminated when the probability model fitted to the data reasonably approximates reality (Table 1; Adams and Birkes 1991; Adams et al. 1992a; Adams and Birkes, in prep.). This assumes some *a priori* information on which model or models are appropriate. In addition, while computer simulations show that with the model approach estimation bias is relatively insensitive to the magnitude of average EP, standard deviations of estimates increase as average EP decreases (Adams and Birkes, in prep.). Thus, EP's must be high for accurate estimation of gene dispersal parameters, regardless of the method employed.

Parentage analysis, or mating models which don't allow for gene flow, strictly apply only to isolated populations where genotypes of all potential parents can be identified. In practice, however, isolation is difficult to achieve in study populations, and sampled offspring resulting from parentage outside the study area are often detected by genetic exclusion. A procedure that can be followed in this situation is to delete the identified contaminants from the data set and apply parentage analysis or the mating model approach to the remaining data, assuming all offspring in the reduced sample come from parents within the study area (Neale 1984; Schnabel 1988; Hamrick and Murawski 1990). The extent to which this procedure leads to valid estimates of gene dispersal within populations depends on the amount of actual gene flow into the study area and the degree to which immigrants differ genetically from offspring produced by parents within the population. Immigrants that cannot be identified as such by genetic exclusion will contribute error in estimating gene dispersal. For example, if actual gene dispersal decreases with distance from parent trees, the inclusion of non-identified immigrants will likely result in estimates of mean gene dispersal within populations that are upwardly biased (Adams et al. 1992a). This is because the likelihood of encountering a potential parent which is genetically compatible with a non-identified immigrant increases with increasing distance away from the physical location of the immigrant in the population (because the number of potential parents increases with increasing distance away). Application of the model approach to computer simulated data sets suggests that the upward bias in estimating mean effective pollen dispersal distance within populations may be relatively modest (i.e., <10%) when gene flow is ignored in mating models, if actual gene flow is low (i.e., $w \leqslant 0.10$) (Adams et al. 1992a). If, however, exclusion probabilities are not high and significant gene flow is suspected, estimates of gene dispersal within populations based on parentage analysis or on mating models which ignore gene flow, should be treated with caution.

Observed patterns of gene dispersal within populations

In this section, I briefly review the results of studies employing isozymes to investigate gene dispersal within natural populations of forest trees. Studies on pollen gene dispersal in seed orchards have recently been reviewed by Adams and Birkes (1989), Wheeler et al. (1992), and Adams et al. (1992a).

Pollen gene dispersal

The rare marker approach was used by Muller (1977) to study effective pollen dispersal in a 120 year-old Scots pine (*Pinus sylvestris* L.) stand (128 trees/ha) in Germany. Seeds were collected from each of 15 mother trees ranging from 5 to 75 m away from a male marker tree and assayed for the presence of a rare isozyme allele. As might be expected from patterns of pollen dispersal from individual trees, the relative frequency of offspring sired by the marker male decreased exponentially with increasing distances of mother trees (Fig. 3). Somewhat unexpected, however, is that the proportion of offspring sired by the marker tree was always low (<11%), even for mother trees less than 10 m away. Similar findings were obtained using rare markers in a 100 year-old Scots pine seed tree stand in northern Sweden, which had a tree density less than one-third (37 tree/ha) that of the German stand (Yazdani et al. 1989). Again the proportion

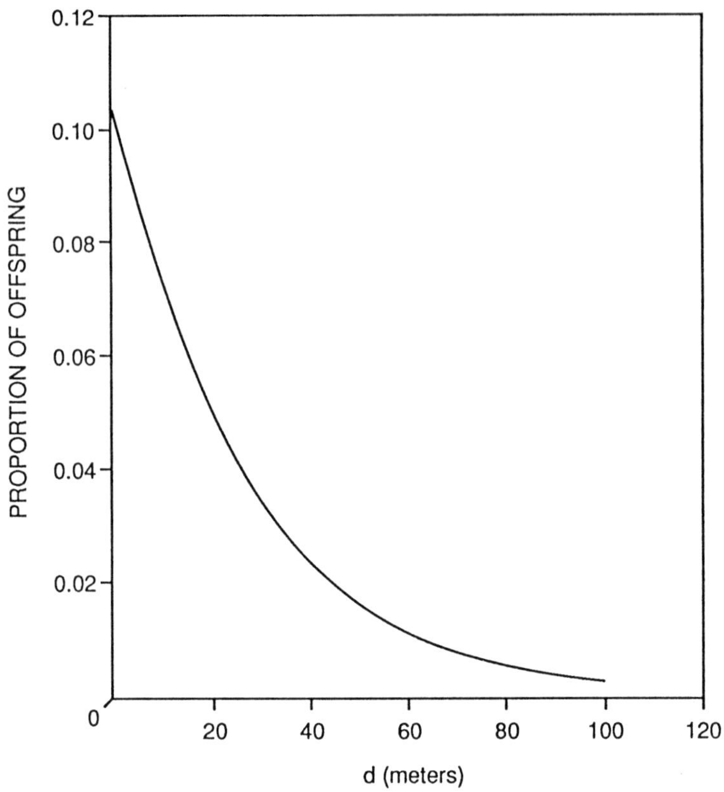

Fig. 3. Relationship between the distance of mother trees from a male marker tree (d) and the proportion of offspring sired by the marker tree in a 120 year-old Scots pine stand (derived from Muller 1977).

of offspring fertilized by the marker tree decreased rapidly with distance, but in this case, the highest proportion of offspring due to the marker was estimated to be no more than 2%. Thus, although the probability of mating between individuals in these Scots pine stands appears to be greatest among near neighbors, the bulk of pollen fertilizing viable offspring comes from other than near neighbor sources.

If near neighbors are not responsible for the majority of effective pollen in these Scots pine stands, what sources of pollen are responsible? The proportion of viable seed due to self-fertilization is expected to be low in this species (i.e., <10%, Muona 1989), thus, the most likely sources are gene flow from nearby stands or the summed accumulations of pollen dispersed from distant individuals within stands. Although large amounts of pollen may come from nearby stands, pollen dispersed by individuals within stands can account for the observations in Scots pine. To illustrate,

I use data from slash pine (Wang et al. 1960), where pollen deposition from an individual tree followed a negative exponential distribution (Fig. 4). As the distance from a mother tree increases, the relative density of pollen from any one male decreases, but, the number of males increases, so that the summed amount of pollen produced by distant males can be substantial. In this example, less than 20% of the pollen cloud around a mother tree would come from the 16 males within 20 m, while more than 40% would come from the 78 males located at distances greater than 40 m. Thus, if fertilization success is proportional to the relative amount of pollen contributed to the pollen cloud around a mother tree, near neighbors would account for only a small percentage of the total offspring.

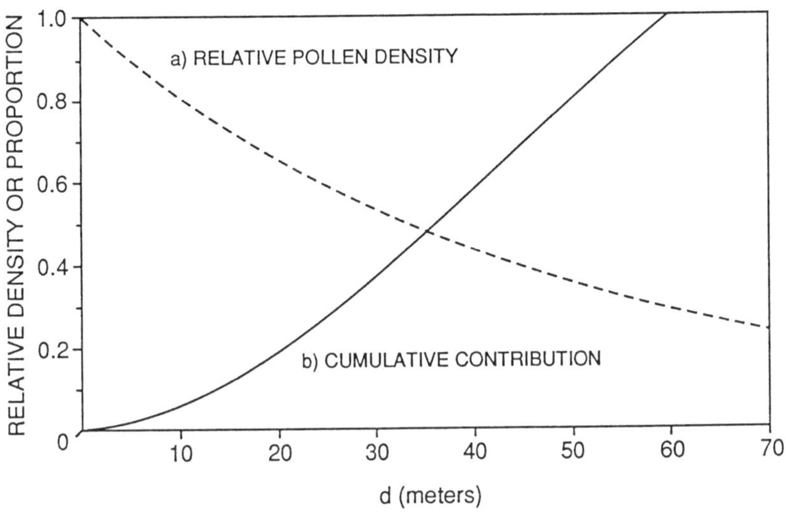

Fig. 4. Curves showing a) the exponential decrease in the density of pollen dispersed by an individual male (relative to source density) with increasing distances (d) from the source, and b) the cumulative proportion of the total pollen cloud around a mother tree contributed by all males at increasing distances away. Curve b assumes that all trees in the stand produce an equal amount of pollen, trees have a uniform density of 124 trees/ha, and for simplicity, only trees within 60 m contribute to the pollen cloud around an individual mother tree.

The results in Scots pine suggest that mean distance between mates and the average number of males contributing to the offspring of any single mother tree can be large in conifers. Estimates of effective pollen dispersal in two old-growth (>200 years) Douglas-fir shelterwood stands (called Benshell and Cutmore) support this notion (Table 2). Seeds sampled from 6 mother trees in each stand were assayed electrophoretically and the

Table 2. Gene dispersal statistics for two old growth Douglas-fir shelterwood stands[1]

	Stand	
	Benshell	Cutmore
Density (trees/ha)	35	15
Average spacing between trees (m)[2]	18.2	27.7
Dispersal by pollen		
Pollen gametes sampled	547	574
Immigrants detected (%)	26.6	20.4
Selfs (%)	6.3	9.2
Outcross dispersal within stands		
Mean dispersal distance (m)	54.8	80.7
Mean squared dispersal distance (m^2)	3776	7663
Effective number of outcross males per female (N_{em})	83	72
Dispersal by seed		
Egg gametes sampled	65	42
Immigrants detected (%)	26.6	16.7
Dispersal within stands:		
Mean dispersal distance (m)	47.3	72.3
Mean squared distance (m^2)	2991	5269
Effective number of Mother trees (N_{ef})	66	50

[1] Data from Neale (1984) and Neale and Adams (unpublished).
[2] Calculated assuming all trees are surrounded by the same sized hexagonal growing area so that the distance to any immediate neighbor, of which there are six, is always the same and equal to the average spacing.

genotypes of their pollen gametes determined at 13 isozyme loci. Since the study plots (2.4 ha in each stand) were not isolated, it is not surprising that the proportion of pollen gametes that could not be assigned parents (i.e., detected immigrants) exceeded 20% in both stands. There is no way to tell whether these immigrants came from nearby trees just outside the study plots or if they were from distant stands (gene flow). Nevertheless, trees near the plot boundaries did not have consistently higher proportions of detected pollen immigrants in their progeny, suggesting that distant pollen sources may have been primarily responsible. The remaining pollen gametes were assumed to come from pollen producing trees within the shelterwoods (36 in Benshell, 43 in Cutmore). The mean exclusion probability estimated from allele frequencies of parent trees in the two shelterwoods was 0.85, so on average, 5—6 potential male parents remained for each pollen gamete after genetic exclusion. Pollen parents were inferred from the non-excluded males using the modified most-likely procedure, where transition probabilities were weighted by probabilities of mating based on distances between mates (Neale 1984). It was assumed that the effective dispersal of pollen gametes from individual fathers could be approximated by a circular normal probability distribution, where the parameter specifying the distribution (i.e., mean squared dispersal distance) was estimated from a small subsample of pollen gamete data using only transition probabilities to assign paternity.

Paternity within shelterwoods was assigned to the mother trees (self-fertilization) in 6.3% of the cases at Benshell and 9.2% at Cutmore. These estimates of the proportion of offspring due to selfing are somewhat larger than estimates obtained using the mixed mating model (2.5 and 3.5%, respectively; Neale and Adams 1985); nevertheless, the proportion of viable selfed offspring appears to be low. Mean effective outcross pollen dispersal was estimated to be 54.8 m at Benshell and 80.7 m at Cutmore, in each case, about three times the average spacing between trees. These values probably overestimate mean outcross pollen dispersal within stands, but not greatly so, since exclusion probabilities were relatively high. Overall effective dispersal from individual trees, however, is probably underestimated, because sampling of dispersal distances was truncated by the plot boundaries; and, significant long-distance dispersal is evident from the substantial gene flow detected in both shelterwoods. The higher mean dispersal in Cutmore may be because the lower tree density in this stand allows for greater air movement. Cutmore, however, has a slightly higher location than Benshell, and thus, is less protected by adjacent stands from wind.

The effective number of outcross males fertilizing offspring of each female (N_{em}) is defined as the number of outcross males it would take to generate an equivalent amount of genetic diversity as observed in the pollen pools of the mother trees, given pollen gametes are drawn at random from trees in a large population. An estimate of N_{em} for the portion of outcross offspring sired by males within shelterwoods, was calculated by

$$N_{em} = 4\pi\sigma_p^2\delta \qquad (2)$$

where σ_p^2 is the axial variance of pollen dispersal distance and δ is the density of pollen producing males (Crawford 1984). Estimates of σ_p^2 for each shelterwood were obtained by $\frac{1}{2}\Sigma^n d_i^2/n$, where d_i is the inferred dispersal distance of the ith pollen gamete in the offspring sample, of which there were a total of n pollen gametes after removing those identified as immigrants or resulting from selfing. Assuming all mature trees produce pollen, N_{em} is estimated to be greater than 70 in both shelterwoods (Table 2). Total N_{em} is expected to be even greater, since gene flow was not accounted for in the calculations.

Estimates of effective pollen dispersal for two insect-pollinated angiosperm tree species have been reported by Hamrick and his students (Schnabel 1988; Hamrick and Murawski 1990; Hamrick, personal communication). Seed crops were sampled from mother trees in each of two years in each of three stands of the dioecious, honeylocust (*Gleditsia triacanthos* L.) in northeastern Kansas. In *Platypodium elegans*, a tropical canopy species with bisexual flowers, seed crops were sampled in each of three years on individuals widely scattered over an 84 ha area (which had a total of 59 adult *P. elegans* trees) on Barro Colorado Island, Republic of Panama. Paternity of offspring was assigned fractionally based on 19 polymorphic isozyme loci in honeylocust and 15 in *P. elegans*. Because findings on effective pollen dispersal were similar in the three honeylocust sites, data from only one site, Nelson, are discussed here. There were 24 male and 16 female trees within the 1.92 ha at Nelson, which was the most isolated of the three study plots, having no other trees of the same species within several hundred meters, accept on the north side of the plot. In addition, the average exclusion probability must have been very high at Nelson, since the number of offspring assigned to a single male within the study plot averaged 44% in the two sample years.

It is generally considered that pollen dispersal in insect-pollinated plants is very limited (Levin and Kerster 1974), but the results in these two species do not support this contention (Table 3). The proportion of offspring identified as having pollen parents outside the study plot exceeded 30% in *P. elegans* and was nearly 50% in honeylocust. In addition, as observed in the Douglas-fir shelterwoods, the proportion of detected immigrants in both species was unrelated to the distance mother trees were from the edge of the study plot. Thus, pollen gene flow in these species may be extensive. The frequency distributions of observed distances between mates within plots were skewed towards shorter-distance matings in both species, with an estimated mean effective pollen dispersal of 39.4 m in honeylocust and 387.6 m in *P. elegans*. The figure for *P. elegans* includes zero distance matings because self-males were included in the fractional paternity analysis. Nevertheless, the estimated proportion of selfed progeny was very low (8%) in the study trees. Thus, insect-mediated cross-pollination appears to very effective in *P. elegans*, despite its low frequency of occurrence in the forest.

Insect-mediated pollen dispersal within populations of honeylocust and *P. elegans* may not be much greater, on average, than the mean distance between nearest flowering trees. In the dioecious honeylocust, mean

Table 3. Gene dispersal in two insect-pollinated angiosperm tree species

	Species	
	Gleditsia triacanthos	*Platypodium elegans*
Density (trees/ha)	21	0.70
Average spacing between trees (m)[1]	23.5	128.4
Dispersal by Pollen[2]		
Offspring sampled	467	198
Immigrants detected (%)	48.0	31.4
Mean dispersal distance within populations (m)	39.4	387.6
Dispersal by seed[3]		
Seeds sampled	310	—
Immigrants detected (%)	1.5	—
Mean dispersal distance within populations (m)	51.5	—

[1] Calculated assuming all trees are surrounded by the same sized hexagonal growing area, so that the distance to any immediate neighbor, of which there are six, is always the same and equal to the average spacing.
[2] Averages over two seed crops at a single site for *G. triacanthos* and over three seed crops from the same area for *P. elegans*.
[3] Based on a sample of young trees (<12 years old) in an old field site adjacent to the study plot of adult trees.

dispersal was less than twice the average spacing between trees (Table 3). In *P. elegans*, less than 20% of the individuals in the study area produced fruit in any one year. If only trees which eventually produced fruit were involved in mating, mean dispersal would be 23% greater than the average spacing between mating individuals (316.2 m) (Table 3). Thus, N_{em} for within-population mating may be relatively limited in these species, but with the extensive gene flow observed, overall N_{em} is still expected to be high.

Gene dispersal by seed

In both the Douglas-fir and honeylocust studies cited above, gene dispersal via seed was also investigated. In Douglas-fir, this was done by collecting seed from traps located in a grid pattern under the two shelterwood stands. In the honeylocust Nelson population, young trees (<12 years-old) were sampled in an old field adjacent to the adult tree study plot. Similar parentage analysis procedures as described for estimating pollen dispersal were used in each case to estimate effective seed movement. In both species, estimates of mean seed dispersal distances within the study areas were of similar magnitude to those observed for mean pollen dispersal (Tables 2 and 3). Seed immigrants, however, were detected at a much higher level in the Douglas-fir stands (26.6 and 16.7%) than in the honeylocust population (1.5%).

Although the frequency distribution of seed dispersal was skewed toward shorter distances in the Douglas-fir shelterwoods, the extent of gene dispersal by seed seems large enough to prevent the formation of tight spatial clustering of family members in offspring. For example, using the same approach employed for calculating N_{em}, the effective number of maternal parents contributing to the proportion of seed at any one location due to within shelterwood sources (N_{ef}), was estimated to exceed 50 (Table 2). When gene flow is also considered, N_{ef} would be expected to be even greater. Tight spatial clustering of siblings was also not evident in the understory of two Scots pine seed stands (Yazdani et al. 1985; Yazdani et al. 1989), where rare markers were employed to estimate the degree to which parent trees contribute (as either males or females) to nearby seedling regeneration. In both stands, marker trees were responsible for less than 10% of the seedlings within 10 m, and for lower percentages further away. The above results are counter to evidence for strong family structure in other conifer species (e.g., Linhart et al. 1981; Park and Fowler 1982; Park et al. 1984; Furnier et al. 1987).

In contrast to Douglas-fir, some family clustering of honeylocust offspring might be expected in the old field adjacent to the Nelson population, because mean seed dispersal was only 1.4 times greater than the mean spacing between female trees (37.2 m), and detected seed immigrants were few (Table 3). Indeed, spatial heterogeneity of isozyme allele frequencies among juvenile trees in the old field suggests that this is the case (Schnabel 1988).

Conclusions

Additional work is needed in developing and testing statistical procedures for estimating gene dispersal parameters. While parentage analysis has the advantage that it can be assumption free, estimation bias may be considerable if exclusion probabilities are not high. Under what conditions can parentage analysis be reliably used? We need a better understanding of how the efficiency of parameter estimation based on parentage analysis is influenced by exclusion probabilities, number of potential parents in the study population, and by the presence of unidentified immigrants. Employing RFLP markers to increase the number of polymorphic loci could substantially increase exclusion probabilities, but their use is not without complications and does not guarantee unambiguous parentage assignment unless the number of potential parents is small (Smouse and Chakraborty 1986). The use of hypervariable molecular markers is complicated by mutation, measurement error, and the inability to discriminate similar sized fragments on gels (Jeffreys et al. 1985; Devlin et al. 1990). Nevertheless, the high allelic diversity of these markers would make them highly desirable for parentage analysis, if inexpensive techniques for their large scale sampling in plants can be developed (Brown 1989).

With the levels of genetic discrimination currently possible in forest trees, the model approach appears to the most reliable means of estimating gene dispersal parameters. This approach, however, can be efficiently employed only when gene dispersal patterns can be approximated by relatively simple relationships. The neighborhood model seems to be a particularly useful alternative in forest trees where it is nearly impossible to find small isolated populations for study. Although the model approach may be superior to parentage analysis in many situations, efficient parameter estimation still requires large numbers of marker loci and large sample sizes (Adams and Birkes, in prep.).

Too few data are available to draw general conclusions about gene dispersal patterns within forest tree populations. In the studies on effective pollen dispersal that have been reported, mating success of males appears to decrease with increasing distance from mother trees as expected from observed patterns of pollen dispersal, but average effective dispersal distance can still be large. In conifers, pollen from distant males within the same stand or from surrounding stands may sire the bulk of offspring of any one mother tree, with near-neighbor mating responsible for a relatively limited proportion. The result is that effective male population sizes can be considerable. In insect-pollinated angiosperms, gene flow may also be strong, but within stand matings may primarily occur between nearest flowering individuals.

The influence of distance on mating success needs to be investigated in more species and for a greater variety of breeding systems. In addition, the influence of ecological factors other than distance need to be understood; such as flower density and pollinator availability in insect-pollinated species, and stand age, density, and structure in both wind- and insect-pollinated species. Differential floral phenology and pollen fecundity of males appear to have a profound influence on mating success in seed orchards (Adams and Birkes 1989; Wheeler et al. 1992). It would be interesting to determine the extent to which these factors affect outcross mating patterns in natural stands.

Because seed dispersal involves diploid genotypes, its variance has twice the influence on effective population size of offspring than an equivalent amount of pollen dispersal variance (Crawford 1984). Thus, patterns of effective seed dispersal may be more relevant to population structure than gene dispersal via pollen. In the two studies discussed above where both effective seed and pollen dispersal were investigated in the same populations, mean gene dispersal distances by seed were similar to those observed for pollen (Tables 2 and 3). If seed dispersal is limited relative to the average spacing between maternal trees, spatial clustering of family members in the offspring generation might be expected. This is apparently the case for honeylocust reproduction in the old field adjacent to the Nelson population. More work is needed to determine under what conditions seed movement may be severely restricted and the extent to which tight clustering of family members is the result (Epperson 1992, this issue, pp. 257—278). In any case, the ecology of effective seed dispersal (a function of fruit and seed morphology, palatability, longevity, etc.) may even be more complex than for pollen dispersal, and thus, generalizations even harder to come by.

References

Adams, W. T. and Birkes, D. S. 1989. Mating patterns in seed orchards. Proc. 20th South. For. Tree Improv. Conf., Charleston, South Carolina. pp. 75—86.

Adams, W. T. and Birkes, D. S. 1991. Estimating mating patterns in forest tree populations, pp. 157—172. In: Fineschi, S., Malvolti, M. E., Cannata, F. and Hattemer, H. H. (Eds) Biochemical Markers In The Population Genetics of Forest Trees. SPB Academic Publishing, The Hague, Netherlands.

Adams, W. T., Birkes, D. S. and Erickson, V. J. 1992a. Using genetic markers to measure gene flow and pollen dispersal in forest tree seed orchards. In: R. Wyatt (Ed) Ecology and Evolution of Plant Reproduction: New Approaches. Chapman and Hall, New York (in press).

Adams, W. T., Griffin, A. R. and Moran, G. F. 1992b. Using paternity analysis to measure effective pollen dispersal in plant populations. Amer. Naturalist (in press).

Bramlett, D. L. 1981. Effectiveness of wind pollination in seed orchards, pp. 10—14. In: Franklin, E. C. (Ed) Pollen Management Handbook. U.S.D.A., Forest Service, Agric. Handbook 587.

Brown, A. H. D. 1989. Genetic characterization of plant mating systems, pp. 145—162. In: Brown, A. H. D., Clegg, M. T., Kahler, A. L. and Weir, B. S. (Eds) Plant Population Genetics, Breeding, and Genetic Resources. Sinauer Assocs. Inc., Sunderland, MA.

Chakraborty, R., Meagher, T. R. and Smouse, P. E. 1988. Parentage analysis with genetic markers in natural populations. 1. The expected proportion of offspring with unambiguous paternity. Genetics 118: 527—536.

Crawford, T. J. 1984. The estimation of neighbourhood parameters for plant populations. Heredity 52: 273—283.

Devlin, B., Risch, N. and Roeder, K. 1990. No excess of homozygosity at loci used for DNA fingerprinting. Science 249: 1416—1420.

Devlin, B., Roeder, K. and Ellstrand, N. C. 1988. Fractional paternity assignment: theoretical development and comparison to other methods. Theor. Appl. Genet. 76: 369—380.

Ellstrand, N. C. 1992. Gene flow among seed plant populations. This issue (pp. 241—256).

Ellstrand, N. C. and Marshall, D. L. 1986. Patterns of multiple paternity in populations of *Raphanus sativus*. Evolution 40: 837—842.

Epperson, B. K. 1992. Spatial structure of genetic variation within populations of forest trees. This issue (pp. 257—278).

Friedman, S. T. and Adams, W. T. 1985. Estimation of gene flow into two seed orchards of loblolly pine (*Pinus taeda* L.). Theor. Appl. Genet. 69: 609—615.

Furnier, G. R., Knowles, P., Clyde, M. A. and Dancik, B. P. 1987. Effects of avian seed dispersal on the genetic structure of whitebark pine populations. Evolution 41: 607—612.

Gregorius, H-R. 1975. A model for the genetic relationship among offspring from open-pollinated plant populations. Theor. Appl. Genet. 46: 109—115.

Hamrick, J. L. and Murawski, D. A. 1990. The breeding structure of tropical tree populations. Plant Species Biol. 5: 157—165.

Hamrick, J. L. and Schnabel, A. 1985. Understanding the genetic structure of plant populations: some old problems and a new approach, pp. 50—70. In: Gregorius, H-R. (Ed) Population Genetics In Forestry. Lecture Notes In Biomathematics 60, Springer—Verlag, New York.

Handel, S. N. 1983. Pollination ecology, plant population structure, and gene flow, pp. 163—211. In: Real, L. (Ed) Pollination Biology. Academic Press, Inc., New York.

Issac, L. A. 1930. Seed flight in the Douglas fir region. J. Forestry 28: 492—499.

Jeffreys, A. L., Wilson, V. and Thein, S. L. 1985. Individual-specific 'fingerprints' of human DNA. Nature 316: 76—79.

Lanner, R. M. 1966. Needed: a new approach to the study of pollen dispersion. Silvae Genet. 15: 50—52.

Ledig, F. T. 1974. An analysis of methods for the selection of trees from wild stands. Forest Sci. 20: 2—16.

Levin, D. A. 1981. Dispersal versus gene flow in plants. Ann. Missouri Bot. Gard. 68: 233—253.

Levin, D. A. 1983. Plant parentage: an alternative view of breeding structure of populations, pp. 171—188. In: King, C. E. and Dawson, P. S. (Eds) Population Biology Retrospect and Prospect. Columbia Univ. Press., New York.

Levin, D. A. and Kerster, H. W. 1974. Gene flow in seed plants, pp. 139—220. In: Dobzhansky, T., Hecht, M. T. and Steere, W. C. (Eds) Evolutionary Biology 7, Plenum Press, New York.

Libby, W. J., Stettler, R. F. and Seitz, F. W. 1969. Forest genetics and forest-tree breeding. Ann. Rev. Genet. 3: 469—494.

Linhart, Y. B., Mitton, J. B., Sturgeon, K. B. and Davis, M. L. 1981. Genetic variation in space and time in a population of ponderosa pine. Heredity 46: 407—426.

Meagher, T. R. and Thompson, E. 1986. The relationship between single parent and parent pair genetic likelihoods in genealogy reconstruction. Theor. Pop. Biol. 29: 87—106.

Meagher, T. R. and Thompson, E. 1987. Analysis of parentage for naturally established seedlings of *Chamaelirium luteum* (Liliaceae). Ecology 68: 803—812.

Mitton, J. B. 1992. The dynamic mating systems of conifers. This issue (pp. 197—216).

Müller, G. 1977. Short note: cross-fertilization in a conifer stand inferred from enzyme gene-markers in seeds. Silvae Genet. 26: 223—226.

Müller-Starck, G. 1982. Reproductive systems in conifer seed orchards. I. Mating probabilities in a seed orchard of *Pinus sylvestris* L. Silvae Genet. 31: 188—197.

Muona, O. 1989. Population genetics in forest tree improvement, pp. 282—298. In: Brown, A. H. D., Clegg, M. T., Kahler, A. L. and Weir, B. S. (Eds) Plant Population Genetics, Breeding, and Genetic Resources. Sinauer Assocs. Inc., Sunderland, MA.

Namkoong, G. 1966. Inbreeding effects on estimation of genetic additive variance. Forest Sci. 66: 8—13.

Neale, D. B. 1984. Population genetic structure of the Douglas-fir shelterwood regeneration system in southwest Oregon. Ph.D. thesis, Oregon State Univ., Corvallis, Oregon, USA.

Neale, D. B. and Adams, W. T. 1985. The mating system in natural and shelterwood stands of Douglas-fir. Theor. Appl. Genet. 71: 201—207.

Nybom, H. and Schaal, B. A. 1990. DNA "fingerprints" applied to paternity analysis in apples (Malus × domestica). Theor. Appl. Genet. 79: 763—768.

Park, Y. S. and Fowler, D. P. 1982. Effects of inbreeding and genetic variances in a natural population of tamarack (*Larix Laricina* (Du Roi) K. Koch) in eastern Canada. Silvae Genet. 31: 21—26.

Park, Y. S., Fowler, D. P. and Coles, J. F. 1984. Population studies of white spruce II. Natural inbreeding and relatedness among neighboring trees. Can. J. For. Res. 14: 909—913.

Ritland, K. 1983. Estimation of mating systems, pp. 289—302. In: Tanksley, S. D. and Orton, T. J. (Eds) Isozymes in Plant Genetics and Breeding, Part A. Elsevier Science Publish. Co. Inc., New York.

Roeder, K., Devlin, B. and Lindsay, B. G. 1989. Application of maximum likelihood methods to population genetic data for the estimation of individual fertilities. Biometrics 45: 363—379.

Schnabel, A. 1988. Population genetic structure and gene flow in *Gleditsia triacanthos* L. Ph.D. thesis, Univ. of Kansas, Lawrence.

Schoen, D. J. and Stewart, S. C. 1986. Variation in male reproductive investment and male reproductive success in white spruce. Evolution 40: 1109—1120.

Shaw, D. V. and Allard, R. W. 1982. Estimation of outcrossing rates in Douglas-fir using isozyme markers. Theor. Appl. Genet. 62: 113—120.

Silen, R. R. 1962. Pollen dispersal considerations for Douglas-fir. J. Forestry 60: 790—795.

Smith, D. B. and Adams, W. T. 1983. Measuring pollen contamination in clonal seed orchards with the aid of genetic markers. Proc. 17th South. For. Tree Improv. Conf., Athens, GA. 69—77.

Smouse, P. E. and Chakraborty, R. 1986. The use of restriction fragment length polymorphisms in paternity analysis. Am. J. Hum. Genet. 38: 918—939.

Sorensen, F. C. and White, T. L. 1988. Effect of natural inbreeding on variance structure in tests of wind-pollination Douglas-fir progenies. Forest Sci. 34: 102—118.

Squillace, A. E. 1974. Average genetic correlations among offspring from open-pollinated forest trees. Silvae Genet. 23: 149—155.

Wang, C-W., Perry, T. O. and Johnson, A. G. 1960. Pollen dispersion of slash pine (*Pinus elliotii* Engelm.) with special reference to seed orchard management. Silvae Genet. 4: 78—86.

Wheeler, N. C., Adams, W. T. and Hamrick, J. L. 1992. Pollen distribution in wind-pollinated seed orchards. In: Bramlett, D. L., Askew, G. R., Blush, T. D., Bridgwater, F. E. and Jett, J. B. (Eds) Pollen Management Handbook, Vol. 2, USDA Forest Service Agric. Bulletin (in press).

Yazdani, R., Lindgren, D. and Rudin, D. 1985. Gene dispersion and selfing frequency in a seed-tree stand of *Pinus sylvestris* (L.), pp. 139—154. In: Gregorius, H-R. (Ed) Population Genetics in Forestry, Lecture Notes in Biomathematics 60, Springer—Verlag, New York.

Yazdani, R., Lindgren, D. and Stewart, S. 1989. Gene dispersion within a population of *Pinus sylvestris*. Scand. J. For. 4: 295—306.

New Forests **6**: 241–256, 1992.

Review paper

Gene flow among seed plant populations

NORMAN C. ELLSTRAND

Department of Botany & Plant Sciences and Program in Genetics, University of California Riverside, CA 92521–0124, USA

Received 9 October 1990; accepted 12 March 1991

Key words: immigration rate, tree breeding, conservation genetics, risks of biotechnology

Application. Gene flow is a critical determinant of population genetic structure, playing an important role in forest tree improvement, conservation genetics, and the potential for the escape of engineered genes. A review of gene flow among plant populations yields the following generalization: although gene flow varies substantially among species, populations, seasons, and even individual plants, at physical isolation distances of hundreds to thousands of meters, gene flow levels are frequently sufficient to counteract genetic drift and moderate levels of directional selection. Gene flow will be hard to predict a priori. Therefore, any project or management plan depending on a given level of gene flow or reproductive isolation should include a genetically-based method to measure the actual gene flow among the populations involved.

Abstract. Gene flow is a critical determinant of population genetic structure, playing an important role in both evolutionary and applied plant population genetics. Four methods have been used to estimate rates of gene flow among plant populations. I review and reconcile the data collected by these methods. The following generalization emerges: although gene flow varies substantially among species, populations, seasons, and even individual plants, at physical isolation distances of hundreds to thousands of meters, gene flow levels are frequently sufficient to counteract genetic drift and moderate levels of directional selection. This pattern suggests the genetic structure of natural plant populations is more dynamic than generally supposed. Furthermore, substantial and variable gene flow has implications for plant breeding, conservation genetics, and the potential for the escape of engineered genes.

Introduction

Gene flow, the movement of genes among populations, can play a potent role in determining the genetic structure of those populations (Slatkin 1985a, 1987). Theoretical studies have demonstrated that relatively low levels of gene flow can swamp the effects of local, directional selection

(Antonovics 1968; Epperson 1990; Gillespie 1975; Haldane 1930; Sanderson 1989). If gene flow and selection work in concert, beneficial alleles can spread rapidly among populations (Wright 1982). Furthermore, rates of gene flow as low as one individual per generation are sufficient to counteract the effects of genetic drift, even in very small populations (Kimura and Weiss 1964; Weiss and Kimura 1965; Wright 1931). On the other hand, if gene flow is highly restricted, microgeographic differentiation is readily established, either by selection (Bradshaw 1972; Endler 1973; Jain and Bradshaw 1966; Levin 1978; Slatkin 1973) or by drift (Levin 1984; Rohlf and Schnell 1971; Turner et al. 1982; Wright 1943, 1946). Finally, if substantial gene flow occurs sporadically over time and space, migration can act in a way similar to drift, creating substantial spatial differentiation (Levin 1988; Slatkin 1987; Wade and McCauley 1988). Because gene flow is critical in determining the distribution of genetic variation within and among populations, its measurement ranks as a high priority among the goals of population geneticists (Grant 1980, 1985; Levin 1984; Slatkin 1985a, 1987).

Furthermore, gene flow can play an important role in applied plant population genetics. In many cases, gene flow could be detrimental. For example, unanticipated long-distance mating ("pollen contamination", Smith and Adams 1983) can disrupt plant breeding programs. Also, gene flow by pollen can provide a conduit for the escape of engineered genes from a transgenic crop into natural, sexually compatible populations (Ellstrand 1988; Ellstrand and Hoffman 1990). In conservation genetics, interspecific mating between an endangered species and a more widespread, sexually compatible congener could lead to extinction via hybridization (e.g., Rieseberg et al. 1989). Alternatively, gene flow could be beneficial. Crop-weed gene flow in primitive agroecosystems may act to increase the in situ gene bank (Plucknett et al. 1987). In endangered species, gene flow among small, semi-isolated populations can prevent the erosion of genetic variation from drift (Allendorf 1983). Clearly, study of gene flow is as important for applied population geneticists as for evolutionary geneticists.

Here I review what is known of gene flow among seed plant populations. I summarize the strengths and weaknesses of the four general approaches used to estimate gene flow and reconcile the datasets resulting from each. The synthetic picture that emerges is a view of gene flow as an often substantial, but highly variable, parameter. I conclude with some implications for evolutionary and applied population genetics.

Methods that estimate gene flow among plant populations

Pollen and seed dispersal from point sources

The most common method of estimating gene flow involves extrapolation from data on pollen and seed dispersal from point sources. Pollen dispersal is inferred from pollinator foraging distances (e.g., Schmitt 1980), or from following pollen analogues (dyes, e.g., Campbell and Waser 1989). Or pollen movement is measured directly by following marked pollen (e.g., Handel 1976), or, rarely, from following naturally polymorphic pollen (Thomson and Thomson 1989). For anemophilous species, dispersed pollen may be collected from samplers set at varying distances from a source (e.g., Tonsor 1985). Similarly, seed dispersal curves are measured indirectly by following animal vectors or directly by collecting dispersed seed at varying distances from source plants (reviewed in Levin and Kerster 1974).

From over a hundred such studies, a common pattern emerges. Most pollen and seed are dispersed very close to the source, their frequency declining rapidly with distance. The skewed, leptokurtic dispersal schedules described by the vast majority of these studies have become almost axiomatic (Levin and Kerster 1974; Richards 1986; Willson 1983). Generally, the mean is from one to about twenty meters (Levin and Kerster 1974; Grant 1980; Hamrick 1982). When both seed and pollen dispersal are measured for a species, seed dispersal is usually more restricted than pollen dispersal (e.g., Bos et al. 1986; Levin and Kerster 1968). Extrapolations from these data have lead to the predication that gene flow among populations should be very restricted, "that the immigration rate is likely to be much less than 1%" (Levin 1984).

These methods have proven valuable for describing dispersal patterns from a source, providing the foundation for understanding some of the ecological factors (vector behavior, plant density, etc.) that determine variation in patterns of dispersal within and among species (Handel 1983a; Howe and Smallwood 1982).

However, data collected by these methods have limited utility for estimating interpopulation gene flow. Measuring dispersal from a point source almost always truncates the actual dispersal curve; "long-range dispersal events are excluded by the methods used in quantitative dispersal studies" (Grant 1985). Surrounding a plant with seed traps at various distances does not ensure that some seed might not fall beyond the most distant trap. Likewise, recording plant-to-plant pollinator foraging bouts excludes the movement of the pollinator from one population to the next; in fact, a fifty meter move may be difficult to observe within a large, continuous population. The problem is exacerbated by the fact that leptokurtic dispersal curves are characterized by tails that contain more long-distance dispersal events than those of an equivalent normal distribution (Hamrick 1987). Unless the researcher can estimate the total number of unrecorded long distance dispersal events (for example, by comparing the total number of seeds on a plant before release to the total number recovered), it would be impossible to judge whether 0.1%, 1%, or 10% of the propagules have dispersed beyond the edges of the experimental site. Also, truncation may occur at the left hand side of the curve, because measures of pollen dispersal will fail to account for self-fertilization rates in self-compatible plants (Hamrick 1987).

Even if dispersal curves are accurate, dispersal is still not necessarily the same as gene flow (Endler 1977). The translation of pollen flow into gene flow depends on many ecological and genetic factors that influence fertilization and seed set (Handel 1983a; Levin 1981; Willson 1984). Likewise, the translation of seed dispersal into gene flow depends on the ecological and genetic factors that influence germination and seedling establishment (Hamrick and Loveless 1986; Levin 1981). Thus, even accurate measurements of dispersal may represent gene flow poorly.

Gene dispersal from point and block sources

The difference between dispersal and gene flow is reduced if the successful dispersal of genes, rather than propagules, is measured. Genetic markers have been used most frequently to measure mating distances within synthetic populations (e.g., Crane and Mather 1943; Schaal 1980), to measure mating rates between contiguous populations (e.g., Handel 1983b; Smyth and Hamrick 1987), or, rarely, to measure gene flow among spatially discrete populations (e.g., Ellstrand et al. 1989). Progeny testing of seed harvested at varying distances from the plant or plants bearing the marker alleles identifies successful fertilization and fruit set. The method has been applied almost exclusively to measure gene dispersal by pollen. Also, studies involving markers in natural populations are exceptionally rare (Levin 1981). However, at least one study has addressed both pollen- and seed-mediated gene dispersal using introduced markers in a natural population (Levin 1983).

When the data are summarized in a single curve, most studies employing genetic markers show gene dispersal within a population follows a skewed and leptokurtic distribution (Levin and Kerster 1974). But detailed maps of the gene dispersal often show patchiness and/or directionality (e.g., Handel 1982; Smyth and Hamrick 1987). Also, when genetic data are contrasted with pollinator foraging patterns for the same sites, the actual gene dispersal is always greater than would be inferred from the vector movements; the difference is attributed to pollen carry-over (Schaal 1980; Levin 1981). Still, the distributions measured by this method are often in accord with those from the dispersal of propagules from point sources (Levin 1981).

The importance of this method is that it measures the movement of genes instead of just pollen or seed. Furthermore, these studies have provided the foundation for understanding some of the ecological factors (population shape, population size, etc.) that determine variation in patterns of dispersal within and among species (Handel 1983a; Howe and Smallwood 1982). A third advantage is that the approach can be used to measure gene flow among discrete populations (e.g., Ellstrand et al. 1989), although it has been used rarely for that purpose.

However, almost all of these experiments are constrained by their artificiality. Relatively uniform plants growing in monospecific experimental plots may be poor analogues for natural populations where variation in plant size, spacing, phenology, and the presence of other species that share pollen or fruit vectors may play a role in determining pollen and seed dispersal (Handel 1983a; Howe and Smallwood 1982).

The approach is most useful for judging gene flow when the data come from experiments involving the exchange of genes from discrete populations. However, gene flow is usually inferred from those experiments studying gene dispersal from a source plant or plants into a contiguous population or populations. Extrapolating from these gene dispersal curves to interpopulation gene flow rates suffers from the same long-distance truncation problems that accrue from extrapolating from propagule dispersal curves because of the difficulty of measuring long-distance events occurring beyond the study plot. Any approach measuring dispersal from a source is inadequate for extrapolation to gene flow rates between populations.

Inferring gene flow from natural population genetic structure

Since gene flow can play an important role in shaping population genetic structure, it follows that the average gene flow should be discernable from the existing population genetic structure. Several statistical methods have been developed to use allele frequency distributions to indirectly estimate the gene flow parameter, Nm, the average number of migrants per generation. One common method is based on the frequencies of "private" or "rare" alleles, those observed in a single population (Barton and Slatkin 1986; Slatkin 1985b). The other common method is based on overall allelic differentiation among populations as measured by Wright's (1951) F_{ST} or Nei's (1973) G_{ST} statistics (Crow 1986). Slatkin and Barton (1989) compared both methods and found they "yield roughly comparable estimates under a wide variety of conditions" under demographic stability but that "F_{ST} is likely to be more useful under realistic conditions". Other similar, less frequently employed, estimates of gene flow include Slatkin's (1981) method based on the conditional average frequencies of alleles, Slatkin and Barton's (1989) maximum likelihood methods and van Dijk's (1987) method based on microspatial differentiation.

Nm estimates have been calculated from the allozyme variation data gleaned from natural plant populations; certain trends have emerged from comparisons of these estimates. First, gene flow estimates vary considerably among species, from very low ($Nm \ll 0.1$) to very high ($Nm \gg 10.0$) (Govindaraju 1988a, 1989; Hamrick 1987). Second, *Nm* varies with the dispersal ecology of the species. In comparisons of pollination syndrome, predominant selfers had typically lower gene flow than outcrossers (Govindaraju 1988a, b; Hamrick 1987; Hamrick et al. 1992). For outcrossers, gene flow varied with pollen vectors, wind-pollinated species generally having more gene flow than animal-pollinated species (Govindaraju 1988a, 1989; Hamrick 1987; Hamrick et al. 1992). *Nm* estimates also varied with seed dispersal characteristics, with low values for gravity dispersal, intermediate values for animal dispersal and high values for wind dispersal (Hamrick et al. 1992).

The primary advantage of the analysis is that gene flow is estimated free of source-sink constraints; the average gene flow is calculated for all of the specific populations under study. In fact, detailed gene flow relationships among populations can be identified by choosing subsets of individual populations for analysis. For example, Govindaraju (1989) calculated *Nm* for subsets of populations of *Pinus rigida* from the data of Guries and Ledig (1982) and of *Eucalyptus caesia* from the data of Moran and Hopper (1983). For *P. rigida*, the detailed analysis revealed two distinct groups of populations, each apparently internally linked by gene flow. For *E. caesia*, one subspecies behaved as if all populations were uniformly joined by moderate levels of gene flow; the other subspecies behaved as if certain populations were tightly linked by gene flow while others were isolated. Given the typical heterogeneity of genetic distances among populations of a single species, it is likely that gene flow heterogeneity will be the rule as more detailed analyses are conducted (cf. Golenberg 1987).

The advantage of the statistical approach is also a limitation. *Nm* estimates are only for the populations sampled. Other populations within the same species may show different patterns of gene flow. The estimates are also constrained by the geographic distribution of those populations. All other things being equal, more widely spaced populations ought to show less gene flow than closely spaced ones (Hamrick 1987). Therefore, generalizations from the populations sampled must be made with caution.

Another limitation is that the method will not differentiate among different kinds of gene flow, such as the relative importance of gene flow by seed vs. gene flow by pollen. Furthermore, indirect estimates of gene flow do not distinguish gene flow by exchange of individuals among populations versus gene flow by extinction and recolonization cycles (Wade and McCauley 1988).

Finally, indirect estimation assumes evolutionary equilibrium. Obviously, if a single population is split into several isolated ones, some time must pass before differentiation reflects the actual cessation of gene flow. For populations with about 100 individuals, the simulations of Slatkin and Barton (1989) demonstrate that 500—1000 generations are necessary to achieve equilibrium values of the appropriate order of magnitude. For long-lived perennials or even annuals with a substantial seed bank, equilibrium will be achieved only after thousands to tens of thousands of years. Thus, major long-term climatic changes would be sufficient to keep the population structure of many species in permanent disequilibrium. For these species, Nm estimates will be higher than actual gene flow rates.

Immigration into isolated sinks

Gene flow estimates can be obtained from estimates of immigration rates into isolated locations or isolated populations. Immigration by dispersed propagules can be measured with artificial samplers, usually for pollen (e.g., Edmonds 1979). Sinks may also be used to measure gene flow directly. The appearance of seedlings at naturally or artificially denuded sites documents gene flow by seed (e.g., Docters van Leeuwen 1936; Waser et al. 1982). Likewise, progeny testing for alien alleles identifies gene flow by pollen (e.g., Squillace and Long 1981).

An extension of this idea has lead to an approach for measuring gene flow by pollen more accurately into unmanipulated populations. The basis of these so-called "paternity" methods is simple (cf. Ellstrand and Marshall 1985; Hamrick and Schnabel 1985). First, multilocus genotypes of all reproductive individuals in population are determined. A comparison of the multilocus genotype of the maternal plant (or megagametophyte, in the case of gymnosperms) with that of its progeny allows for reconstruction of the paternal gametic contribution. If that gamete could not have arisen from any local reproductive plant, it is classed as an immigrant. The approach just outlined is called simple paternity exclusion.

Simple paternity exclusion gives a minimum estimate of gene flow because alien gametes that mimic local ones are misidentified ("detectable" vs. "ambiguous" *sensu* Adams and Birkes 1990; "apparent" vs. "cryptic" *sensu* Devlin and Ellstrand 1990). Refined methods have been developed to compensate for these gametes. For gymnosperms, these methods take advantage of the fact that the precise genotype of the maternal gamete can be assayed directly from the megagametophyte. Consequently, the genotype of the successful paternal gamete can also be inferred. Allele frequencies in surrounding source populations can be used to infer the frequency of detectable gametes relative to cryptic gametes, and consequently, total gene flow (Smith and Adams 1983; El-Kassaby and Ritland 1986; Adams and Birkes 1990). The situation is more complex for angiosperms in which the female gamete's genotype is rarely obtainable. A Monte Carlo method has been developed that provides an approximate maximum likelihood estimate of gene flow for angiosperms (Devlin and Ellstrand 1990). A bonus of this method is that it can be used to estimate the amount of gene flow received by individual plants. Similar multilocus approaches could be developed to measure gene flow by seed as well (cf. Meagher and Thompson 1986, 1987).

Studies measuring gene flow into sinks usually show that plant populations spatially isolated by hundreds to thousands of meters are not genetically isolated. In many cases, the rate of gene flow is moderate to high (Ellstrand and Hoffman 1990). Also, they have revealed gene flow by pollen to be highly variable. At one extreme, Schoen and Stewart (1987) estimated gene flow into a *Picea glauca* orchard isolated by 1 km to be on the order of 0.1 to 0.7%. At the other extreme, Squillace and Long (1981) found 88% of seeds sired in an orchard of *Pinus elliotii* came from non-orchard sources at least 122 m away.

Nm can be crudely estimated from such data (Table 1). First, the estimated gene flow rate (m) must be halved since a gamete is a haploid immigrant. Next, the rate is multiplied by the number of individuals in the target population (N). Although the seed pool should be included, it is virtually never available and is not accounted for in this analysis. Also, since *Nm* is a per-generation estimate, these per-year data should be corrected by an estimate of the number of years in a generation. In Table 1, the values are uncorrected because it is difficult to estimate how many of these interpopulation hybrids would enter the next generation. Both of these omissions, plus the fact that seed dispersal rates are unknown, suggest that this procedure may underestimate *Nm* by as much as an order of magnitude. Therefore, it is interesting that the haphazard sample

Table 1. Some crude direct estimates of *Nm* from interpopulation mating rates

Species (Reference)	Isolation	Estimated interpopulation mating rate	Direct *Nm* estimate (See text)	Indirect *Nm* estimate (**)
Cynosurus cristatus (Ennos and Dodson 1987)	"well-isolated"	0.4%	1.9	—
Gleditsia triacanthos (Hamrick and Schnabel 1985)	200 m	6.0%	1.5	—
Picea glauca (Schoen and Stewart 1987) I	1000 m	0.1%	0.3	—
Picea glauca II (Schoen and Stewart 1987)	1000 m	0.7%	2.3	—
*Pinus taeda** (Friedman and Adams 1985)	122 m	36.0%	54.0	2.8
*Pseudotsuga menziesii** (El-Kassaby and Ritland 1986)	8 m	0.2%	1.3	3.0
Raphanus sativus I (Devlin and Ellstrand 1990)	150 m	6.3%	1.3	—
Raphanus sativus II (Devlin and Ellstrand 1990)	150 m	7.0%	1.4	—
Tachigalia versicolor (Hamrick and Loveless 1989)	500 m	26.0%	0.8	3.3

* Based on mean over multiple estimates
** (Govindaraju 1988a; Hamrick and Loveless 1989)

of *Nm* estimates calculated (Table 1) are almost uniformly high ($Nm > 1$), despite the wide range of inter-population mating rates among species. I have compared these direct estimates with indirect ones, when available, but the comparisons are too few to draw any conclusions.

While gene flow differences among species are pronounced, variation among populations within a species may be more than an order of magnitude as well. For example, paternity studies in orchards and natural stands of *Pseudotsuga menziesii* have measured inter-population mating at rates from as low as 0.2% to as high as 52% (Adams and Birkes 1990; El-Kassaby and Ritland 1986; Neale 1983). Natural populations of wild radish isolated by 100 to 1000 m were found to have inter-population mating rates from 3% to 18% (Ellstrand et al. 1989). Interestingly, in both species, the distance from the nearest pollen source did not vary closely with rate of gene flow. Also, these paternity studies have shown that gene flow receipt may vary greatly among individuals. For example, in one wild radish population, certain individuals set only seeds sired by local fathers, while others had as many as 20% of their seeds fertilized by gametes coming from at least 150 m beyond the edges of the population (Devlin and Ellstrand 1990).

The advantages of using sinks is that they give an estimate of all gene flow arriving in a site or population. The spatial truncation encountered from using sources is never a problem. However, the multilocus methods operate under certain assumptions:

— that sources are homogeneous in their allele frequencies,
— that successful foreign gametes are a random sample from the sources, and
— that all populations are in linkage equilibrium.

Finally, paternity methods work best in relatively small populations that are highly polymorphic (Adams and Birkes 1990; Chakraborty et al. 1988; Devlin et al. 1988; Smith and Adams 1983). Because theory predicts that gene flow receipt should increase as population size decreases (Handel 1983a), the gene flow rates in large continuous populations might be different from those detected by these methods which use relatively small populations (generally 40—1000 individuals).

Reconcilation of the data

Contemporary plant population geneticists vary in their views of gene flow. At one extreme, gene flow is viewed as highly restricted (Levin 1981); at the other, extensive (Muona 1990). A third view is emerging that gene flow in plants is idiosyncratic, ranging from very low to very high, and varying among species, populations, individual plants, and even over a season (Ellstrand and Marshall 1985; Ellstrand et al. 1989; Grant 1985; Hamrick 1987; Slatkin 1987). The first view is so strong among certain evolutionary biologists that, after calculating Nm values greater than one, they have still concluded gene flow "is likely to be rather small" (e.g., Bos et al. 1986; Knight and Waller 1987)! Given the strengths and the weaknesses of the four general approaches to measuring gene flow in plant populations, certain trends emerge from the gene flow database that support the latter two of these views.

The two methods that measure gene flow without the constraints of spatial truncation show that gene flow rates for most species studied (particularly outcrossing species) range from moderate ($0.5 < Nm < 1.0$) to extensive ($Nm \gg 1.0$). As a general rule, gene flow appears to be low only in highly autogamous species (Govindaraju 1988a; Hamrick 1987). In addition to gene flow variation among species, when the data are available, it is clear that gene flow varies among populations, both from statistical analysis of population structure (Govindaraju 1989) and from paternity methods (Ellstrand et al. 1989). Data from both sources and sinks have indicated that many ecological and genetic factors may influence gene flow variation, even at the level of the individual. Individual variation in gene flow receipt has been well-documented from both studies using genetic markers in adjacent synthetic populations (Smyth and Hamrick 1987) and multilocus paternity methods in natural populations (Devlin and Ellstrand 1990). And when sought, gene flow was found to vary within individuals over time (Ellstrand and Marshall 1985; Ellstrand et al. 1989). These data make it clear that, as a force that structures populations, gene flow may be as variable and idiosyncratic as natural selection.

Is it possible to reconcile data from the first two approaches which have been used to support the popular conclusion that gene flow is limited among plant populations with the latter two which yield data suggesting that gene flow often occurs at levels high enough to be important? The datasets are, in fact, compatible. The apparent conflict arises because attention has been focused on the wrong part of the source dispersal curve, that is, the peak and mean instead of the tail. Because dispersal from a source generally follows a leptokurtic distribution, many gene flow events are expected to occur in the long tail. Even if individuals disperse a few seeds or pollen grains at a great distance, when the dispersal distribution tails of thousands, hundreds of thousands, or millions of plants are summed, then substantial interpopulation gene flow is not so surprising, especially because evolutionarily substantial gene flow is one immigrant per generation in the case of drift or a few percent per generation in the case of moderate selection (Ellstrand et al. 1989; Hamrick 1987; Slatkin 1985). Although measuring dispersal out of a source almost always truncates the true dispersal curve, the real truncation is an intellectual one that plant evolutionary biologists (myself included; Ellstrand et al. 1978) have suffered regarding the significance of gene flow as an evolutionary force.

Implications for forest tree genetics

Gene flow cannot be ignored. Data collected by both population genetic structure methods and paternity methods suggests that most tree species have moderate to high levels of gene flow (cf. Govindaraju 1989; Adams and Birkes 1990). These levels are sufficient to counteract genetic drift and moderate levels of directional selection. Because the majority of tree species are probably highly outcrossed (e.g., Fryxell 1957; Levin 1975;

Muona 1990), forest geneticists should operate under the assumption that gene flow is probably present and should be monitored if it is expected to have an impact on a management program. For example, pollen contamination may erode the efforts of breeders. In fact, Muona (1990) suggested "pollen contamination may remain the most serious single problem in wind-pollinated seed orchards". Likewise, conservation geneticists concerned that an endangered species might be hybridizing with a sympatric, common one (e.g., Rieseberg et al. 1989; Ellstrand 1991) may be justified. Furthermore, if a genetically engineered tree is to be field tested, planning for field release should include special strategies to reduce or eliminate gene flow (Ellstrand 1988; Ellstrand and Hoffman 1990).

The heterogeneity of gene flow that occurs among individuals and populations within species (e.g., *Pseudotsuga menziesii*, see above) suggests that high gene flow cannot necessarily be assumed. However, gene flow can often be managed; certain combinations of ecological and/or genetic factors can be manipulated to enhance or reduce local gene flow. For example, cooling treatments of trees may alter their phenology, increasing their isolation from conspecifics (e.g., El-Kassaby and Ritland 1986). Because spatial distance and gene flow might not be as closely correlated as previously thought (e.g., Ellstrand et al. 1989), spatial isolation may prove a less effective method of isolating populations. In any forest genetic management program where a certain level of gene flow would help or hinder progress, measurements of background gene flow (particularly using paternity methods) coupled with strategies for modifying it would be well worth the trouble.

Acknowledgments

Bill Libby and members of the Department of Forest Genetics at the UDSA Pacific Southwest Forest Station at Berkeley provided my first education in forest population genetics during my sabbatical year in Berkeley 1987—88. My views of gene flow have been enhanced by discussions with Jim Hamrick, Tracy Kahn, and Monty Slatkin. I am grateful to Aileen Wietstruk for her careful preparation of this manuscript.

References

Adams, W. T. and Birkes, D. S. 1990. Estimating mating patterns in forest tree populations. In: Hattemer, H. H. and Fineschi, S. (Eds) Biochemical markers in the population genetics of forest trees. S.P.B. Academic Publishing, BV. The Hague (in press).

Allendorf, F. W. 1983. Isolation, gene flow, and genetic differentiation among populations.

In: Schonewald-Cox, C. M., Chambers, S. M., MacBryde, B. and Thomas, W. L. (Eds) Genetics and conservation: a reference for managing wild animal and plant populations. Benjamin Cummings Publishing, Menlo Park. pp. 51—65.

Antonovics, J. 1968. Evolution in closely adjacent plant populations. VI. Manifold effects of gene flow. Heredity 32: 507—524.

Barton, N. H. and Slatkin, M. 1986. A quasi-equilibrium theory of the distribution of rare alleles in a subdivided population. Heredity 56: 409—415.

Bos, M., Harmens, H. and Vrieling, K. 1986. Gene flow in *Plantago*. I. Gene flow and neighborhood size in *P. lanceolata*. Heredity 56: 43—54.

Bradshaw, A. D. 1972. Some evolutionary consequences of being a plant. Evol. Biol. 5: 25—47.

Campbell, D. R. and Waser, N. M. 1989. Variation in pollen flow within and among populations of *Ipomopsis aggregata*. Evolution 43: 1444—1455.

Chakraborty, R., Meagher, T. R., and Smouse, P. E. 1988. Parentage analysis with genetic markers in natural populations. I. The expected proportion of offspring with unambiguous paternity. Genetics 118: 527—536.

Crane, M. B. and Mather, K. 1943. The natural cross-pollination of crop plants with particular reference to the radish. An. Appl. Biol. 30: 301—308.

Crow, J. F. 1986. Basic concepts in population, quantitative, and evolutionary genetics. New York, W. H. Freeman.

Devlin, B. and Ellstrand, N. C. 1990. The development and application of a refined method for estimating gene flow from angiosperm paternity analysis. Evolution 44: 248—259.

Devlin, B., Roeder, K., and Ellstrand, N. C. 1988. Fractional paternity assignment: theoretical development and comparison to other methods. Theor. App. Genet. 76: 369—380.

Docters van Leeuwen, W. M. 1936. Krakatau, 1883 to 1933. Ann. Jard. Bot. Buitenzorg. 56—57: 1—506.

Edmonds, R. L. 1979. Aerobiology: the ecological systems approach. Dowden, Hutchinson, and Ross. Stroudsburg, PA.

El-Kassaby, Y. A. and Ritland, K. 1986. Low levels of pollen contamination in a Douglas-fir seed orchard as detected by allozyme markers. Silv. Genet. 35: 224—228.

Ellstrand, N. C. 1988. Pollen as a vehicle for the escape of engineered genes? In: Hodgson, J. and Sugden, A. M. (Eds) Planned Release of Genetically Engineered Organisms (Trends in Biotechnology/Trends in Ecology and Evolution Special Publication) Elsevier Publications, Cambridge. pp. S30—S32.

Ellstrand, N. C. 1991. Gene flow by pollen: implications for plant conservation genetics. Oikos, (in press).

Ellstrand, N. C., Devlin, B. and Marshall, D. L. 1989. Gene flow by pollen into small populations: data from experimental and natural stands of wild radish. Proc. Nat. Acad. Sci., U.S.A. 86: 9044—9047.

Ellstrand, N. C. and Hoffman. C. A. 1990. Hybridization as an avenue for escape of engineered genes. BioScience 40: 438—442.

Ellstrand, N. C. and Marshall, D. L. 1985. Interpopulation gene flow by pollen in wild radish, *Raphanus sativus*. Am. Nat. 126: 606—616.

Ellstrand, N. C., Torres, A. M., and Levin, D. A. 1978. Density and the rate of apparent outcrossing in *Helianthus annuus* (Asteraceae). Syst. Bot. 3: 403—407.

Endler, J. A. 1973. Gene flow and population differentiation. Science 179: 243—250.

Endler, J. A. 1977. Geographic variation, speciation, and clines. Princeton University Press, Princeton.

Ennos, R. A. and Dodson, R. K. 1987. Pollen success, functional gender, and assortative mating in an experimental plant population. Heredity 58:119—128.

Epperson, B. K. 1990. Spatial patterns of genetic variation within plant populations. In:

Brown, A. H. D., Clegg, M. T., Kahler, A. L., and Weir, B. S. (Eds) Plant population genetics, breeding and resources. Sinauer Associates, Sunderland. pp. 229—253.

Friedman, S. T. and Adams, W. T. 1985. Estimation of gene flow into two seed orchards of loblolly pine (*Pinus taeda* L.). Theor. Appl. Genet. 69:609—615.

Fryxell, P. A. 1957. Mode of reproduction in higher plants. Bot. Rev. 23: 135—233.

Gillespie, J. H. 1975. The role of migration in the genetic structure of populations in temporally and spatially varying environments. I. Conditions for polymorphism. Am. Nat. 109: 127—135.

Golenberg, E. M. 1987. Estimation of gene flow and genetic neighborhood size by indirect methods in a selfing annual, *Triticum dicoccoides*. Evolution 41:1326—1334.

Govindaraju, D. R. 1988a. Relationship between dispersal ability and levels of gene flow in plants. Oikos 52: 31—35.

——. 1988b. A note on the relationship between outcrossing rate and gene flow in plants. Heredity 61: 401—404.

——. 1989. Estimates of gene flow in forest trees. Biol. J. Linn. Soc. 37: 345—357.

Grant, V. 1980. Gene flow and the homogeneity of species populations. Biol. Zbl. 99: 157—169.

Grant, V. 1985. The problem of gene flow on a geographical scale. Zhurnal Obschei Biologii 46: 20—31.

Guries, R. R. and Ledig, F. T. 1982. Genetic diversity and population structure in pitch pine (*Pinus rigida* Mill.). Evolution 36: 387—402.

Haldane, J. B. S. 1930. A mathematical theory of natural and artificial selection. Part IV. Isolation. Proc. Cambridge Philos. Soc. 26: 220—230.

Hamrick, J. L. 1982. Plant population genetics and evolution. Am. J. Bot. 69: 1685—1693.

Hamrick, J. L. 1987. Gene flow and distribution of genetic variation in plant populations. In: Urbanska, K. (Ed) Differentiation in higher plants. Academic Press, New York, pp. 53—67.

Hamrick, J. L., Griswold, G. B., and Godt, M. J. 1992. Association between Slatkin's measure of gene flow and the dispersal ability of plant species. Am. Nat. (in press).

Hamrick, J. L. and Loveless, M. D. 1986. The influence of seed dispersal mechanisms on the genetic structure of plant populations. In: Estrada, A. and Fleming, T. H. (Eds) Frugivores and seed dispersal. Dr W Junk, Publishers, Dordrecht, pp. 211—223.

Hamrick, J. L. and Loveless, M. D. 1989. The genetic structure of tropical tree populations, associations with reproductive biology. In: Bock, J. H. and Linhart, Y. B. (Eds) The evolutionary ecology of plants. Westview Press, Boulder, pp. 129—146.

Hamrick, J. L. and Schnabel, A. 1985. Understanding the genetic structure of plant populations: some old problems and a new approach. In: Gregorius, H. R. (Ed.) Population genetics in forestry. Springer-Verlag, Berlin. pp. 50—70.

Handel, S. N. 1976. Restricted pollen flow of two woodland herbs by neutron-activation analysis. Nature 260: 422—423.

——. 1982. Dynamics of gene flow in an experimental population of *Cucumis melo* (Cucurbitaceae). Am. J. Bot. 69: 1538—1546.

——. 1983a. Pollination ecology, plant population structure, and gene flow. In: Real, L. (Ed) Pollination biology. Academic Press, Orlando. pp. 163—211.

——. 1983b. Contrasting gene flow patterns and genetic subdivision in adjacent populations of *Cucumis sativus* (Cucurbitaceae). Evolution 37: 760—771.

Howe, H. F. and Smallwood, J. 1982. Ecology of seed dispersal. Ann. Rev. Ecol. Syst. 13: 201—228.

Jain, S. K. and Bradshaw, A. D. 1966. Evolutionary divergence among adjacent plant populations. I. The evidence and its theoretical analysis. Heredity 21: 407—441.

Kimura, M. and Weiss, G. 1964. The stepping-stone model of population structure and the decrease of genetic correlation with distance. Genetics 49: 561—576.

Knight, S. E. and Waller, D. M. 1987. Genetic consequences of outcrossing in the cleistogamous annual, *Impatiens capensis*. I. Population-genetic structure. Evolution 41: 969—978.

Levin, D. A. 1975. Pest pressure and recombination systems in plants. Am. Nat. 109: 437—451.

——. 1978. Some genetic consequences of being a plant. In Brussard, P. F. (Ed) Ecological genetics: the interface. Springer-Verlag, New York. pp. 189—212.

——. 1981. Dispersal versus gene flow in plants. Ann. Mo. Bot. Gard. 68: 233—253.

——. 1983. An immigration-hybridization episode in *Phlox*. Evolution 37: 575—582.

——. 1984. Immigration in plants: an exercise in the subjunctive. In: Dirzo, R. and Sarukhan, J. (Eds) Perspectives on plant population ecology. Sinauer, Sunderland. pp. 242—260.

——. 1988. Consequences of stochastic elements in plant migration. Am. Nat. 132: 643—651.

Levin, D. A. and Kerster, H. W. 1968. Local gene dispersal in *Phlox*. Evolution 22: 130—139.

Levin, D. A. and Kerster, H. W. 1974. Gene flow in seed plants. Evol. Biol. 7: 139—220.

Meagher, T. R. and Thomson, E. 1986. The relationship between single-parent and parent-pair likelihoods in genealogy reconstruction. Theor. Pop. Biol. 29: 87—106.

Meagher, T. R. and Thomson, E. 1987. Analysis of parentage for naturally established seedlings of *Chamaelirium luteum* (Liliaceae). Ecology 68: 803—812.

Moran, G. F. and Hopper, S. D. 1983. Genetic diversity and the insular population structure of the rare granite rock species, *Eucalyptus caesia* Benth. Austral. J. Bot. 31: 161—172.

Muona, O. 1990. Population genetics in forestry. In: Brown, A. H. D., Clegg, M. T., Kahler, A. L., and Weir, B. S. (Eds) Plant population genetics, breeding, and genetic resources. Sinauer Associates, Sunderland, pp. 282—298.

Neale, D. B. 1983. Population genetic structure of the shelterwood regeneration system in southwest Oregon. Ph.D. Diss. Oregon State Univ., Corvallis.

Nei, M. 1973. Analysis of gene diversity in subdivided populations. Proc. Natl. Acad. Sci. USA, 70: 3321—3323.

Plucknett, D. L., Smith, N. J. H., Williams, J. H., and Anishetty, N. M. 1987. Gene Banks and the World's Food. Princeton University Press, Princeton, NJ.

Richards, A. J. 1986. Plant Breeding Systems. Allen and Unwin, London.

Rieseberg, L. H., Zona, S., Aberbom, L. and Martin, T. D. 1989. Hybridization in the island endemic, Catalina mahogany. Conservation Biol. 3: 52—58.

Rohlf, F. J. and Schnell, G.D. 1971. An investigation of the isolation by distance model. Am. Nat. 105: 295—324.

Sanderson, N. 1989. Can gene flow prevent reinforcement? Evolution 43: 1223—1235.

Schaal, B. A. 1980. Measurement of gene flow in *Lupinus texensis*. Nature 284: 450—451.

Schmitt, J. 1980. Pollinator foraging behavior and gene dispersal in *Senecio* (Compositae). Evolution 34: 934—943.

Schoen, D. J. and Stewart, S. C. 1987. Variation in male fertilities and pairwise mating probabilities in *Picea glauca*. Genetics 116:141—152.

Slatkin, M. 1973. Gene flow and selection in a cline. Genetics 75: 733—756.

——. 1981. Estimating levels of gene flow in natural populations. Genetics 99:323—335.

——. 1985a. Gene flow in natural populations. Ann. Rev. Ecol. Syst. 16: 393—430.

——. 1985b. Rare alleles as indicators of gene flow. Evolution 39: 53—65.

——. 1987. Gene flow and the geographic structure of natural populations. Science 236: 787—792.

Slatkin, M. and Barton, N. 1989. A comparison of three indirect methods for estimating average levels of gene flow. Evolution 43: 1349—1368.

Smith, D. B. and Adams, W. T. 1983. Measuring pollen contamination in clonal seed orchards with the aid of genetic markers. In: Proc. 17th Southern Forest Tree Improvement Conference. Univ. Georgia, Athens pp. 69—77.

Smyth, C. A. and Hamrick, J. L. 1987. Realized gene flow via pollen in artificial populations of musk thistle, *Carduus nutans* L. Evolution 41: 613—619.

Squillace, A. E. and Long, E. M. 1981. Proportion of pollen from nonorchard sources. In: Franklin, E. C. (Ed) Pollen management handbook. USDA Agricultural Handbook # 587, Washington DC, pp. 15—19.

Thomson, J. D. and Thomson, B. A. 1989. Dispersal of *Erythronium grandiflorum* pollen by bumblebees: implications for gene flow and reproductive success. Evolution 43: 657—661.

Tonsor, S. J. 1985. Intrapopulational variation in pollen-mediated gene flow in *Plantago lanceolata* L. Evolution 39: 775—782.

Turner, M. E., Stephens, J. C., and Anderson, W. W. 1982. Homozygosity and patch structure in plant populations as a result of nearest-neighbor pollinations. Proc. Nat. Acad. Sci., U.S.A. 79: 203—207.

van Dijk, H. 1987. A method for the estimation of gene flow parameters from a population structure caused by restricted gene flow and genetic drift. Theor. Appl. Genet. 73: 724—736.

Wade, M. J. and D. E. McCauley. 1988. Extinction and recolonization: their effects on the genetic differentiation of local populations. Evolution 42: 995—1005.

Waser, N. M., Vickery, R. K., and Price, M. V. 1982. Patterns of seed dispersal and population differentiation in *Mimulus guttatus*. Evolution 36: 753—761.

Weiss, G. and Kimura, M. 1965. A mathematical analysis of the stepping stone model of genetic correlation. J. Appl. Prob. 2: 129—149.

Willson, M. F. 1983. Plant Reproductive Ecology. John Wiley and Sons, New York.

Willson, M. F. 1984. Mating patterns in plants pp. 261—276. In: Dirzo, R. and Sarukhan, J. (Eds) Perspectives on plant population ecology. Sinauer, Sunderland.

Wright, S. 1931. Evolution in Mendelian populations. Genetics 16: 97—159.

——. 1943. Isolation by distance. Genetics 28: 114—138.

——. 1946. Isolation by distance under diverse systems of mating. Genetics 31: 39—59.

——. 1951. The genetical structure of populations. Ann. Eugenics 15: 323—354.

——. 1982. Character change, speciation, and the higher taxa. Evolution 36: 427—443.

New Forests **6**: 257–278, 1992.

Review paper

Spatial structure of genetic variation within populations of forest trees

B. K. EPPERSON

Department of Botany and Plant Sciences, University of California, Riverside, CA 92521, USA

Received 8 October 1990; accepted 24 May 1991

Key words: genetic structure, natural selection, population genetics, autocorrelation

Application. New spatial statistical methods are introduced for studies of genetic structure within stands. These methods measure levels of inbreeding and can detect natural selection. This information may be useful for improving the genetic quality of stand regeneration.

Abstract. The spatial pattern and structure of genetic variation are important aspects of the population genetics of forest stands. Combined with limits to seed and pollen dispersal, spatial structure affects the level of inbreeding and the action of natural selection. The genetic constitution of stand regeneration, following different forestry practices, is also influenced by spatial structure. For example, natural regeneration with seed trees involves sampling seed trees from a stand that may be genetically nonhomogeneous. This paper reviews theoretical and empirical results on spatial patterns of genetic variation, produced under limited gene flow and selection, in terms of recently developed spatial statistics (e.g., spatial autocorrelation). Genetic correlations in samples from spatially structured populations are also described, as well as how spatial samples can be used to characterize the structure of genetic variation, and how inferences can be made about (spatially distributed) components of fitness and yield.

Introduction – the nature of spatial structure

Knowledge of structure of genetic variation is fundamental to understanding the population genetics of forest trees and other plants. Structure influences and is influenced by virtually every aspect of the population genetics of forest trees, including mating system and the action of natural selection. *Spatial* structure is the distribution of genotypes over the two dimensional space of a stand. It can be characterized through the physical locations and the genetic or genealogical relationships between individual trees. Summary measures of spatial structure may be used to improve our understanding of the population genetics of forest trees. In addition, forest management and gene conservation practices usually include a component of spatial sampling procedures. For example, the spatial distribution of genotypes of seed trees following logging affects the genetic constitution of regeneration. Thus spatial structure is relevant to the measurement and management of the genetic composition of stands. Considerable advances have recently been made in relating spatial patterns of genetic variation within populations to different types of natural selection and amounts of dispersal (Sokal and Wartenberg 1983; Sokal, Jacquez and Wooten 1989; Epperson 1989, 1990).

Statistical measurements of genetic structure within stands have primarily involved either measures of polymorphism and heterozygosity or labor intensive multilocus parentage analysis. Parent-offspring data or similar genealogical data can reveal some but not all of the kinship relationships among individuals (Morton 1973a, b). More generally, in a spatial context, kinship is averaged among individuals or groups of individuals, based on their orientation in space, usually in terms of distances of separation. For selectively neutral loci, spatial patterns of genetic variation reflect the average total kinship between individuals, which usually is a decreasing function of distances of separation (Malecot 1948; Morton 1973a, b; Barbujani 1987). Spatial structure combined with limits to pollen dispersal results in consanguinous matings and local inbreeding. Moreover, where selection is operating, the population dynamics of genetic variation may depend importantly on the spatial distribution of genotypes. Conversely, spatial pattern analyses can be useful in detecting and characterizing natural selection, especially microenvironmental selection.

There have recently been major advances in spatial statistics that can be used to measure the important features of spatial patterns of genetic variation precisely (Sokal and Oden 1978; Cliff and Ord 1981; Upton and Fingleton 1985). Few studies have applied spatial correlation statistics to the population genetics of forest trees (Epperson 1983; Epperson and Allard 1989; Wagner et al. 1989) or to other plant species (Epperson and Clegg 1986; Dewey and Heywood 1988; Schoen and Latta 1989).

The present paper examines the utility of various spatial statistics for the study of *spatial structure* of genetic variation for allozyme and other loci, within forest stands. First, some general features of genetic structure within stands of forest trees and microenvironmental selection are presented. Following sections discuss spatial statistics and how measures of spatial structure are influenced by dispersal and selection.

Spatial structure of genetic variation, life history, and mating system

Spatial patterns of genetic variation within stands are expected to vary widely among forest tree species, which have a wide variety of dispersal mechanisms, reproductive biologies, and mating systems. Conifers generally exhibit wind dispersal of pollen and seed, high rates of outcrossing, and high levels of inbreeding depression. Many angiosperm trees, including some economically important trees in tropical regions, have quite different pollination mechanisms and other mating system features. The effects of mating system on levels of heterozygosity within populations are well established (see Wright 1946; Hamrick and Godt 1989). Plant populations with low to moderate distances of dispersal of pollen and seed (relative to tree density) are generally expected to build up substantial genetic isolation by distance and spatial autocorrelations of genotypes for selectively neutral loci (Wright 1943; Malecot 1948; Sokal and Wartenberg 1983; Sokal, Jacquez and Wooten 1989; Epperson 1989). In such populations there develops a distinctive structure of patches, or large areas where one homozygous genotype predominates. Moreover, many forests, particularly tropical forests, have a high level of species diversity (e.g., Sakai 1985). Wherever a population has low densities, pollen and/or seed may disperse long physical distances, yet there could be considerable isolation by distance and spatial correlations of genotypes. There is some evidence of marked isolation by distance in temperate hardwoods (e.g., Merzeau et al. 1989).

Spatial structure of genetic variation has been implicated in many conifer populations. "Structure-sensitive" single locus estimates of outcrossing rates are commonly lower than the relatively structure-insensitive multilocus measures (Shaw, Kahler and Allard 1981; Ritland 1985). A rare exception occurred in my own studies of populations of lodgepole pine (*Pinus contorta* ssp. *latifolia* Dougl. ex Loud.), which in fact had little or no spatial structuring (Epperson and Allard 1984). It is worth noting that spatial and/or temporal heterogeneity of gene frequencies in pollen are the proximal causes of deflated single locus estimates. Such heterogeneity could occur even if the spatial distribution of reproducing individuals is random, but individuals differ in timing or amount of pollen production (Epperson and Allard 1989).

Studies of spatial distributions and genealogical relationships between individual progeny in forest stands have revealed considerable family clumping in both temperate conifers and tropical angiosperms (Mitton 1983; Sakai 1985). The degree of family clumping varies in studies of progeny of remaining trees in seed tree or shelterwood stands (Neale and

Adams 1985; Yazdoni, Lindgren and Rudin 1985). Much of the spatial correlations caused by familial clumping of seedlings may be subsequently removed by intense competition that usually leaves only one or a few survivors per clump at reproductive maturity. It is worth noting that the structure among the genotypes of adult trees should continue to be governed primarily by the amount of dispersal per generation, as long as the survival of seedlings is random. However, when self progeny and other highly inbred progeny are less likely to survive (Sorensen and Miles 1982), the development of local patches of homozygotes may be retarded, because inbred progeny are more likely to be homozygous than are outbreds (Bennet and Binet 1956).

Theoretical studies of structure under different kinds of selection may help to resolve the controversy surrounding interpretations of the common observation that deficits of heterozygosity (which may be caused by spatial structure and/or selfing) for allozyme and other loci tend to disappear as regeneration ages (Phillips and Brown 1977; Brown and Albrecht 1980; Moran and Brown 1980; Farris and Mitton 1984; Yazdoni, Muona, Rudin and Szmidt 1985, Plessas and Strauss 1986; Muona et al. 1987). Although some recent studies reveal a lack of heterozygote advantage among outbred progeny, suggesting that inbreeding per se is the cause (Strauss 1986; Strauss and Libby 1987), recent review articles continue to call for more work in this important area (Muona 1989). In general, it is not clear how low levels of inbreeding in progeny from mildly consanguinous mating affects survival. A potentially great problem in forest regeneration practices is that slightly inbred seedlings may survive but show inferior qualities as mature trees. It is important to study how inbreeding caused by spatial structure behaves in natural stands, and in manipulated stands and samples.

Experimental and theoretical evidence indicates that marked spatial patterns are produced in areas of species hybridization and introgression. Spatial autocorrelation analysis may improve our understanding of hybrid viability and introgression. Wagner et al. (1989) used spatial autocorrelation statistics to describe the distributions of chloroplast DNA genotypes in large samples from a hybrid swarm of lodgepole pine and jack pine (*Pinus banksiana* Lamb.). Many cpDNA variations showed strong autocorrelations across the contact region. Similar patterns may be found in zones of recent contact and admixture between populations of the same species.

Population biology and microenvironmental selection in forest trees

The potential for selection in conifers and many other forest trees is great.

Populations can tolerate tremendous genetic loads because of a great overabundance of pollen, seed, seedlings and finally saplings. For example, Campbell (1979) estimates that some 20,000 seeds are produced to replace one mature Douglas Fir. Moreover, there are remarkable levels of genetic variation maintained within a conifer population, even within a stand. Individual forest trees must tolerate a wide range of environmental conditions during a lifetime (Mitton and Grant 1984). Conditions can also differ greatly over very small distances. Differences in soil depth, various soil qualities, water availability, exposure, depth of winter snow pack, light availability, competing species, all can exist on a spatial scale as small as a few meters. On a somewhat larger scale are the classic contrasts, of conditions and often species compositions, between north and south facing slopes. Large intrapopulational differences in allozyme allele frequencies have been found between stands on north and south facing slopes (Mitton et al. 1977; Beckman and Mitton 1984), and between wet and dry sites (Mitton et al. 1989).

An understanding of microenvironmental selection would be of considerable direct importance to forest management. In order to utilize potential local adaptation of stands, regeneration methods must use materials that are derived from source materials from nearby areas (Brown and Moran 1981). However, important genetic variation is contained within populations and there is often little differentiation among populations (Hamrick and Godt 1989; Muona 1989). Given the potential for microenvironmental selection, it may be that populations have already arrived at near-optimal local solutions to microenvironmental heterogeneity within stands areas, the limits to dispersal, and constraints from avoidance of inbreeding depression. This could result in a near-optimal genotype for each potential tree site. In general, if microenvironmental selection within a stand is strong, then plus tree selection greatly loses its appeal. Moreover, planting of genetically superior seedlings becomes impractical if not impossible because it would require matching favorable genotypes of each planted seedling to identified microenvironments within the stand. It is important to characterize the theoretical structures within stands that result from different space-time or spatial evolutionary models of dispersal and selection. In general terms, each generation gene flow tends to reduce correlations, which are then reinforced by another round of selection favoring different genotypes in different microenvironments. With the high levels of dispersal for most conifers, selection must be strong in order for correlations to build up over time. More detailed discussions are presented below.

Microenvironmental selection may be detected from strong spatial correlations of genetic variation, and strong associations of genotypic frequencies with microenvironments (Bradshaw, 1984). Our ability to attribute genetic differentiation to microenvironmental selection is complicated by the fact that several other factors, including limits to seed and pollen dispersal can also create differentiation. Spatial statistics may improve interpretations of associations of genetic variation with microenvironments. These methods measure spatial relations of genetic variation with microenvironmental factors more precisely. They may be modified to account properly for spatial correlations of genetic variation due to dispersal alone, and to the spatial patterns of the microenvironmental factors themselves. Discussions of various spatial statistics are presented in more detail below.

The most detailed study of spatial structure within stands is on populations of lodgepole pine (Epperson 1983; Epperson and Allard 1987, 1989). These populations are essentially 100 percent outcrossing and estimated values of Wright's neighborhood sizes exceed 1,000. However, two populations, separated by only 11 km, differed in allele frequencies for some allozyme loci as much as populations at different ends of the vast geographical range of the species (Yeh and Layton 1979; Wheeler and Guries 1982). Within each study population, there were spatial autocorrelations of genotypes for some loci but not others, possibly indicating microenvironmental selection for some loci.

Spatial autocorrelation statistics for genetic variation

In this section, two important spatial statistical methods appropriate for genetic data are outlined. More detailed discussions can be found in Epperson (1989). The first method considered below is for point samples of genetic values that are approximately continuously varying. For example, allele frequencies, p_i, in a collection of n quadrat subsamples ($i = 1, \ldots, n$) are mapped so that each subsample is assigned to a point location. The second method is for data of nominal types, in this case genotypes, where a single nominal type can be assigned to a sample point location. Spatial autocorrelation statistics for both types of traits have better properties where the sample point locations are both fairly large in number and regularly spaced. Ideally, sample points are located on a regularly spaced sample grid or lattice (Epperson 1989).

Spatial statistical analyses of allele frequencies in quadrat subsamples proceeds through establishing distance measures between the pairs of quadrats. Usually this measure is simply the physical distance between the centerpoints of quadrats, but other distance measures may better reflect long term gene flow (Gabriel and Sokal 1969; Cliff and Ord 1981).

One important measure of correlation is based on the unweighted

Moran's I-statistic. First, the pairs of quadrats are classified into distance range classes. For each distance class k Moran's I statistic is calculated by: $I = n\ \Sigma\Sigma_{i \neq j}^{i,j \in k}\ Z_i Z_j / W_k\ \Sigma_{i=1}^{n}\ Z_i^2$, where $Z_i = p_i - \bar{p}$, and $\bar{p}$ is the mean allele frequency of all n quadrats and W_k equals twice the number of pairs of quadrats in the distance class k. Under the random hypothesis, I has expected value $u_1 = -1/(n-1)$. The variance, u_2, is given for example in Sokal and Oden (1978) and Cliff and Ord (1981). If the number of quadrats is fairly large and the number of genotypes per quadrat is moderate, then the statistic $(I - u_1)/\sqrt{u_2}$ has an approximate standard normal distribution under the random hypothesis (Cliff and Ord 1981).

A set of unweighted I statistics for mutually exclusive distance classes is known as an I-correlogram. Thus I-correlograms measure relative correlations in allele frequencies as a function of the distance measure. I-correlograms can be tested as a whole for significant deviation from the random hypothesis (Oden 1984), but exact tests are lacking for differences between correlograms from different data sets, for example, frequency distributions for different loci (Sokal and Wartenberg 1983).

An alternative method is to calculate separate weighted I statistics, in which the weights, w_{ij}, between locations i and j, can be specified based on any independent information on the relative strengths of correlations among pairs of subsample quadrats. In general, $I = n\ \Sigma\Sigma_{i \neq j}\ w_{ij} Z_i Z_j / W\ \Sigma_{i=1}^{n}\ Z_i^2$, where W is the sum of w_{ij} over all i, j such that $j \neq i$ ($w_{ii} = 0$). In addition, modified I statistics can be used to test the fit of a pattern to a theoretical (determined) map (Cliff and Ord 1981).

A more general method employs the calculation of a single statistic, the Mantel statistic (Cliff and Ord 1981), which tests the independence of two matrices. In the present context, one matrix contains values of a measure of differences in allele frequencies [for example Nei's genetic distance (Nei 1973)], and the other contains values of a physical distance measure.

Directionality in a spatial data set, such as may be caused by directional dispersal, maybe detected by classifying pairs of sample points by their respective locations in subareas defined by sectored concentric rings of an encircled total sample area (Oden and Sokal 1986). Alternatively, directionality along a predetermined spatial axis may be included in the distance measure in each of the statistical models (Upton and Fingleton 1985).

The second method involves using join counts to analyze a spatial map of n point sample genotypes. Two criteria are used to form subsets of the $n(n-1)/2$ total number of pairs of points, or joins. The first is based on some distance measure of the spatial relationship between pairs of points, and distance classes k are formed. The joins are further defined by the two genotypes for a pair of points. Thus, $n_{ij}(k)$ is the number of joins between genotypes i and j for distance class k (i.e., the number of pairs of points which have genotype i at one point and j at the other and are separated by distances that fall within distance class k). For example, each distance class k may contain all pairs of sample points separated by d sample lattice units, where $k - 0.5 \leqslant d < k + 0.5$. Figure 1 shows the distribution of single locus heterozygotes for one of three rare alleles, 1a, 1b, 1c (denoted A, B, C, respectively in Fig. 1) and for one of two common alleles (alleles 2 and 3) of locus GOT I in a sample of trees from a stand of lodgepole pine (Epperson and Allard 1989). In this example the total sample size is 204, and the total number of joins is 204 × 203/2 = 20,706. For the first distance class, $0.5 \leqslant d < 1.5$ lattice units (note that this includes strict nearest neighbors and "diagonal" neighbors), there are 15, 5, 1, 3, 0, 0 joins of A × A, A × B, A × C, B × B, B × C, and C × C, respectively out of a total number of 731 joins for distance class one. Other joins would also be calculated which included pairs in combination with the genotypes with alleles 2 and 3.

Test statistics can be calculated for the null hypothesis, H_0, that the sampling distribution of the numbers of joins is "random", i.e., that produced by sampling pairs *without replacement* from the total sample of genotypes. Effectively, H_0 purports that the locations of sample genotypes are randomized. Under H_0 the expected number of joins between genotypes i and j for any distance class k is $u_{ii} = Wn_i(n_i - 1)/2n(n - 1)$ and $u_{ij} = Wn_in_j/n(n - 1)$ for $j \neq i$. Here n_i is the number of times that

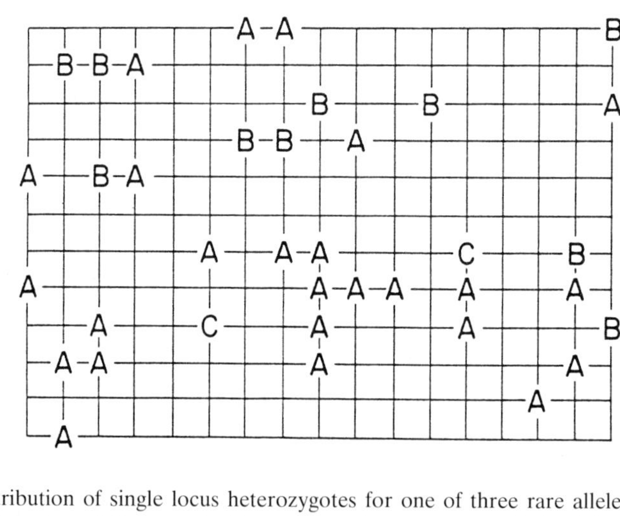

Fig. 1. Distribution of single locus heterozygotes for one of three rare alleles, 1a, 1b, and 1c (denoted A, B, and C respectively) and one of two common alleles (alleles 2 or 3) for locus GOT I in a sample of trees on a regularly spaced lattice from a population of lodgepole pine. Lattice points are separated by about 16 m. (50 feet).

genotype i occurs in the sample, and (suppressing k) W is twice the number of joins in total for class k. [Alternatively w_{ij} can be assigned to pairs of points as described above, and W will equal the sum of the w_{ij}]. The standard errors, $SE_{ij}(k)$, under H_0 can be found (Sokal and Oden 1978; Cliff and Ord 1981), and under H_0 the test statistic $SND_{ij} = (n_{ij} - u_{ij})/SE_{ij}$ (k suppressed) has an asymptotic standard normal distribution (SND) (Cliff and Ord 1981). These significance tests generally have high statistical power (Cliff and Ord 1981), and SND test statistics for short distances are particularly sensitive to most forms of structure (Epperson 1989). For selectively neutral genes, isolation by distance generally causes large positive SNDs for joins between like homozygotes at short distances, whereas the signs of SNDs between heterozygotes may be unpredictable and dependent on details of dispersal (Epperson 1989). SNDs are also useful as measures of spatial correlations. Spatial patterns of populations may be directly inferred from sample distributions (Ord 1980) to the resolution of the sample lattice. In addition, join counts can be used to test other null hypotheses (Epperson 1989).

Correlograms of SND statistics for mutually exclusive distance classes can be formed, analogous to I-correlograms. Like I-correlograms the distances at which SND correlograms intercept zero, known as the X-intercepts, provide measures of the spatial scales of sample autocorrelations (Sokal and Wartenberg 1983; Epperson 1990).

In some cases, it may be desirable to calculate other types of joins. One alternative is to simply combine genotypes into nominal classes. For example, more significant positive correlations were found between rare heterozygotes of the GOT I locus in the lodgepole pine data (see Fig. 1) after combining all rare heterozygotes into one nominal type, and then calculating the SND statistics for all joins between rare heterozygotes (Epperson and Allard 1989). In contrast, SND statistics for other sums of types of joins cannot be calculated by simply combining nominal types, as for example all joins between like homozygotes (Epperson 1989). Explicit formulae for the expected value and variance under H_0 have been given for one particularly informative summary measure, the total number of joins between unlike nominal types (Sokal and Oden 1978; Cliff and Ord 1981). This statistic is closely allied to measures of diversity as functions of distance. For diploid genotypes, it is closely related to measures of genotypic distance such as the probability of individuals having different genotypes as a function of distance. For haploid data it is inversely related to probabilities of gene identity (Epperson 1989). Multilocus data can be handled in a variety of ways (Epperson and Allard 1989), but the simplest is to place all multilocus genotypes into separate nominal classes.

Dispersal and spatial autocorrelations in populations and samples under isolation by distance for neutral genes

Limited seed dispersal creates kinship among adjacent individuals, and subsequent limits to pollen dispersal results in consanguinous matings and inbred progeny. The classical isolation by distance models of Wright (1943) and Malecot (1948) incorporate analogous measures of dispersal (Crawford 1984), e.g., marital distances or parent-offspring distances. *Spatial averaging* and calculations from dispersal parameters, produce a function of the average a priori coefficient of kinship on distance of separation, $\phi(d)$ [the probability of identity by descent for two genes, one from each of two individuals or alternatively two subpopulations, that are separated by a given distance, d (Malecot 1948, 1973)].

If an outside systematic pressure (of strength m), which may represent either mutation or immigration, is added to keep the population from becoming fixed for one gene, then for long distances (at least) an equilibrium kinship function on distance (d) is obtained of the general form: $\phi(d) \cong \mathrm{a}d^{-c}\mathrm{e}^{-\mathrm{b}d}$. The constants a and b are positive and are controlled by the dispersal parameters and the value of m. There has been some controversy over how c may depend on the number of spatial dimensions, especially for the case generally of most interest, i.e., for short distances in a model with two spatial dimensions (Imaizumi et al. 1970; Morton 1973a, b). The utility of the a priori coefficient of kinship is limited because it is defined with respect to an ancestral population, and can be determined only by tracing probabilities of descent through genealogies back to the founding population.

A measure which is directly estimable from genetic survey data is the *conditional kinship*, r_{ij}, or kinship between i and j relative to the existing population (when i and j represent two subpopulations rather than two individuals, r_{ij} is essentially the covariance in gene frequencies; Malecot 1973). It has been suggested that the spatially averaged conditional kinship function on distance is, $r(d) \cong (1 - \mathrm{L})\phi(d) + \mathrm{L}$ (Morton 1973a, b). However, the validity of this equation and the genetic meaning of L (a negative constant) have been questioned (reviewed in Epperson 1989). Unweighted Moran's I statistics are closely related to conditional kinship. If both are defined using the same distance classes and the number of points (subpopulations) is large, then $\mathrm{I}(d) \approx r(d)\, F_{st}$, where F_{st} is Wright's measure of variation in gene frequencies among all subpopulations (Barbujani 1987). There is no corresponding direct theoretical relationship between individual kinship coefficients and genotypic join counts. As a complete set, join counts contain more information which corresponds roughly to the information contained in additional descent measures (Epperson 1989).

Direct results for spatial autocorrelation of genetic variation within populations in two spatial dimensions were obtained in several Monte Carlo simulation studies (Sokal and Wartenberg 1983; Sokal, Jacquez and Wooten 1989; Epperson 1990). In each study, simulated populations consisted of 10,000 individual genotypes at a selectively neutral diallelic locus, where the individuals were continuously distributed over a regularly spaced lattice of 100 × 100 points. Distances and rates of dispersal of individual propagules ranged from very low to moderate (see Epperson 1989). Spatial correlations were characterized by first partitioning a simulated population into 400 nonoverlapping 5 by 5 quadrats and then calculating I-correlograms on quadrat allele frequencies. Generally, the values of I statistics for short distances increased rapidly as patch structures (large areas where several hundred individuals mostly have one homozygous genotype) built up from the initially random distributions during the first 30 to 50 generations. After 50 generations the I-correlograms changed little. Moreover, for the wide range of dispersal parameters, I-correlograms varied only slightly (Sokal and Wartenberg 1983; Sokal, Jacquez and Wooten 1989; Epperson 1990). A typical correlogram is shown in Fig. 2. The salient features are:

(1) a steep decrease in values of I as distance increases from 1 to 4 or 5 quadrat units (or from 5 to 20 or 25 "interplant" units);
(2) X-intercepts in the range of 4 to 6 quadrat units; and
(3) very little change in the values of I at greater distances.

Correlograms for join count statistics also reflect the patch structure (Epperson 1989). SNDs for joins between like homozygotes are in excess at distances up to about 20 to 25 interplant units (4 to 5 quadrat units). In addition, similar patch structures were observed in the simulated populations with highly limited dispersal in a study by Turner et al. (1982).

The amount of genetic correlation expected in samples of genotypes for a neutral locus collected from populations with low to moderate dispersal could be predicted by imposing a sampling grid over the typical patch structures, or more roughly by rescaling the typical I-correlogram. For example, suppose that a set of sample quadrats, with nearest neighbor quadrats having centers 15 interplant units (3 simulation quadrat units) apart, were collected from a frequency surface with a I-correlogram like that in Fig. 2. Then nearest neighbor pairs of these sample quadrats should have values of I near 0.10, and sample quadrats twice as far apart should have negligible correlation. In addition, the predicted covariance or

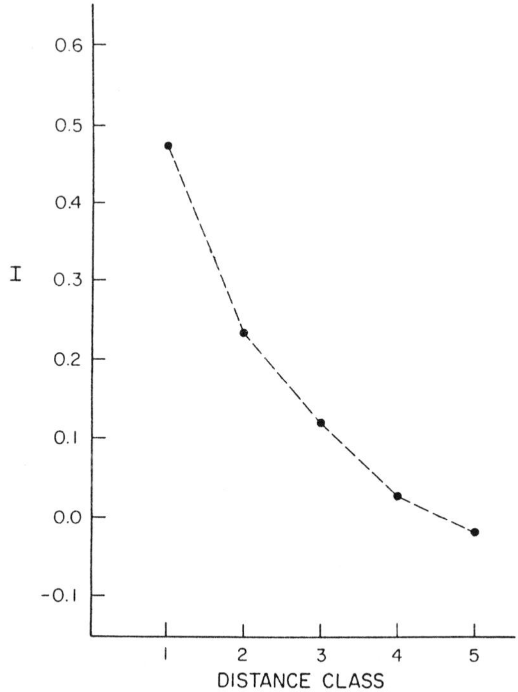

Fig. 2. Typical correlogram for simulated populations with low amounts of dispersal. Distance classes are in quadrat units, and each quadrat unit equals five times the distance between nearest neighbor individuals (i.e., five "interplant" units).

conditional kinship could be calculated by multiplying values of I by estimates of F_{st} from the sample quadrats. Moreover, sample quadrats located within the same patch should have high genetic correlations and kinship. These theoretical results are supported by several studies of plant populations with low to moderate levels of dispersal (Epperson and Clegg 1986; Schoen and Latta 1989). Another example of sampling illustrates the effects of seed tree logging from such populations. Suppose that prior to logging there are 1,000 mature trees per hectare, and after logging 40 well-spread out seed trees per hectare. Then average genetic correlations among nearest neighbor seed trees would approximate that for distance class one in Fig. 2. For theoretical purposes suppose that there were only 10 seed trees per hectare, then correlations would be closer to that for distance class two.

Many conifer populations have greater distances of dispersal (relative to density) than those in the simulations described above, and the theoretical spatial correlations in such populations are not well studied. However, for populations with very high levels of dispersal (e.g., Wright's neighborhood size >1000), there is virtually random mating (among outcrosses), little isolation by distance and thus little spatial autocorrelation (Wright 1943).

Spatial patterns under (non-microenvironmental) natural selection and immigration

Within populations that have limited dispersal, spatial distributions of genotypes for a locus under natural selection differ markedly from those for loci that are selectively neutral. This is true even where there are no components of microenvironmental selection. I conducted a series of Monte Carlo simulations of a diallelic locus under additive directional selection with intensity s, in populations with highly limited dispersal (Epperson 1990). In these simulations, selectively removed individuals were replaced by neighbors. Thus selection operated like local competition. (The simulations also featured an outside systematic pressure which could represent either mutation or immigration from outside populations with constant allele frequencies). The I-correlograms for populations with $s = 0.1$ were greatly reduced in comparison to the neutral case, and selection eroded patches of deleterious genotypes down to small average sizes. Differences in patch structures and I-correlograms caused by selection were established within 30 to 50 generations, and were unaffected by either initial or equilibrium allele frequencies. In contrast, selection had very little effect where $s = 0.01$.

Contrasts of spatial patterns in populations of morning glory confirm these results. Dispersal in populations of *I. purpurea* is very limited. Genotypes for loci that are selectively neutral (as based on independent experiments; Epperson and Clegg 1987) are distributed in large patches (Epperson and Clegg 1986), in accordance with neutral theory. In contrast, large consistent reductions of patch sizes occur for deleterious white-flowered genotypes.

To my knowledge there are no theoretical results on the spatial distributions of genotypes within populations when fitnesses are nonadditive. In populations with low seed and/or pollen dispersal, individuals with deleterious genotypes (via either viability or fecundity deficits) are "replaced" or outcompeted by their neighbors. In such cases, it may be expected that patches of recessive deleterious homozygotes would have larger average sizes than for the additive case, for the same intensity of selection. This follows from the likelihood that heterozygotes near the boundary of a patch, although having full fitness, also contribute homozygous progeny to the patch through local inbreeding. Thus such structures may be expected for cases where selection acts primarily through inbreeding depression, which is a common feature of many populations of forest trees (Sorensen and Miles 1982). (Conversely, patches of dominant deleterious homozygotes might be somewhat smaller). In contrast, heterozygote advantage might cause some reductions in patch sizes for both homozygotes but an increase in sizes of areas containing mostly heterozygotes. This is an important area for further theoretical studies. Contrasts of spatial correlations (especially join counts) could be useful in distinguishing between heterozygote advantage and inbreeding depression (caused by recessive deleterious genes).

Random immigration and mutation alone, with rates lower than 0.01, have little effect on spatial correlations (Epperson 1990). Slight changes in structure and I-correlograms can be caused by certain other types of immigration when rates are on the order of 0.01. In particular, when the immigrant source population has very different allele frequencies, when immigrants arrive at only one boundary of the recipient population, and when dispersal within the recipient population falls within a certain intermediate range, then immigration results in a weak cline of allele frequencies superimposed onto the patch structure (Sokal, Jacquez and Wooten 1989).

The independence of structure from most mutation and immigration processes, and independence from initial conditions and allele frequencies, strengthen our ability to make inferences based on contrasting patterns for loci that are selected with those for loci that are neutral. In addition, for pairs of unlinked neutral loci, there should be little correspondence or correlations over large areas (Sokal and Wartenberg 1983). Thus, contrasts and correlations among spatial patterns for different loci in multiple-locus studies provides a very powerful framework for studying natural selection.

Multilocus selection may be expected to interact complexly with the spatial structure of genetic variation. Even with nonepistatic selection, the average fitness of single locus genotypes may depend on linkage disequilibrium created from correlations between loci over space (Prout 1973). With nonadditive or epistatic selection, and linkage, the interplay between marginal fitnesses and multilocus spatial structure must be very complex.

In a detailed spatial autocorrelation analysis of two study populations of lodgepole pine, the effects of selection and spatial structure could be disentangled. Multiallelic genotypes at eleven of fourteen allozyme loci showed no evidence of spatial structure or convincing evidence of linkage disequilibrium (Epperson and Allard 1987, 1989). This accords with the high recombination rates among these eleven loci, combined with the facts that the populations are nearly 100 percent outcrossing, and have Wright's neighborhood sizes exceeding 1,000. There was significant disequilibrium between some pairs of alleles of the three other, tightly linked, loci (GOT I, PER I, and PER II). Single locus spatial correlations were small but statistically significant for some genotypes of GOT I, PER I and PER II. Genotypes for different loci were not significantly correlated spatially. Thus structure could not be responsible for the observed disequilibrium (Epperson and Allard 1989), which was consistent only with epistatic selection. Both epistatic and nonepistatic (but microenvironmental) components of selection may be acting on these loci themselves or on adjacent chromosomal segments.

Microenvironmental selection–theory and statistical measures

As mentioned briefly above, most forest tree populations are prolific enough to tolerate tremendous genetic loads, and they contain high levels of genetic variation for both allozymes and morphological traits. Microenvironmental heterogeneity in numerous important abiotic and biotic factors can exist on spatial scales ranging down to a few meters. Thus there is considerable potential for microenvironmental selection. Differences in the genotypes that are favored between different microenvironments could cause a variety of changes in the spatial patterns of genetic variation within populations. Microenvironmental selection can also result in correlations between genetic variation and spatially distributed parameters of small scale microenvironmental heterogeneities. Both types of correlations will depend on several features of microenvironmental selection and the stochastic events inherent in local systems of mating system and seed and pollen dispersal patterns. Correlation analysis may be used to detect important factors of microenvironmental selection (Epperson 1989). Such information points to genetic and microenvironmental factors that may warrant further studies to measure fitness directly, possibly including the use of genealogical survey data to directly measure the reproductive success of individual mature trees. However, spurious correlations for the genotypes of a neutral locus can occur simply because of the patch structures created by limited dispersal (Epperson 1989). Spatial structure and correlations can be predicted for some cases in which the scales of both dispersal and population densities, and the grains of microenvironmental heterogeneities, take certain extreme values. For intermediate values of these factors, more complex spatial statistical methods may be used.

First, consider cases where microenvironmental heterogeneity is scaled much larger than the average spacing between (mature) individuals, where microenvironmental selection strongly favors opposite genotypes is different microenvironmental zones, and where dispersal is extremely limited. This situation would create strong associations between genotypic frequencies and the type of zone. In each zone, gene frequencies should simply evolve to values near the equilibrium values predicted from the relative fitnesses within a zone. In addition, if instead dispersal is very great, strong microenvironmental selection could create predictable local deviations in gene frequencies as a stand matures; however, dispersal would erase the spatial correlations during reproduction. Gene frequencies in different zones in a mature stand could be near those predicted after one generation of selection from the fitness differentials between genotypes, for a zone. These patterns would generally exhibit less sharp features than where gene flow is highly limited. Spatial patterns of microenvironmental factors may take on any form, including very irregularly shaped zones, and thus may produce irregular patterns of gene frequencies. It is useful to consider theoretical results for autocorrelations calculated for artificially generated irregular patterns of values (e.g., gene frequencies). Sokal and Oden (1978) and Sokal (1979) found that:

(1) X-intercepts of I-correlograms are nearly equal to the average size of zones of spatial data where different zones vary in size;
(2) X-intercepts are closer to the smaller dimension of zones that are rectangular or irregular in shape; and
(3) helical and "ridge" clines of spatial data both result in large positive correlations at short distances and negative correlations at longer distances.

Naturally spatial associations of genotypic frequencies with microenvironmental zones would be essentially complete, and correlations between allele frequencies and microenvironmental values should reach values near 1.0 in the low dispersal case.

Where environmental heterogeneity is scaled much smaller than both the spacing of (adult) individuals and the dispersal distances, then the progeny from each tree genotype will fall into zones with probabilities that are nearly independent of the parents' location. Thus overall average fitnesses of genotypes may be essentially independent of spatial structure, even though the average fitnesses of genotypes may depend on the total amounts of each microenvironmental type. In mature stands the spatial correlations of genotypes with microenvironments would be near zero.

Where dispersal distances are intermediate, the spatial autocorrelations of genetic variation depend complexly on how the genotypic fitness differentials vary in strength and direction over space. Some theoretical results are available from Monte Carlo simulations. Sokal, Jacquez and Wooten

(1989) simulated populations with low to moderate dispersal and with several patterns of microenvironmental selection. A gradient of strengths of directional selection (i.e., always against the same allele of a diallelic locus) produced I-correlograms that differed little from the neutral case, except that I-statistics for long distances were much more negative. It appears that a cline is superimposed onto patch structure. Discrete microenvironmental zones that were similar in size to patches produced in the neutral case, accordingly produced I-correlograms that were very similar to those for the neutral case. More simulation studies need to be done, with different spatial scales and patterns, and different forms of within zone selection.

Correlations *between* genetic variation *and* microenvironmental factors also are confounded with a genetic patch structure overlay supported by limited dispersal. Patch structure alone could cause spurious correlations between genotypes for a neutral locus and microenvironmental factors in small samples (Epperson 1989). Conversely, gene flow also may blur spatial differences caused by microenvironmental selection, especially near the boundaries of microenvironmental zones.

A number of spatial statistical methods can be used for partitioning autocorrelations due to limits to gene flow from correlations caused by location (i.e., local selection) in known microenvironments. Important considerations of sampling design and statistical details are discussed in a previous paper (Epperson 1989). In cases where microenvironmental heterogeneities are represented as classification variables and the spatial point data are genotypic, then loglinear and similar statistical models, modified with spatial autocorrelation parameters, can be employed. If instead, genetic data are allele frequencies, then modified analysis of variance would be appropriate.

Using multiple regression methods we can model a spatially distributed set of n subsamples of allele frequencies, Y_i, $i = 1, n$ as the observed vector Y of the dependent variable, and X as an $n \times m$ matrix of values of m different microenvironmental factors (independent variables) for each of the n sample locations. The usual multiple regression model is then: $Y = Xb + e$, where b is a vector of coefficients of the strengths of the effects of the microenvironmental selection factors, and e is a vector of independent and identically distributed error terms. Two methods of incorporating spatial autocorrelations are: 1. include a spatial autoregressive component so that $Y = pWY + Xb + e$; or 2. incorporate interactions into the error term so that $Y = Xb + pWe + u$ (u is a vector of independent identically distributed error terms). Here W is a matrix of relative weights of interactions, and p is a scalar which measures the overall strength of interactions, caused by proximity and gene flow. It is anticipated that even very simple forms of spatial weighting in W (i.e., with only low order spatial lags) may remove most of the spatial autocorrelation due to limits to dispersal. Statistical procedures exist for obtaining estimates and tests of significance for b and p, as well as tests for model fit. Details and examples are presented in Cliff and Ord (1981) and Upton and Fingleton (1985). A preliminary indication of whether these more complex models are required, or whether the usual multiple regression model is sufficient, can be gained by testing the residuals of the multiple regression model, $\hat{e} = Y - Y\hat{b}$ (where $\hat{b}$ is the vector of the estimates of the b_j, $j = 1, m$), for autocorrelations, using a modified I statistic (see Cliff and Ord 1981).

Summary

Spatial analyses of the distributions of genetic variation provide important information on the population genetics of plant species. High levels of variation for allozyme loci are particularly useful fore forest trees. In addition, spatial statistics can be used to relate spatial structure to consanguinity relationships such as parent-offspring data. Spatial autocorrelation statistics can be used to detect natural selection, because patterns produced under different types of natural selection are distinct from those for neutral loci, at least in populations with low to moderate dispersal distances. Such distinctions between patterns appear to be maintained under many conditions, and in part emerge from the fact that patterns "capture" the cumulative effects of a number of generations.

The amount of spatial structure of genetic variation within a stand is itself important to forest genetics. In general, many forestry practices contain a component of spatial sampling. Amounts of variation, levels of inbreeding, and local adaptedness of genotypes in spatial samples depend on the amounts of isolation by distance and natural selection operating within stands. This affects regeneration methods and collections for germplasm. High levels of genetic variation and great tolerance for genetic loads are present within stands. In addition, forest stands contain considerable microenvironmental heterogeneity. Thus, the opportunity for microenvironmental selection is great. The presence of strong genotype by microenvironment interactions would have important implications for the genetic quality of forest regeneration. Further theoretical and experimental studies are needed to more fully understand the amounts and causes of spatial structure of genetic variation within stands.

Acknowledgments

This investigation was supported in part by National Science Foundation grants BSR-8614608 and BSR-9006984 to M. T. Clegg. The author thanks Jim Hamraick for stimulating discussions, and Mike Clegg and two anonymous reviewers for helpful comments on earlier drafts of the manuscript.

References

Barbujani, G. 1987. Autocorrelation of gene frequencies under isolation by distance. Genetics 117: 777—782.

Beckman, J. S. and Mitton, J. B. 1984. Peroxidase allozyme differentiation among successional stands of ponderosa pine. Amer. Midl. Nat. 112: 43—49.

Bennett, J. H. and Binet, F. E. 1956. Association between Mendelian factors with mixed selfing and random mating. Heredity 10: 51—56.

Bradshaw, A. D. 1984. Ecological significance of genetic variation between populations, pp. 213—228. In: Dirzo, R. and Sarukhan, J. (Eds) Perspectives on Plant Population Ecology. Sinauer, Sunderland, Mass.

Brown, A. H. D. and Albrecht, L. 1980. Variable outcrossing and the genetic structure of predominantly self-pollinated species. J. Theor. Biol. 82: 591—606.

Brown, A. H. D. and Moran, G. F. 1981. Isozymes and the genetic resources of forest trees, pp. 1—10. In: Conkle, M. T. (Ed) Isozymes in North American Forest Trees and Forest Insects. Pac. SW Forest and Range Expt. Station Tech. Rep. # 48.

Campbell, R. K. 1979. Genecology of Douglas-fir in a watershed in the Oregon cascades. Ecology 60: 1036—1050.

Cliff, A. D. and Ord, J. K. 1981. Spatial Processes. Pion, London.

Crawford, T. J. 1984. The estimation of neighborhood parameters for plant populations. Heredity 52: 273—283.

Dewey, S. E. and Heywood, J. S. 1988. Spatial genetic structure in a population of *Psychotria nervosa*. I. Distribution of genotypes. Evolution 42: 834—838.

Epperson, B. K. 1983. Multilocus Genetic Structure of Natural Populations of Lodgepole Pine. Ph.D. dissertation, Department of Genetics, University of California, Davis.

Epperson, B. K. 1989. Spatial patterns of genetic variation within plant populations, pp. 229—253. In: Brown, A. H. D., Clegg, M. T., Kahler, A. L. and Weir, B. S. (Eds) Plant Population Genetics, Breeding, and Genetic Resources. Sinauer Associates, Sunderland, Mass.

Epperson, B. K. 1990. Spatial autocorrelation of genotypes under directional selection. Genetics 124: 757—771.

Epperson, B. K. and Allard, R. W. 1984. Allozyme analysis of the mating system in lodgepole pine populations. J. Hered. 75: 212—214.

Epperson, B. K. and Allard, R. W. 1987. Linkage disequilibrium between allozymes in natural populations of lodgepole pine. Genetics 115: 341—352.

Epperson, B. K. and Allard, R. W. 1989. Spatial autocorrelation analysis of the distribution of genotypes within populations of lodgepole pine. Genetics 121: 369—377.

Epperson, B. K. and Clegg, M. T. 1986. Spatial autocorrelation analysis of flower color polymorphisms within substructured populations of morning glory (*Ipomoea purpurea*). Am. Nat. 128: 840—858.

Epperson, B. K. and Clegg, M. T. 1987. Frequency-dependent variation for outcrossing rate among flower color morphs of *Ipomoea purpurea*. Evolution 41: 1302—1311.

Farris, M. S. and Mitton, J. B. 1984. Population density, outcrossing rate and heterozygote superiority in ponderosa pine. Evolution 38: 1151—1154.

Gabriel, K. R. and Sokal, R. R. 1969. A new statistical approach to geographic variation analysis. Syst. Zool. 18: 259—270.

Hamrick, J. L. and Godt, M. J. 1989. Allozyme diversity in plant species, pp. 43—63. In: Brown, A. H. D., Clegg, M. T., Kahler, A. L. and Weir, B. S. (Eds) Plant Population Genetics, Breeding, and Genetic Resources. Sinauer Associates, Sunderland, Mass.

Imaizumi, Y., Morton, N. E. and Harris, D. E. 1970. Isolation by distance in artificial populations. Genetics 66: 569—582.

Malecot, G. 1948. Les Mathématiques de l'Hérédité. Masson, Paris.

Malecot, G. 1973. Isolation by distance, pp. 72—75. In: Morton, N. E. (Ed) Genetic Structure of Populations. University of Hawaii Press, Honolulu.

Merzeau, D., DiGiusto, F., Comps, B., Thiebaut, B., Letouzey, J. and Cuguen, J. 1989. Genetic control of isozyme systems and heterogeneity of pollen contribution in Beech (*Fagus sylvatica* L.). Silvae Genetica 38: 195—201.

Mitton, J. B. 1983. Conifers, pp. 443—472. In: Tanksley, S. D. and Orton, T. J. (Eds) Isozymes in Plant Genetics and Breeding. Elsevier, Amsterdam.

Mitton, J. B. and Grant, M. C. 1984. Associations among protein heterozygosity, growth rate and developmental homeostasis. Ann. Rev. Ecol. Syst. 15:479—499.

Mitton, J. B., Linhart, Y. B., Hamrick, J. L. and Beckman, J. S. 1977. Observations on the genetic structure and mating system of ponderosa pine in the Colorado Front Range. Theor. Appl. Genet. 51: 5—13.

Mitton, J. B., Stutz, H. P., Schuster, W. S. and Shea, K. L. 1989. Genotypic differentiation at PGM in Engelmann spruce from wet and dry sites. Silvae Genetica 38: 217—221.

Moran, G. F. and Brown, A. H. D. 1980. Temporal heterogeneity of outcrossing rates in alpine ash (*Eucalyptus delagatensis* R. T. Bak.). Theor. Appl. Gent. 57: 101—105.

Morton, N. E. 1973a. Kinship and population structure, pp. 66— 69. In: Morton, N. E. (Ed) Genetic Structure of Populations. University of Hawaii Press, Honolulu.

Morton, N. E. 1973b. Kinship bioassay, pp. 158—163. In: Morton, N. E. (Ed) Genetic Structure of Populations. University of Hawaii Press, Honolulu.

Muona, O. 1989. Population genetics in forest tree improvement, pp. 282—298. In: Brown, A. H. D., Clegg, M. T., Kahler, A. L. and Weir, B. S. (Eds) Plant Population Genetics, Breeding, and Genetic Resources. Sinauer Associates, Sunderland, Mass.

Muona, O., Yazdani, R. and Rudin, D. 1987. Genetic change between life stages in *Pinus sylvestris*: allozyme variation in seeds and planted seedlings. Silv. Genet. 36: 39—42.

Neale, D. B. and Adams W. T. 1985. The mating system in natural and shelterwood stands of Douglas-fir. Theor. Appl. Genet. 71: 201—207.

Nei, M. 1973. Analysis of gene diversity in subdivided populations. Proc. Nat. Acad. Sci. USA 70: 3321—3323.

Oden, N. L. 1984. Assessing the significance of a spatial correlogram. Geographical Analysis 16: 1—16.

Oden, N. L. and Sokal, R. R. 1986. Directional autocorrelation: an extension of spatial correlograms to two dimensions. Syst. Zool. 35: 608—617.

Ord, J. K. 1980. Tests of significance using non-normal data. Geographical Analysis 12: 387—392.

Phillips, M. A. and Brown, A. H. D. 1977. Mating system and hybridity in *Eucalyptus pauciflora*. Aust. J. Biol. Sci. 30: 337—344.

Plessas, M. E. and Strauss, S. H. 1986. Allozyme differentiation among populations, stands and cohorts in Monterey pine. Can. J. For. Res. 16: 1155—1164.

Prout, T. 1973. Appendix to Mitton, J. B. and Koehn, R. K. Population genetics of marine pelecypods. III. Epistasis between functionally related isozymes in *Mytilas edulis*. Genetics 73: 487—496.

Ritland, K. 1985. The genetic mating structure of subdivided populations. I. Open-mating model. Theor. Pop. Biol. 27: 51—74.

Sakai, K. 1985. Studies on breeding structure in two tropical tree species, pp. 212—225. In: Gregorius, H. R. (Ed) Population Genetics in Forestry. Lecture Notes In Biomathematics 60. Springer-Verlag, Berlin.

Schoen, D. J. and Latta, R. G. 1989. Spatial autocorrelation of genotypes in populations of *Impatiens pallida* and *Impatiens capensis*. Heredity 63: 181—189.

Shaw, D. V., Kahler, A. L. and Allard, R. W. 1981. A multilocus estimator of mating system parameters in plant populations. Proc. Natl. Acad. Sci. USA 78: 1298—1302.

Sokal, R. R. 1979. Ecological parameters inferred from spatial correlograms, pp. 167—196. In: Patil, G. P. and Rosenweig, M. L. (Eds) Contemporary Quantitative Ecology and Related Econometrics. International Cooperative Publishing House, Fairland, Md.

Sokal, R. R., Jacquez, G. M. and Wooten, M. C. 1989. Spatial autocorrelation analysis of migration and selection. Genetics 121: 845—855.

Sokal, R. R. and Oden, N. L. 1978. Spatial autocorrelation in biology. 1. Methodology. Biol. J. Linn. Soc. 10: 199—228.

Sokal, R. R. and Wartenberg, D. E. 1983. A test of spatial autocorrelation analysis using an isolation-by-distance model. Genetics 105: 219—237.

Sorensen, F. C. and Miles, R. S. 1982. Inbreeding depression in height, height growth, and survival of Douglas-fir, ponderosa pine and noble fir to 10 years of age. For. Sci. 28: 283—292.

Strauss, S. H. 1986. Heterosis at allozyme loci under inbreeding and cross-breeding in *Pinus attenuata*. Genetics 113: 115—134.

Strauss, S. H. and Libby, W. J. 1987. Allozyme heterosis in radiata pine is poorly explained by overdominance. Am. Nat. 130: 879—890.

Turner, M. E., Stephens, J. C. and Anderson, W. W. 1982. Homozygosity and patch structure in plant populations as a result of nearest-neighbor pollination. Proc. Natl. Acad. Sci. USA 79: 203—207.

Upton, G. J. G. and Fingleton, B. 1985. Spatial Data Analysis by Example. Vol. 1. Point Pattern and Quantitative Data. Wiley, New York.

Wagner, D. B., Sun, Z.-X., Govindaraju, D. R. and Dancik, B. P. 1989. The population genetic structure of chloroplast DNA polymorphism in a *Pinus banksiana-P. contorta* sympatric region: spatial autocorrelation, pp. 19—32. In: Hallgren, J. E. (Ed) Molecular Genetics of Forest Trees. Proc. Frans Kempe Symp. Swedish Univ., Agric. Sci., Umea, Rep. 8.

Wheeler, N. C. and Guries, R. P. 1982. Population structure, genic diversity and morphological variation in *Pinus contorta* Dougl. Can. J. For. Res. 12: 595—606.

Wright, S. 1943. Isolation by distance. Genetics 28: 114—138.

Wright, S. 1946. Isolation by distance under diverse systems of mating. Genetics 31: 39—59.

Yazdani, R., Lingren, D. and Rudin, D. 1985. Gene dispersion and selfing frequency in a seed tree stand of *Pinus sylvestris* (L.), pp. 1139—1154. In: Gregorius, H. R. (Ed)

Population Genetics in Forestry. Lecture Notes In Biomathematics 60, Springer-Verlag, Berlin.

Yazdani, R., Muona, O., Rudin, D. and Szmidt, A. E. 1985. Genetic structure of a *Pinus sylvestris* L. seed-tree stand and naturally regenerated understory. For. Sci. 31: 430—436.

Yeh, F. C. and Layton, C. 1979. The organization of genetic variability in central and marginal populations of lodgepole pine *Pinus contorta* spp. *latifolia*. Can. J. Genet. Cytol. 21: 487—503.

New Forests **6**: 279—309, 1992.

Review paper

Allozyme markers in breeding zone designation

R. D. WESTFALL[1, 2] and M. T. CONKLE[1]

[1] *Institute of Forest Genetics, USDA Forest Service, Pacific SW Forest & Range Experiment Station, Box 245, Berkeley, CA 94701, USA;* [2] *National Forest Genetic Electrophoresis Laboratory, USDA Forest Service, 2345 Fruitridge Road, Camino, CA 95709, USA*

Received 7 December 1990; accepted 29 August 1991

Key words: geographic variation, multivariate analysis, trend-surface analysis, transfer risk

Application. Multi-locus analyses of allozymes indicate significant geographic variation in widespread forest trees. Under these conditions, allozymes in tandem with other traits, are useful in the development of breeding zones.

Abstract. Early studies of allozyme variation in plant populations suggested that allelic frequencies in some loci vary by geography. Since then, the expectation that allozymes might be useful in describing geographic patterns has generally not been borne out by single locus analyses, except on the broadest scale. Multi-locus analyses reveal the converse: canonical correlation analysis of individual, uniformly-spaced genotypes describe statistically-significant, complex patterns with geography. Multi-locus scores in four major species, *Abies concolor, Pinus lambertiana, P. ponderosa*, and *Pseudotsuga menziesii*, of the mixed conifer forest in the Sierra Nevada correlate 0.40 or greater with the first canonical vector of a geographical trend surface equation. The different species follow similar patterns by latitude and elevation. In contrast with patterns in the Sierra Nevada, large-scale differentiation is weak ($R^2 < 0.20$) among populations of *Pseudotsuga menziesii* in the Coast Ranges and Siskyou Mountains of northern California and southern Oregon, where differentiation may be local. For the purpose of forming zones, we subdivided scores of the first two to four canonical vectors into groups and plotted them as multidimensional contour intervals. Reclassification by discriminant analysis serves as an approximate guide to transfer risks within and among these groups.

Introduction

Early studies of the variation in allozyme markers in plant populations suggested strong associations with geography (Allard and others 1972). Similar results in tree species (Bergmann 1975; Tigerstedt 1973) led some forest geneticists to speculate that allozymes might contribute towards the development of breeding zones (Conkle 1974; Feret and Bergmann 1976). But subsequent results of studies have so diminished such expectations that the present conventional wisdom asserts that allozymes offer little potential usefulness, except in the broadest sense (Adams and Campbell 1981; Falkenhagen 1985; Muona 1990). While mean genetic diversity in many forest tree species is usually high, little of this variation is distributed among populations. Mean heterozygosity in polymorphic loci for both gymnosperms and perennial woody plants is 0.30, while the proportion of electrophoretic variation among populations (G_{st}) is about 0.07 (Hamrick and Godt 1990). Moreover, disequilibria between pairs of loci often do not deviate significantly from zero (Brown 1984; W. T. Adams, pers. comm.). Although patterns have been found by latitude (Bergmann 1975), elevation (Lundkvist 1979), aspect (Mitton and others 1977), and climatic measures (Guries and Ledig 1981), most studies fail to detect ordered geographic differentiation (Moran and Adams 1989; Neale and Adams 1985).

Smouse and others (1982) recognized that the cumulative effects of small differences among allelic frequencies, when summed across loci, can increase differentiation among populations so that individuals can be reliably allocated to biological groups on the basis of their multilocus genotypes. Later, Conkle and Westfall (1984), Guries (1984), and Yeh and others (1985) showed that geographic patterns in trees could be detected with a variety of multivariate statistical techniques. Moreover, individual genotypes in classes formed by multidimensional contour intervals could be reclassified by discriminant analysis into their original groups in higher frequencies than than those randomly expected (Conkle and Westfall 1984). Thus multivariate descriptions of allozyme patterns can contribute to developing and monitoring breeding zones.

The discussion that follows will be limited to allozyme variation. Allozymes, compared with biochemical markers, have found the greatest use in forest population genetics, because of their ease of analysis and their Mendelian basis. Monoterpenes, of the other markers, also have found extensive application to population studies in forest trees (Hanover 1992, this issue, pp. 159—178), but have not been widely applied to the development of breeding zones. Monoterpenes can contribute to zone design, but their greatest use might be in gene conservation (Millar and Westfall 1992, this issue, pp. 347—371).

In the sections that follow, we compare the relative utility of allozyme versus metric traits for describing geographic patterns. Next we will outline the procedural steps used in describing geographic variation and applying the results to the formation of breeding zones. Because the ability to detect pattern is fundamentally a matter of the analytical methods used, we focus on multivariate techniques. We include a summary of our results in four conifer species in California, present a case example of the contribution of allozyme patterns to developing zones for one of these species in the Sierra Nevada Mountains, review unresolved issues, and point to further developments.

Our intention in this review is to demonstrate that significant, regular geographic patterns exist in allozyme loci in forest tree species. Depending on the objectives of the breeder, information on allozyme patterns can therefore augment that of other traits in constructing breeding zones.

The utility of allozymes in developing breeding zones

Zone designations for breeding programs are made on the basis of two perspectives: the source populations and the distribution of planting sites (Namkoong and Kang 1990). When local populations are confirmed or assumed to be optimal (Rehfeldt 1990a), then source populations, which define the seed zone, and the breeding zone, defined by the planting environments, occupy the same zone. But when non-local sources are optimal (Namkoong 1969), planting site zones and the commercially important traits determine the appropriate zones in the source populations (Namkoong and others 1988: 52—53). Once a breeding program advances beyond the establishment of a base population, zones lose their utility. Even in later generations, zones have relevance for defining variation in potential commercial traits and managing gene conservation programs.

Zone designation under the local-is-best assumption has been based on geoclimatic and ecological data (Buck and others 1970) or descriptions of metric traits from short-term tests (Campbell 1984; Rehfeldt 1986). But under the non-local assumption, zones are based on long-term tests (Dorman 1976). Performance of metric traits can be attributed to ecological factors and, if commercially important traits are measured, optimum germplasm source locations for these traits can be determined. But at least 3 years are needed to gather data after the dissemination of propagules, low heritabilities reduce the precision of estimating geographic patterns, and genotype-environment (G x E) interactions complicate the development of seed transfer guidelines. In contrast, allozyme data can be amassed within a few months after collection and processing plant material. Also, multilocus genotypes can be inferred directly and are not subject to G x E interactions.

The utility of geographic patterns in allozymes depends on the objectives of a breeding program and the underlying assumptions. If non-local sources are optimal, only differences among source populations in the economically important traits are relevant (Namkoong 1969). Although breeding for single, broadly-adapted varieties in an attractive goal for breeding programs, multiple local breeding populations may be more stable and allow for greater flexibility over the long run (Namkoong and others 1988: 51–52, 112–115). Also, current forest practices are shifting to those that minimally alter the genetic structure in natural populations: practices that conserve temporal and spatial heterogeneity (Namkoong 1985; Namkoong 1990) and minimize gene flow from plantations to native stands (Ledig 1988). Allozymes, along with other genetic markers, are applicable under this conservative management philosophy by providing data on natural population genetic structures and measuring gene flow (Loveless; Epperson; Mitton; Adams; and Ellstrand 1992, this issue), although consistent associations between allozyme genotypes and ecological factors have not been established for forest tree species (Hamrick and Godt 1990; Hamrick and others 1992, this issue, pp. 95–124). Moreover, the precision of methods for detecting geographic trends in single loci is low and patterns in allozymes have not consistently correlated with those in morphological traits (Hamrick and Godt 1990). Yet, patterns in some morphological and physiological traits are not necessarily correlated among one another (Dickinson and others 1988; Rehfeldt 1986), and it is these independent patterns in sets of traits that contribute to zonation (Campbell 1986). Nevertheless, metric traits are more highly weighted than isozymes in defining source populations for breeding zones.

Methodological considerations

Single-locus methods

In forest genetics, the method of choice for inferring geographic patterns from distance measures computed from single loci is cluster analysis. Almost without exception, the average linkage (UPGMA) clustering algorithm (Dunn and Everitt 1982) has been applied to matrices of Nei's distance (Nei 1978). However, the little-used Cavalli-Sforza and Edwards arc and chord distances (Cavalli-Sforza and Edwards 1967), which weight differences in allelic frequencies by their sampling errors, are better suited for determining the existence of geographic pattern. Additionally, the average linkage algorithm assumes equal rates of divergence. However, the distance-Wagner procedure is more appropriate for visualizing clines (e.g., Millar and others 1988), as it is not constrained by the assumption of equal rates of divergence (Swofford 1981).

A serious drawback with clustering algorithms is that they are sensitive to error variation, diminishing their ability to detect population differences (Archie and others 1989; Dunn and Everitt 1982). When population samples encompass relatively large areas and contain large numbers of individuals per population (greater than 50), means estimated from these populations will reduce sampling errors and clines will be more easily detected than with means from smaller sample sizes (Millar and others 1988). However, clines may be highly nonlinear and therefore not readily detectable in the cluster analysis.

Multi-locus methods

Multivariate analyses are preferable to univariate because single loci are not apt to reveal patterns well, even when patterns are significant (Thorpe 1985). Secondly, multivariate analyses are more efficient: only one analysis is needed for all loci simultaneously. If a pattern exists for a single locus that is independent of all others, it will be clearly shown in any multivariate analysis.

Below we discuss techniques for the multilocus analysis of allozyme data starting with a scoring procedure that transforms these data to a form that is suitable for multivariate analytical methods. Then we discuss four methods that, with the exception of principal components analysis, are special cases of the multivariate general linear model.

Transforming genotypic data. Because multivariate statistical methods require that linear combinations of traits must be normally distributed, haploid or diploid genotypes must be transformed. An appropriate and elegant transformation for diploid genotypes is outlined by Smouse and Williams (1982). For each allele at a locus minus one, assign a value of 0.5 when the allele is present and 0.0 when the allele is absent. Scores for an allele in a homozygote would be $0.5 + 0.5 = 1.0$, for a heterozygote would be $0.5 + 0.0 = 0.5$, and for homozygotes and heterozygotes of other alleles would be 0.0. A similar scoring method for haploid genotypes is given in Weir (1990: 143). Diploid scores for a locus with three alleles are as follows:

	Genotype					
	11	12	13	22	23	33
Allele 1	1.0	0.5	0.5	0.0	0.0	0.0
Allele 2	0.0	0.5	0.0	1.0	0.5	0.0

Other additive scoring systems, such as that in Spielman and Smouse (1976), are equivalent to the above, except that the Smouse and Williams'

(1982) score represents the frequency of an allele at a given level of sampling. For a diploid individual, the score is the frequency of that allele in the individual and for a population, the score is the frequency for the population. According to the expectations of quantitative genetic theory (Kempthorne 1969: 18—19), linear combinations of such allelic scores over ten or more loci are normally distributed (Smouse and others 1982). Such expectations are confirmed by our analyses: residuals in canonical models fit the normal distribution. Although scores for codominant alleles are additive, the infrequent presence of null alleles will generate some small proportion of dominance variance.

Principal components analysis (PCA). PCA is a method used to describe patterns among populations in multidimensional space whereby principal axes in this space are aligned sequentially in the direction of greatest variance and covariance among traits (Morrison 1990). When all traits are uncorrelated, there will be as many principal components as there are traits. But if all traits are highly correlated with one another, their variation will collapse onto one or two principal components. PCA is sensitive to error variation resulting from sampling and genetic recombination (Dunn and Everitt 1982; Gittins 1985; Wartenberg 1985). Hence, analyses are based on population means, which will increase linkage disequilibrium among loci and reduce the number of significant principal components. Used for the analysis of geographic patterns among populations, PCA is hypothesis-free; no *a priori* hypotheses are imposed. Instead, the intent of PCA is to generate hypotheses about population structure (Guries 1984). Although PCA is poorly suited to describing complex clines, principal component scores of populations can be regressed against measures of their geographic location (cf Campbell 1986).

Canonical variate analysis (CVA). CVA, also known as canonical discriminant analysis, is the multivariate equivalent of a one-way analysis of variance (Gittins 1985). Allelic scores are weighted by their ratios of between-to-within group variation, thus reducing error variances. Consequently, CVA is usually better suited than PCA for describing stepped clines or ecotypic patterns (Gittins 1985). Because Mahalanobis distances from CVA (Morrison 1990) have properties similar to the Cavalli-Sforza and Edwards arc distances, they can also be applied to hierarchical cluster analyses (cf Merkle and others 1988). CVA is also hypothesis free when based on samples stratified by populations and individuals or subpopulations and when prior hypotheses about associations have not been imposed on the groups of populations (Kinloch and others 1986; Knowles 1985; Yeh and others 1985). CVA is also sensitive to sampling effects: the first few canonical vectors will focus on alleles that are uniquely represented in one or a few populations. As with PCA, CVA is limited in its ability to detect nonlinear patterns, although canonical scores can be regressed against geographic measures (cf Merkle and others 1988).

Discriminant analysis. Discriminant analysis is related to CVA except that it functions in classification, whereby individuals are assigned into one of several groups (Morrison 1990: 269—287). The method was the first multilocus procedure to be applied to allozymes by Smouse and others (1982), who used the method to allocate Amarindians to populations. The utility of discriminant analysis in describing geographic differences is not so much in classification per se, but in the comparative classification errors among candidate groups. Relative percentages of classified and misclassified individuals indicate similarities among groups in comparison with their geographic distances. For reliable results, sample sizes must be very large or the classification must be validated by an independent dataset. The quadratic function (Morrison 1990: 275—278), used when the within-group variances and covariances are heterogeneous, has limited use in allozyme data because some loci are fixed in one or more groups and not in others, resulting in unequal numbers of variables among groups.

Canonical correlation analysis (CCA). CCA is the multivariate equivalent of multiple regression, but with more than one dependent variable, and the procedure is related to CVA in the same way that regression is related to ANOVA. CCA partitions the coefficient of multiple determination (R^2) of the multiple regression of single variables into separate, independent models, which are shared among dependent variables in the analysis. Although CCA has been applied to ecological studies (Gittins 1985), it's use in population genetics has been only recent (Wartenberg 1985). With the exception of sensitivity to outliers, CCA is less sensitive to errors than PCA and CVA, and errors that occur usually reside in the residual variation. CCA is also superior to PCA for detecting irregular geographic structure (Wartenberg 1985).

Multi-locus geographic patterns and the development of breeding zones

Zone delineation depends on the assumptions behind the choice of optimal source populations. These assumptions in turn determine procedures that are appropriate for sampling, data analysis, and subdivision of source populations and planting sites into zones. Because premises that underly using data from allozyme loci or other genetic markers are similar to those associated with short-term test data, methods that have been used with short-term tests also can be applied to those from allozyme allelic scores (Campbell 1986; Rehfeldt 1990b).

Sample sizes

Multivariate techniques require large sample sizes and complete sets of data. As to the latter issue, each observation with an unresolved locus is deleted from the analysis, thus effectively reducing sample size. With the exception of PCA, all analytical methods require that the sample size to be larger than the number of parameters. A general rule developed from simulation studies of CVA, is that the sample size per group, N_i, should be at least equal to the number of dependent variables, p, in the model (Williams and Titus 1988). In allozyme data, the number of dependent variables is equal to the number of allelic variables. Thus, the minimum sample size is p times the number of groups, k, in CVA, or the number of independent variables in CCA. This rule appears to be quite conservative: Williams and Titus (1988) showed that because errors stabilize more rapidly with increasing sample sizes as the number of dependent variables increase, N_i can be less than p for $p > 30$. Our experience with opportunistically jacknifing data having more than ten loci indicates that samples with three times the total number of variables ($p + k$) are sufficient for stable estimates of the first vector. Thus El-Kassaby (1990) errs in stating that the sample size in any group cannot be smaller than the number of dependent variables. The parameters in canonical analyses or the discriminant function can be estimated as long as $N - p - k$ is greater than zero (preferably much greater), where N is the total sample size.

To meet the requirement for multivariate normality, our experience suggests that data from ten polymorphic loci are sufficient. When each locus averages three alleles, or two allelic variables, p equals 20. With nine independent variables in the model, the minimum sample would be 180.

Sampling methods

If irregular or stepped-cline patterns exist, then CVA is applicable and a hierarchical sampling structure can be used, as in Merkle and others (1988). But when patterns are clinal, sampling should be systematically made as in a factorial design, on a regular grid or set of transects (Box and others 1978; cf Campbell 1986; Campbell and others 1989). Pairs of trees should be sampled at about two thirds of the intersections to provide samples for estimating the variance within locations and to test lack-of-fit in the model (Box and Draper 1987).

In practice, rigid grids are not feasible because species' ranges are not arrayed on a regular grid. Alternatively, a uniform distribution of samples over the range of interest would be suitable. Or when precise estimates of genetic diversity statistics are desired, 20 to 50 trees could be sampled at strategically-spaced locations, and single and paired samples collected at the remaining grid intersections within practical limits of the study.

Analytical procedures

Initially, an exploratory analysis might be desired and for this, PCA and CVA are appropriate. Even when geographic patterns are not obvious from the exploratory analysis, the existence of higher-order patterns and their statistical signifiance should be tested by regression, either by regressing measures of location on principal or canonical scores (Campbell 1984), or on allelic scores by CCA. Because principal and canonical vectors might not be maximally alligned with geography in multi-locus space, CCA is the preferred method. The simplest models should be tested first, beginning with the trend-surface analysis of principal or canonical scores (Gittins 1968) or on allelic variables with its multivariate extension, canonical trend-surface analysis (CTSA) (Lee 1969; Wartenberg 1985). The model used in trend-surface analysis is a second-order polymomial of the measures of geographic location and can be applied by such statistical procedures as SAS' PROC GLM in the univariate condition and PROC CANCORR in the multivariate (SAS Institute Inc. 1985). The importance of additional measures of geography or of higher order terms can be tested by their "extra sums of squares" (Box and Draper 1987), using the Type I sums of squares of canonical scores in PROC GLM, for example (SAS Institute Inc. 1985). To reduce the number of allelic variables in the model, we drop alleles that are near zero for both the standardized canonical weights (< 0.4 approx.) and structure-correlations between the canonical score and the trend surface equation (approx. < 0.05).

For describing pattern, the important statistical data from CCA are the R^2s, the unbiased R^2s, and the statistical significance of the canonical vectors (PROC CANCORR, SAS Institute Inc. 1985). Also useful are the structure correlations between the dependent variables and the scores of the canonical vectors formed by the independent variables. Such structure correlations are the multivariate equivalent of the correlation between the dependent variable and the model described by the independent variables in multiple regression. The square of this structure correlation is the proportion of variance in the dependent variable described by the canonical model, and when summed over k canonical vectors, is the total proportion of variance described by the vectors, also called the redundancies (Gittins 1985).

Final choice of a model depends in part on its adequacy. In addition, we rely on various descriptions of the fitted surface to subdivide that surface into zones. We test lack-of-fit in the model and the relative importance of linear and quadratic terms of the trend-surface model, using a program such as PROC RSREG (SAS Institute Inc. 1985). Using this same SAS procedure, we also describe the shape of the surface of canonical scores from CTSA by canonical analysis (Box and Draper 1987). To aid in the visual description of the patterns, we compute predicted canonical scores and their residuals using the same model from CANCORR in GLM. Contour plots of the predicted scores show the pattern in geographic space (e.g., by latitude and longitude), whereas contour plots of the residuals will suggest discontinuities in the pattern. The latter plots, along with a plot of the residuals against the canonical scores can show biases in the model and suggest additional terms or transformations of the data. (Draper and Smith 1966). To evaluate the normality of the residuals we use both the direct statistical test and the normal probability plot in PROC UNIVARIATE (SAS Institute Inc. 1985). In our experience with allozyme allelic score data in various tree species, residuals do not significantly deviate from normality, or when they do, normal probability plots indicate no important deviations.

Subdivision into zones

Zone formation is a matter of balancing operational needs for a minimal number of units with the biological necessity to minimize risk of transfer within each unit. To apply allozyme data to this process, we determine the pattern and amount of multi-locus geographic variation and integrate this information with data from other traits and with the objectives of a breeding program.

Contour plots of predicted canonical scores indicate regions of similarity in multilocus frequencies in geographic space. Canonical analysis (Box and Draper 1987; PROC RSREG, SAS Institute Inc. 1985) also estimates rates of change in canonical scores in directions within geographic space and provides an indicator of number of subdivisions in these directions. To aid in visualizing patterns in more than one canonical vector, we partition the scores of each vector into contour intervals, assign a letter to each combined interval, and map the multi-vector groups by latitude and longitude. Mapping these interval groups over topographic maps helps to visualize associations with elevation, aspect, watersheds, and other physiographic features.

Differentiation between and within interval groups can be evaluated by discriminant analysis and transfer risk, respectively. Reallocation of individuals into interval groups by discriminant analysis not only indicates similarities by geographic distance, but also suggests pattern unrelated to distance. Such patterns then can be used to develop alternative divisions of the data. Groups having a large proportion of individuals classified into neighboring groups suggest that these groups might be consolidated into a single one. Alternative groups can be evaluated by omitting individuals in the test alternatives from the analysis, recomputing the discriminant functions, and classifying the test individuals into the remaining groups (SAS Institute Inc. 1985: 321—322). For example, this procedure can be used to test subdivisions or consolidations of geographically distant individuals that are grouped into the same contour interval.

The latter method, transfer risk, estimates the proportion of genotypes at one location that does not match those at another location (Campbell and Sugano 1987; Westfall 1991). In metric traits, the number of subdivisions is based on a predetermined minimum risk within groups or zones (Campbell and Sugano 1987). An application of minimum transfer risk in multi-locus allozyme genotypes to the location of gene reserves is illustrated in Millar and Westfall (1992, this issue, pp. 347—371).

Multi-locus patterns in four species of the mixed conifer zone in the Sierra Nevada

With support from many individuals in the Pacific Southwest Region Tree Improvement Program, we have analyzed geographic patterns in allozyme data from four species of the mixed conifer zone in California's Sierra Nevada: White fir (*Abies concolor* Lindl.) (Conkle and Westfall 1988), sugar pine (*Pinus lambertiana* Dougl.), ponderosa pine (*Pinus ponderosa* Laws.), and Douglas-fir (*Pseudotsuga menziesii* Franco). We will also summarize results from analyses of Douglas-fir from the northern Coast Range in California (Conkle and Westfall 1987). The seed parents genotyped to provide diploid data were superior trees that were well distributed throughout the commercial range of each species.

The latitudinal range for the white fir sample is from south of Mt. Lassen to south of Yosemite National Park (Fig. 1a). The ponderosa pine sample has the widest range, from Mt. Shasta in the north to the Greenhorn Mountains in the south (Fig. 1c). Northernmost samples for Douglas-fir and sugar pine are near Mt. Lassen (Figs. 1b and d). The southermost samples of Douglas-fir are near the southern limits of the species' distribution in the Sierra Nevada (Griffin and Critchfield 1972; Fig. 1d).

Individual trees were genotyped using megagametophytes. Diploid genotypes were transformed to allelic scores and analyzed by CTSA and discriminant analysis. Seed parent origins were identified by latitude and longitude (nearest minute), and elevation (nearest 30 m). These three

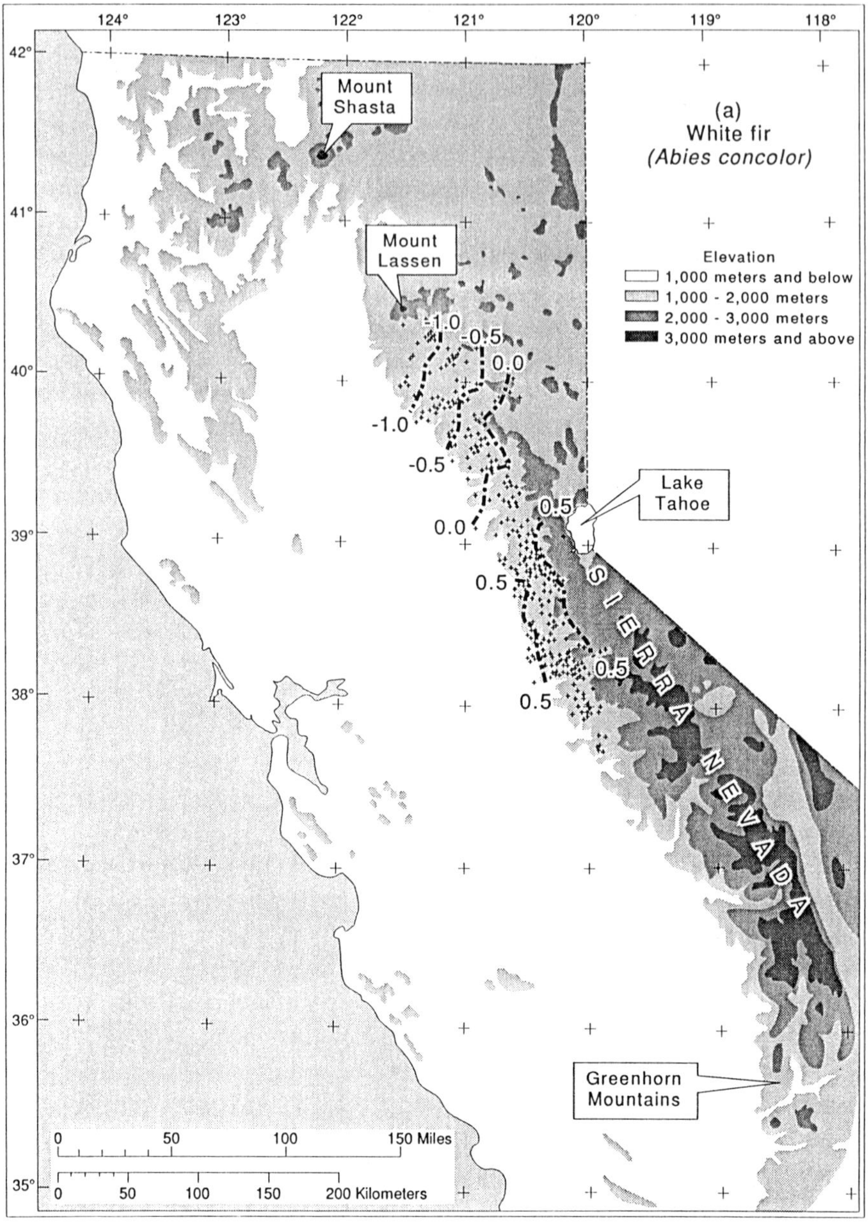
124°
123°
122°
121°
120°
119°
118°
42°
41°
40°
39°
38°
37°
36°
35°
Mount Shasta
Mount Lassen
(a)
White fir
(Abies concolor)
Elevation
1,000 meters and below
1,000 - 2,000 meters
2,000 - 3,000 meters
3,000 meters and above
-1.0
-0.5
0.0
-1.0
-0.5
0.0
0.5
0.5
0.5
0.5
Lake Tahoe
SIERRA NEVADA
Greenhorn Mountains
0 50 100 150 Miles
0 50 100 150 200 Kilometers

Fig. 1a

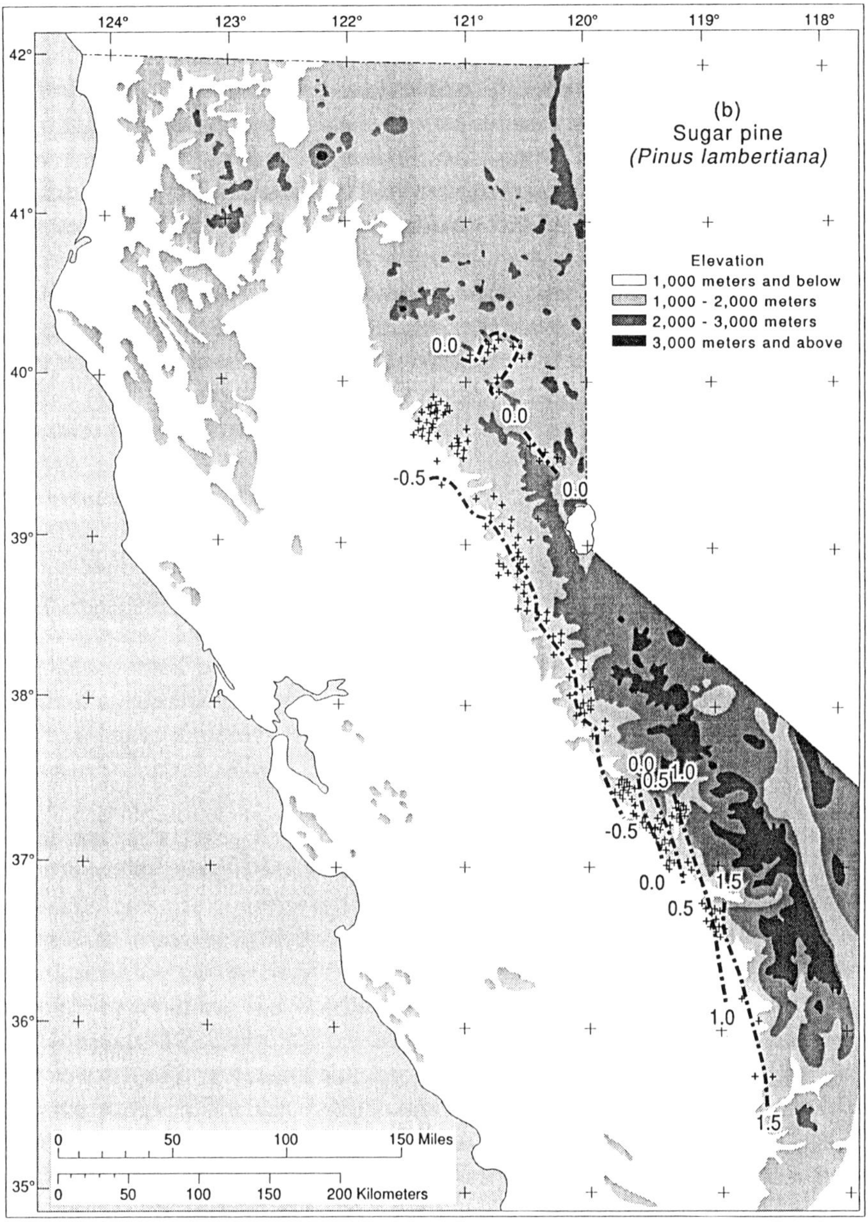
(b)
Sugar pine
(Pinus lambertiana)
Elevation
1,000 meters and below
1,000 - 2,000 meters
2,000 - 3,000 meters
3,000 meters and above
124°
123°
122°
121°
120°
119°
118°
42°
41°
40°
39°
38°
37°
36°
35°
0.0
-0.5
0.5
1.0
1.5
0 50 100 150 Miles
0 50 100 150 200 Kilometers

Fig. 1b

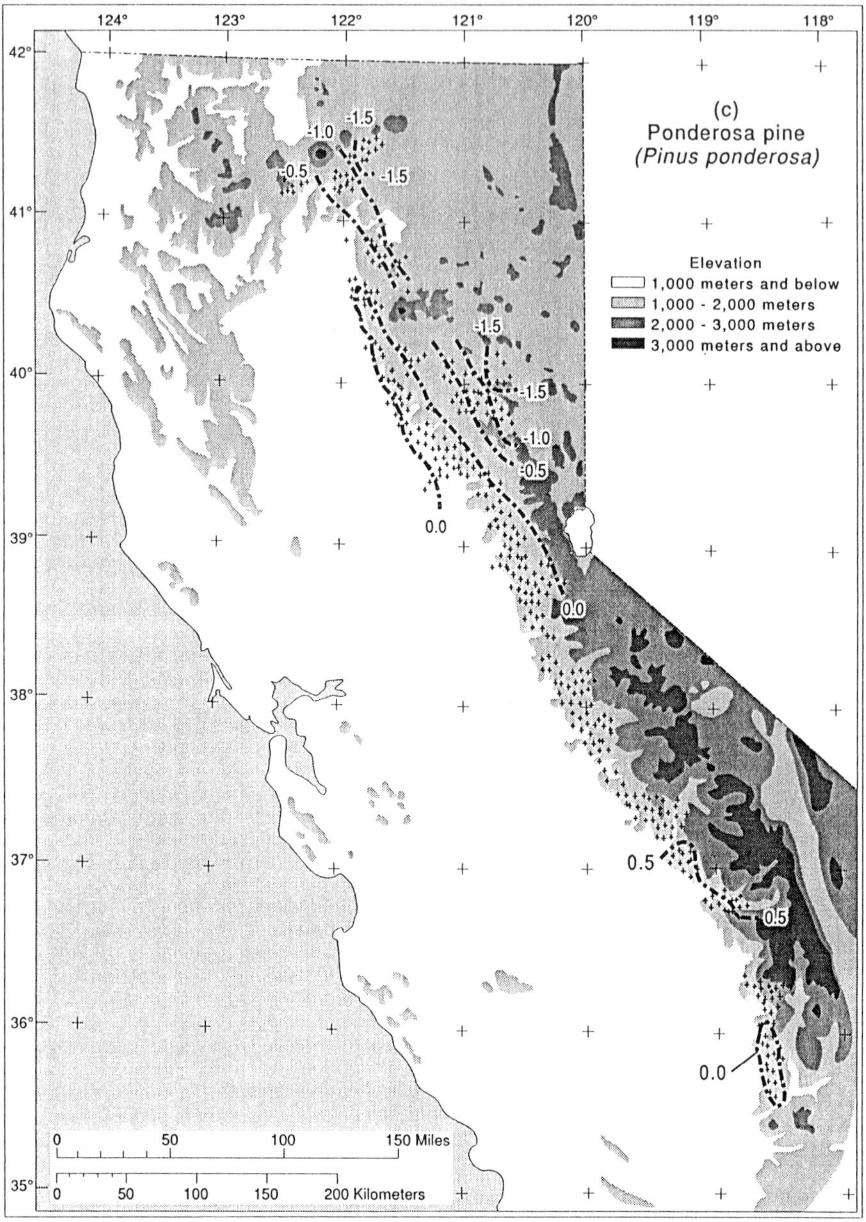
(c)
Ponderosa pine
(Pinus ponderosa)
Elevation
1,000 meters and below
1,000 - 2,000 meters
2,000 - 3,000 meters
3,000 meters and above
124°
123°
122°
121°
120°
119°
118°
42°
41°
40°
39°
38°
37°
36°
35°
-1.5
-1.0
-0.5
0.0
0.5
0 50 100 150 Miles
0 50 100 150 200 Kilometers

Fig. 1c

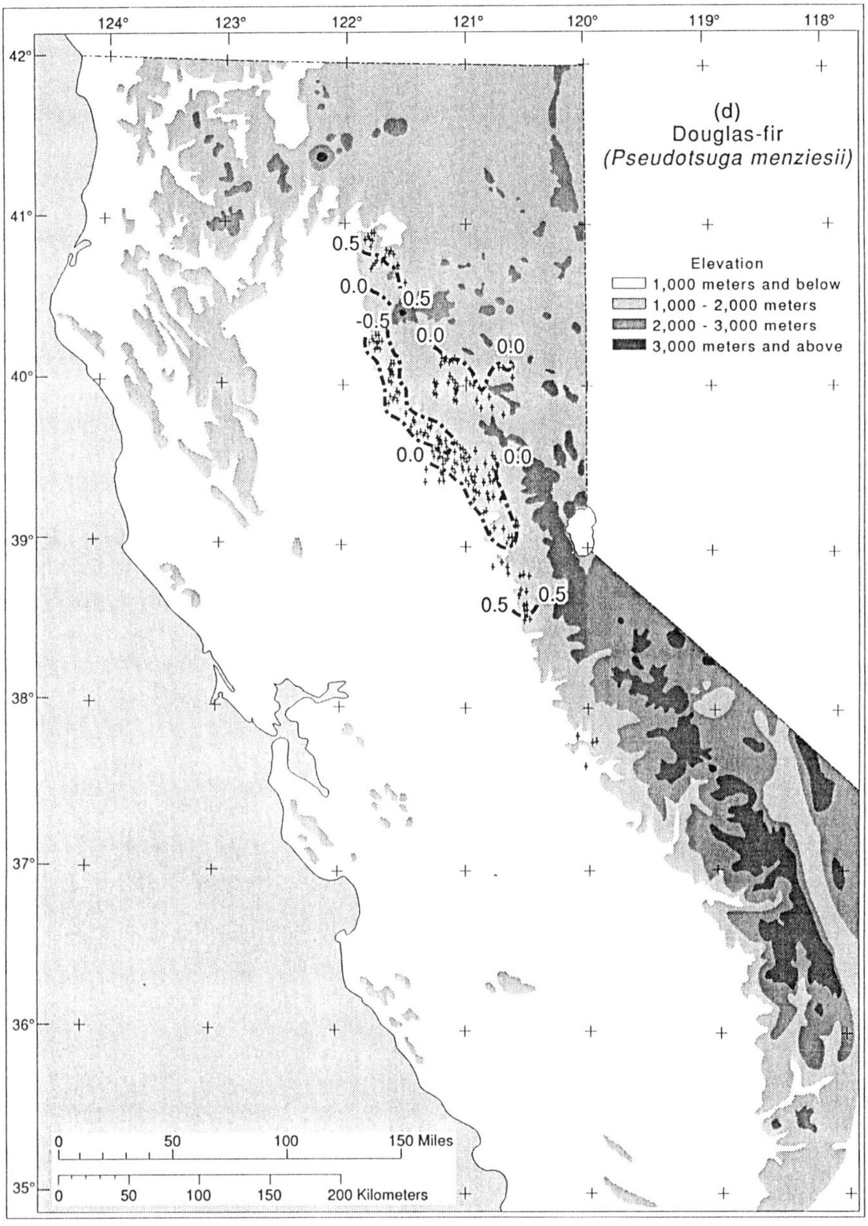

Fig. 1d

Fig. 1. Multi-locus contour plots of the first vector from canonical trend-surface models in four conifer species of the mixed conifer zone in the Sierra Nevada Mountains: *a*) white fir; *b*) sugar pine; *c*) ponderosa pine; and *d*) Douglas-fir. Contour intervals are in 0.50 standard deviation units.

measures of location formed the basis for the second-order trend-surface equations.

Geographic patterning was strongest for Douglas-fir and weakest for white-fir, both on the basis of the amount of variation described by the model in the first vector (43% vs 14%) and the amount of variation described by the first three vectors (63% vs. 28%) (Table 1). In spite of the weak patterning in white fir, a multivariate test for departures from equilibrium (Morrison 1990: 292) was highly significant even though a multiple locus test of linkage disequilibrium (Brown 1984) was not. There is striking similarity in pattern for the first vectors among all four species (Fig. 1a–d). Canonical analyses of the trend-surface equations indicated that all are saddle-shaped surfaces (Box and Draper 1987: 346–350),

Table 1. Single- and multi-locus characteristics of variation in four conifer species in California.

Region/species	N	Nr loci	H_T	Diff. lat.	R_1^2	Signif. vectors	R_T^2	% Correct classif.
Sierra Nevada								
Abies concolor	373	10	0.20	2.8	0.14	2	0.28	0.34
Pinus lambertiana	167	12	0.42	4.8	0.31	2	0.44	0.53
Pinus ponderosa	516	32	0.19	6.1	0.25	4	0.40	0.50
Pseudotsuga menziesii	158	11	0.37	3.5	0.43	2	0.63	0.53
North coast								
Pseudotsuga menziesii[1]	315	27	0.21	3.8	0.17	0	0.35	0.52
Pseudotsuga menziesii[2]	374	11	0.34	2.0	0.14	2	0.24	0.37

Shown are the total sample size (N), number of loci in the analysis, expected total heterozygosities (H_T) of the loci in the analysis, latitudinal differences between the northern- and southern-most samples (in degrees), the proportion of variation in a linear combination of alleles described the the trend-surface model in the first canonical vector R_1^2, the total amount of variation described the first three canonical vectors R_T^2, the number of statistically significant vectors, and the percent of individuals correctly classified to multi-vector contour intervals by discriminant analysis.

R_1^2 = the proportion of variation described by the trend-surface model in the first canonical vector. These are the adjusted (unbiased) estimates (SAS Institute Inc. 1985) and are presented so that the data of varying sample sizes and number of parameters in the models can be compared directly.

$R_T^2 = \Sigma E_i/(1 + \Sigma E_i)$, where E_i is the i'th eigenvalue and $E_i = R_i^2/(1 - R_i^2)$ and R_i^2 = the unbiased R^2 for i'th vector. Eigenvalues are the multivariate equivalent of the sums of squares attributed to the model and the error sums of squares. These are summed over the first three vectors, assuming that the error terms are the same.

[1] Klamath National Forest samples (also includes samples from the Mendocino and Six Rivers NF).

[2] Mendocino, Trinity, and Shasta NF samples.

with the long axis of the saddle oriented in a northwest/southeast direction. In all species, the greatest rate of change in multi-locus frequencies in the first vector is along the northeast/southwest axis and the next largest on the northwest/southeast axis (cf Fig. 1a—d), whereas directions of greatest change in the second or third vectors usually include elevation. Such saddle-shaped forms of the allozyme patterns were also shared with those of growth traits from progeny tests in California (Kitzmiller 1990) whereby growth decreases along the Sierras to the northwest (increasing latitude and longitude) from the location of the growth maximum at a site, but increases to the southwest (decreasing latitude, increasing longitude, and decreasing elevation).

Multi-vector contour interval groups were constructed by dividing each the first three vectors into intervals above and below the mean score, which forms eight contour intervals in six-dimensional space (three geographic and three multi-locus dimensions). These groups, mapped over the sample ranges, show similarity in form among the four species (Fig. 2a—d). The proportions of correctly classified individuals into the eight groups by discriminant analysis are high relative to the random expectation of the percentage correctly classified, or 1/g, where g is the number of groups (Table 1). Because seven or eight groups are designated in each analysis, the random classification rate is thus 0.14 or less and the observed rates are substantially above this level. Although these proportions are biased upwards because the size of each group is small compared to the number of allelic variables, they do represent the degree of multi-locus distinctiveness in each group and do not necessarily follow trends in clinal differentiation among groups.

Patterns are also similar in all species near Mt. Lassen and southward to Lake Almanor (approx. 100 km NW of Lake Tahoe, Fig. 2a—d). The Sierra crest is more diffuse around 40 °N latitude, and the region is in a transition zone between the marine climatic influences of the Sierra Nevada west slope and the continental climate of the Great Basin; allozyme patterns tend to parallel that transition.

In contrast to the Sierra Nevada populations, multi-locus geographic variation in Coast Range Douglas-fir was weak, even though the sample spanned the Kalamath and Mendocino National Forests, ranging from the Oregon border to Clear Lake (which is approximately 130 km north of San Francisco) (Conkle and Westfall 1987; Kitzmiller 1990). The R^2 for the first vector was 0.43 in the Sierra Nevada populations and was 0.17 for populations in the Coast Range; R^2's for the first three vectors were 0.63 and 0.35, respectively (Table 1). Moreover, none of the vectors in the North Coast sample were statistically significant. These results parallel those of Merkle and others (1988) who described statistically significant,

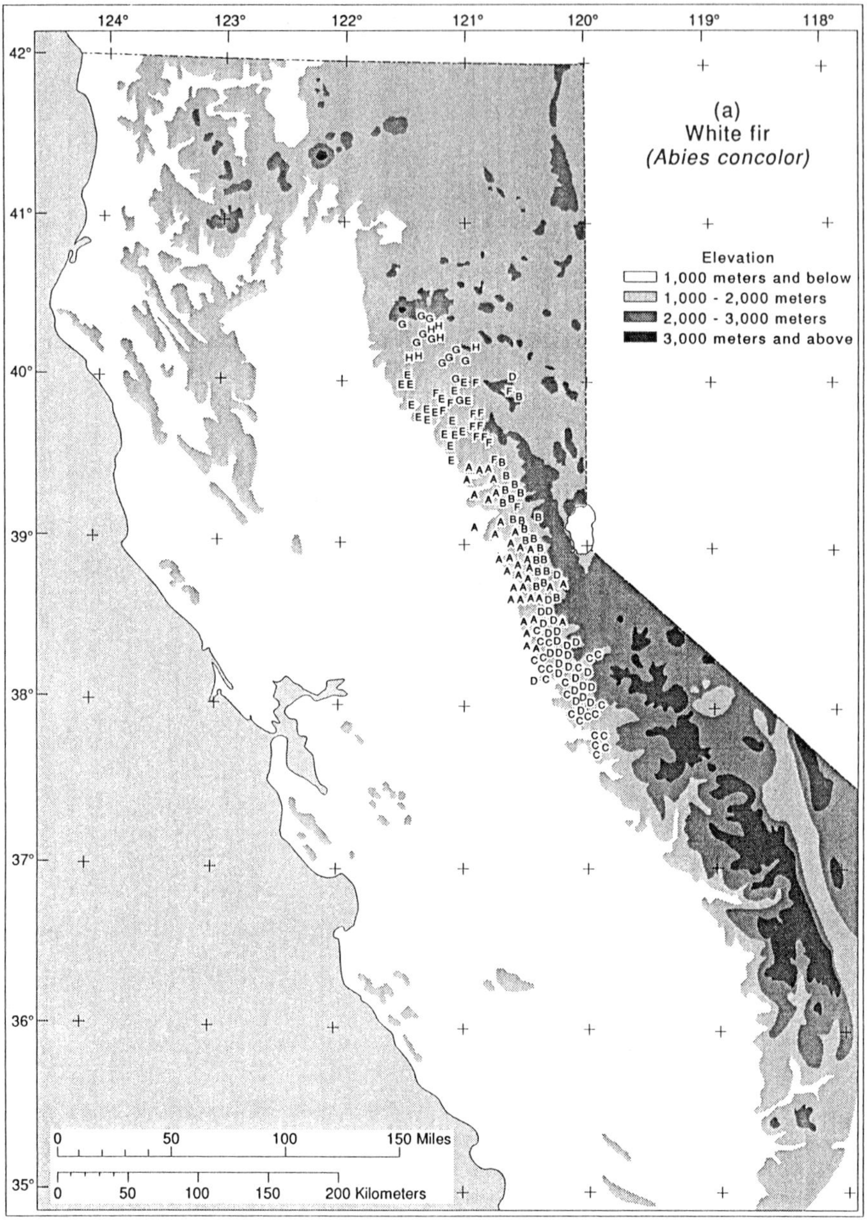

(a)
White fir
(Abies concolor)
Elevation
1,000 meters and below
1,000 - 2,000 meters
2,000 - 3,000 meters
3,000 meters and above
124°
123°
122°
121°
120°
119°
118°
42°
41°
40°
39°
38°
37°
36°
35°
0 50 100 150 Miles
0 50 100 150 200 Kilometers

Fig. 2a

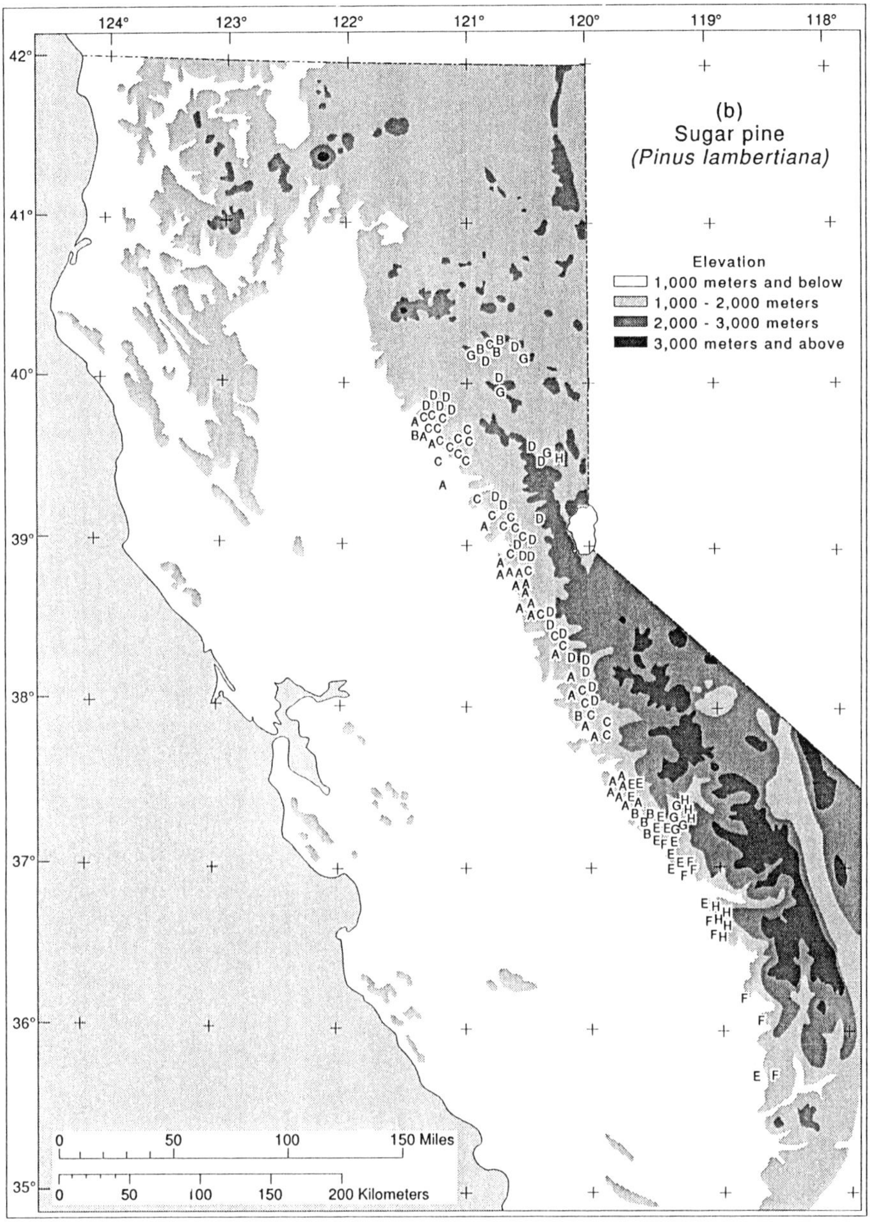

(b)
Sugar pine
(Pinus lambertiana)
Elevation
1,000 meters and below
1,000 - 2,000 meters
2,000 - 3,000 meters
3,000 meters and above
124°
123°
122°
121°
120°
119°
118°
42°
41°
40°
39°
38°
37°
36°
35°
0 50 100 150 Miles
0 50 100 150 200 Kilometers

Fig. 2b

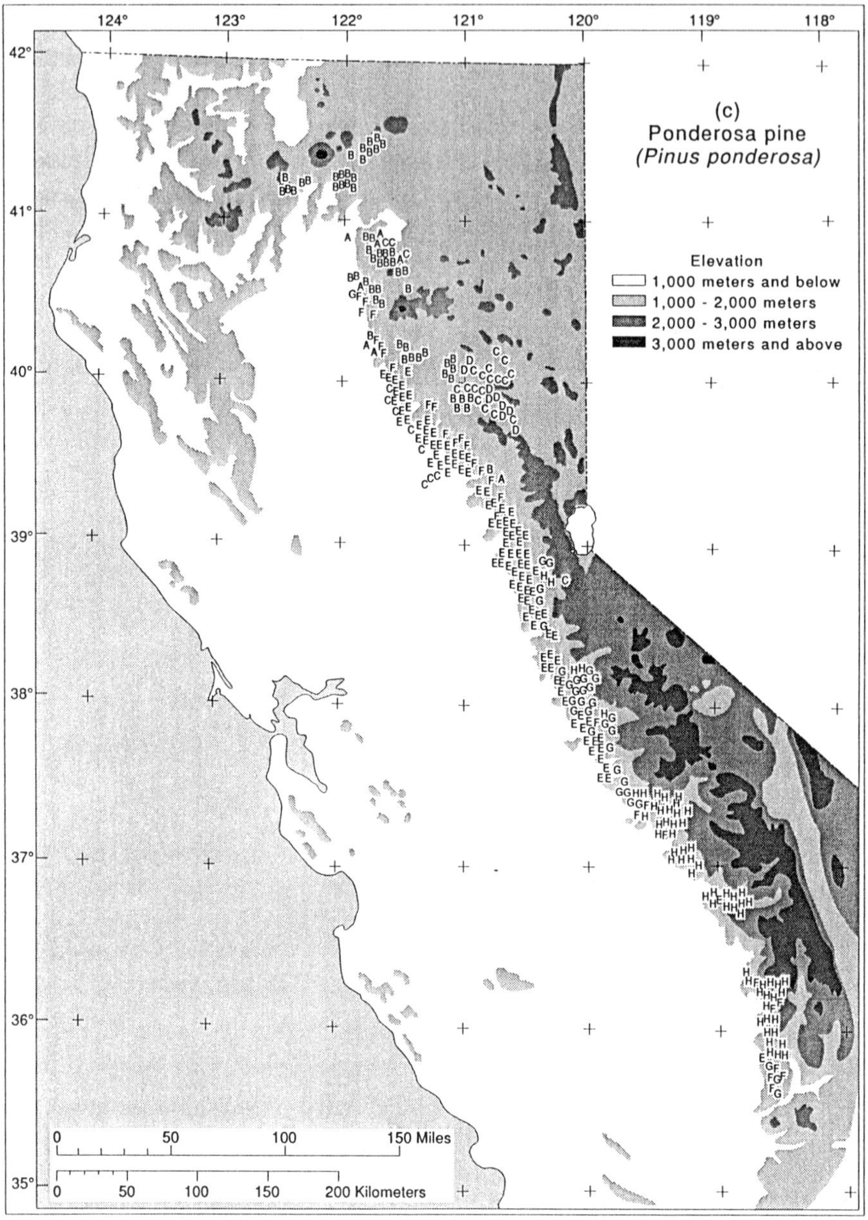

Fig. 2c

Fig. 2. Multi-locus contour class intervals of the first three vectors from canonical trend-surface models in four conifer species of the mixed conifer zone in the Sierra Nevada Mountains: *a*) white fir; *b*) sugar pine; *c*) ponderosa pine; and *d*) Douglas-fir. Contour intervals for each vector are above and below the mean. Class A contains those individuals

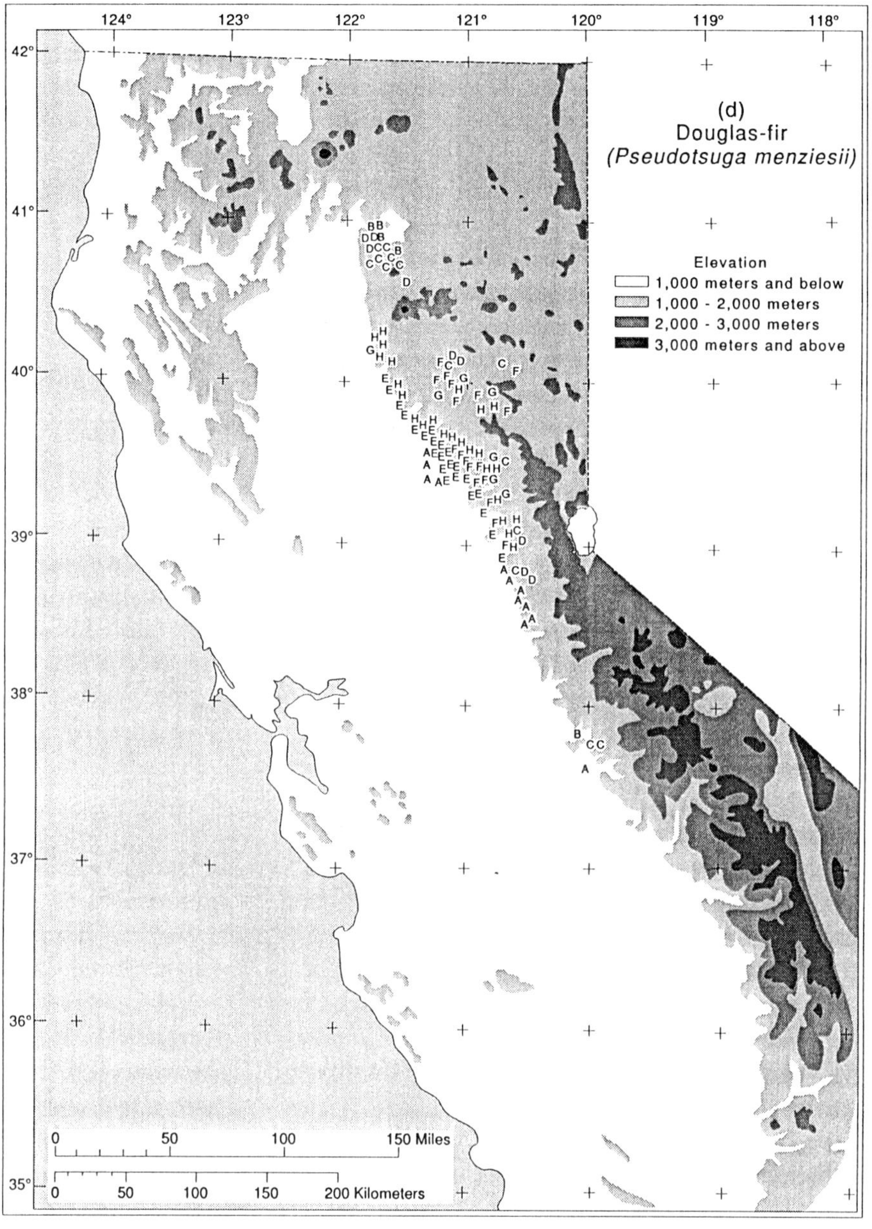

Fig. 2d

---

below the mean for all three vectors, class B individuals below the mean for the first two vectors and above the mean in the third, and so on to class H, which contains individuals above the mean in all three vectors.

but weak patterning in populations from Southern Oregon. The strongest pattern in multi-locus frequencies in the California North Coast data was east/west with some elevational differentiation, which follow existing breeding zone subdivisions for Douglas-fir in the region (Kitzmiller 1976; Kitzmiller 1990). We also analyzed patterns in Douglas-fir bud tissue allozymes of parent trees from the Trinity and Mendocino National Forests, plus those of the Shasta NF to the east. Levels of differentiation were similar to those in the Klamath data: the R^2 for the first vector was 0.14 and for the first three combined, 0.24. Moreover, geographic patterns in the area of overlap between the two datasets were nearly identical.

In these allozyme data, the correlation between an allele and the canonical model only occasionally exceeds 0.25. So, although very small proportions of the variability in any one allele is associated with geography, the aggregate pattern is much stronger (Table 1). Therefore, these loci and their alleles behave much in the way as that expected for quantitative trait loci.

Because large numbers of the resolved loci are not common to all of the analyses, we cannot easily test frequencies of associations and make generalizations regarding loci that pattern among all four species and regions. However, some of the enzymes associated with glycolysis and the Krebs cycle (ACO, IDH, MDH, 6PGDH, and PGM), and LAP2 and GOT pattern in most of the samples.

Geographic patterns in allozymes approximate those patterns for morphological traits, but matched comparisons are difficult to make. Even though progenies from the trees subjected to allozyme analyses are planted in tests, they are split into subsets and planting sites, and genotype-by-environment interactions and the allocation of families among sets make comparisons difficult. In spite of these encumbrances, current progeny test results in Klamath Mountains sites suggest that geographic differentiation in most family sets is on the same order as that suggested by the allozyme data (Kitzmiller 1990).

We have also compared allozyme and morphological patterns in ponderosa pine, where parent trees in the allozyme analysis described above were assessed in a three-year nursery test. Open-pollinated family means of 19 morphological and growth traits were analyzed by the same procedures used in the allozyme study. The trend-surface model in the first vector described 26% of the additive genetic variance in a linear combination of nursery traits, and the first three vectors described 36% of the variation, which are very similar in magnitude to proportions found in the allozyme data (Table 1). Three statistically-significant, saddle-shaped vectors in the nursery data described 44% of the variation in the four allozyme vectors, whereas four allozyme vectors accounted for 62% of the variation in the nursery vectors. Contour intervals, classification percentages from discriminant analysis, and transfer risks were similar in both data sets and together suggest that California Seed Zones in the central Sierra (Buck and others 1970) are conservative for ponderosa pine and could be consolidated.

An example: breeding zone designation in Sierra Nevada Douglas-fir

Results from the analyses of the Sierra Nevada Douglas-fir sample described above contributed to the development of breeding zones, which embrace much of the commercial range of the species in this region of California (Kitzmiller 1976). Allozyme variation in these data is among the greatest of the four Sierran species: total heterozygosity is 0.37, and R^2's for the first and the first three canonical vectors are 0.37 and 0.63, respectively. The rate of change in multi-locus frequencies with geographic distance is largest in the northeast/southwest direction and, at right angles to this direction, is much lower (Fig. 1d). The direction of maximum change in the second vector is similar to that in the first, but the second largest rate of change is by elevation.

The initial step in applying these results to the formation of zones was to subdivide the first two statistically significant vectors each into theoretical thirds: 0.5 standard deviations (SD) below, ± 0.5 SD around, and 0.5 SD above the mean. This subdivision resulted in nine classes or joint contour intervals in five-dimensional space (Table 2 and Fig 3; group G contained no trees within its interval). Maximum differences in multi-locus genotypes were predicted to be between groups A and I.

Subdivisions among the groups are largely by latitude and elevation. Because of the saddle-shaped structure of the data, groups A and B are at both the northern and southern extremes of the sample (Figs. 1d and 3). However, these two groups are below 1470 m in the north and above that elevation in the south. Rates of change in pattern are most rapid between the mid elevations on the west slope of the northern Sierras (Group I) to the southern Cascades (A—C; Fig. 3). Groups D through F subdivided by elevation: D and E divides into contour intervals at 1070 m, and E and F at 1470 m. Groups H and I are located in the elevation band described by E. A heterogeneous assemblage of groups are found in seed zone 523, again suggesting a transition between the northern Sierra to the west and the Cascades to the northwest. Discriminant analysis shows similarity between groups D and E, E and F, and E and H (Table 2).

Breeding zones were developed from these allozyme patterns, ecological and geoclimatic information, and seed production needs. The major demand for seed is anticipated in California Seed Zones 524, 525, and

Table 2. Classification percentages for eight multi-locus allozyme groups in Sierra Nevada Douglas-fir

From class	Percent classified into class								Total number
	A	B	C	D	E	F	H	I	
A	43	14	29		14				7
B		62	12	12	12			3	8
C	8		62	8	15		8		13
D	4	4	7	56	11		15	4	27
E			3	14	52	9	21	2	58
F		8	8		17	33	8	25	14
H				11	30		52	7	27
I						17	17	67	6
Total									158
Percent correctly classified									

Table 3. Classification percentages for five candidate seed zones in Sierra Nevada Douglas-fir

From class	Percent classified into class					Total number
	522	523	524/526L	524/526M	524/526H	
522	53	13	3	13	17	30
523		63		26	11	19
524/526L	10	14	52	19	5	21
524/526M	3	15	10	69	3	71
524/526H	10	10		30	50	10
Total						151
Percent correctly classified						62

Classes 522 and 523 refer to the respective California Seed Zones. Classes 524/526L, M, and H refer to the low (<1070 m), middle (>1070 <1470 m), and high (>14870 m) elevation subdivisions of the pooled California Seed Zones 524, 525, and 526.

526 (Fig. 3; Buck and others 1970), which compose most of an existing breeding zone (Kitzmiller 1976). The allozyme results support that grouping, but also suggest three subdivisions by elevation: below 1070 m, between 1070 and 1470 m, and above 1470 m. A fourth zone was set to encompass Seed Zone 523, with no current subdivisions.

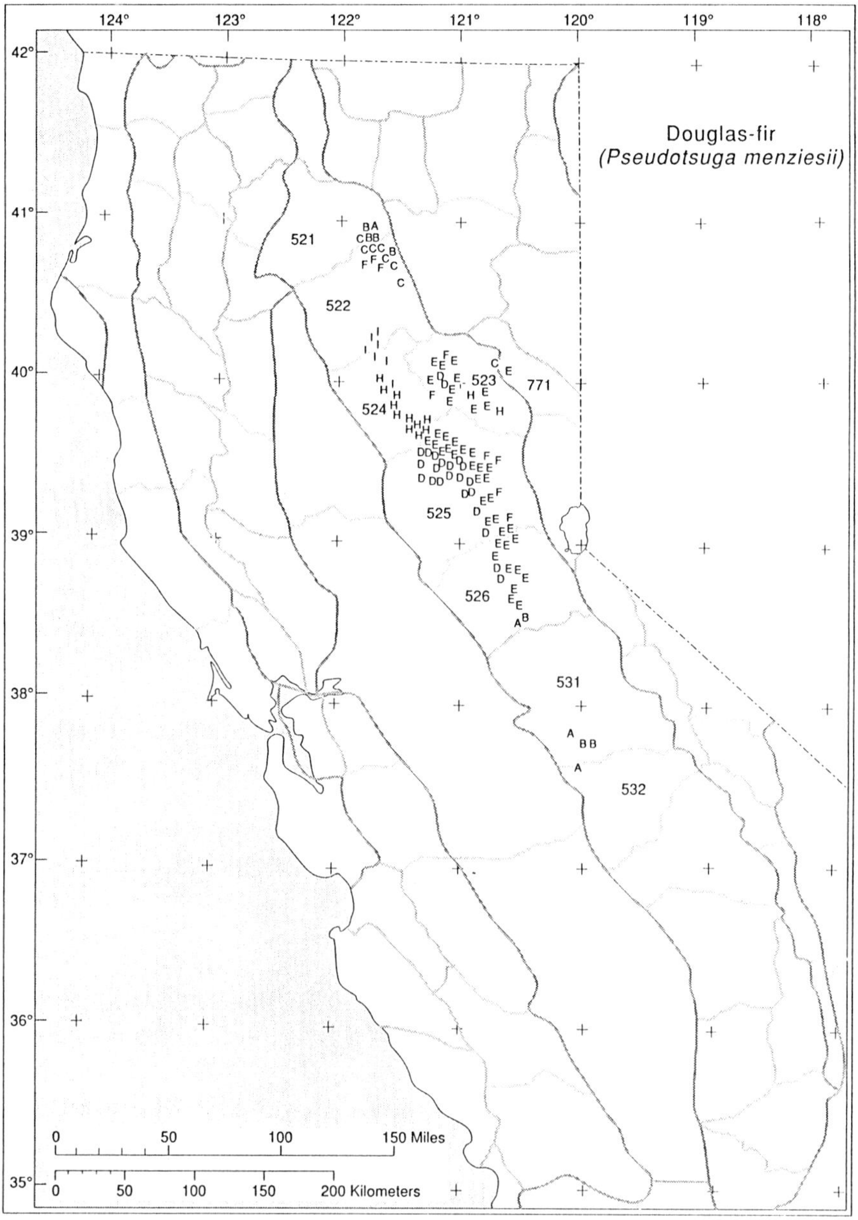

Fig. 3. Multi-locus classes in Sierra Nevada Douglas-fir, developed from intervals of the first two canonical vectors. Groups A-C, D-F, and G-I represent contour intervals in the first vector and the three subdivisions in each (e.g., A, B, and C) represent subdivisions in the second (see the text).

Discriminant allocation results for these zones plus that for Seed Zone 522 are shown in Table 3. The proposed zones formed moderately cohesive classes in comparison to the groups formed by contour intervals (Tables 2 and 3). For the proposed central Sierra zone (Zones 524—526), misclassifications were highest among adjacent elevational groups. Moreover, individuals from the northermost zone, 522, tended to classify into the middle to high elevation classes in the central Sierra group and in the eastern zone, 523.

Future prospects

Statistical stability of the analyses

An important issue to resolve is the minimum number of loci necessary to represent patterns. A procedure similar to that used by Thorpe (1985, 1987) could be applied to establishing minimum numbers. Thorpe determined optimum numbers of traits by bootstrap selection of varying numbers of traits (Efron 1979) and then correlating patterns at the differing trait numbers with the pattern generated from all the traits. In this work, using independent metric traits that were significantly correlated with geography, he found that about 10 randomly selected traits were sufficient for simple clines and 20 were needed to describe complex patterns.

An additional issue is the stability of the patterns, which again can be addressed by adapting procedures used in Thorpe (1987). A measure of stability can be assessed by repetitively resampling the data by bootstrapping, approximately 100 times. Then, for each sample, we compute predicted scores of the canonical trend-surface model and the correlation between predicted scores for the original model and those of the bootstrapped sample. From these bootstrapped samples, we compute a mean correlation with the pattern in the original dataset and 95% confidence interval. Limitations on the widespread use of this method are the daunting demands on time and computing resources.

Further analytical approaches

Spatial autocorrelation methods have received limited attention in forest genetics (Epperson 1990; Epperson 1992, this issue, pp. 257—278). A potentially useful approach is the Mantel test (Manly 1986) of the correlation between genetic and geographic distance matrices, using the residuals from trend surfaces (Bocquet-Appel and Sokal 1989). Significant correlations would suggest localized population structure, perhaps related to factors not considered in the original trend surface models.

DNA polymorphisms

Recent surveys of restriction fragment length polymorphisms (RFLPs) of nuclear and plastid DNA suggest that a greater number of polymorphic loci can be obtained than with allozymes (Neale 1992, this issue, pp. 391—407). If so, then additional loci, some representing different parts of the genome than allozyme loci, would be available for studies of population structure. However, if such polymorphisms only result from mutations in the third codon position or from non-coded regions, their utility in describing geographic patterns will be limited.

Conclusions

In contrast to single locus results, multi-locus analyses reveal statistically significant geographic patterns among tree populations. These multi-locus surfaces tend to follow major geoclimatic patterns in California and also parallel patterns in metric traits. Consequently, these allozyme data can contribute useful information to the formation of breeding zones.

Perhaps the most important questions raised by our results and those from CVA pertain to the causes of the patterns. The geographic patterns may result from chance and historical events. But if they result from selection or selection and drift, can we make global statements about genetic variation in populations within a specific geographic region? If so, can mechanistic models be constructed to explain the patterns, rather than the empirically based trend-surface models?

Acknowledgments

J. H. Kitzmiller enthusiastically supported the work described here. G. Parks, P. Stover, and S. Samman supplied information about the parent trees. We thank P. Hodgskiss, C. Niebling, and S. T. Friedman who supplied the electrophoretic data. And we thank R. K. Scagel and J. Baldwin for helpful suggestions and F. T. Ledig, C. I. Millar, the associate editor, and two anonymous reviewers for constructive comments.

References

Adams, W. T. 1992. Gene dispersal within forest tree populations. This issue (pp. 217—240).

Adams, T. and Campbell, R. K. 1981. Genetic adaptation and seed source specificity, pp. 78—85. In: Hobbs, S. D. and Helgerson, O. T, (Eds) Reforestation of Skeletal Soils, Forest Res. Lab., Oregon State Univ., Corvallis, OR.

Allard, R. W., Babbel, G. R., Clegg, M. T. and Kahler, A. L. 1972. Evidence for coadaptation in *Avena barbata*. Proc. Nat. Acad. Sci. USA 69: 3043—3048.

Archie, J. W., Simon, C. and Martin, A. 1989. Small sample size does decrease the stability of dendrograms calculated from allozyme-frequency data. Evolution 43(3): 678—683.

Bergmann, F. 1975. Adaptive acid posphatase polymorpism in conifer seeds. Silvae Genet. 24(5—6): 175—177.

Bocquet-Appel, J.-P. and Sokal, R. R. 1989. Spatial autocorrelation analysis of trend residuals in biological data. Syst. Zool. 38(4): 333—341.

Box, G. E. P. and Draper, N. R. 1987. Empirical Model-building and Response Surfaces. J. Wiley & Sons, New York.

Box, G. E. P., Hunter, W. G. and Hunter, J. S. 1978. Statistics for Experimenters. An Introduction to Design, Data Analysis, and Model Building. J. Wiley & Sons, New York.

Brown, A. D. H. 1984. Multilocus organization of plant populations, pp. 159—169. In: Wöhrmann, K. and Loeschcke, V. (Eds) Population Biology and Evolution. Springer-Verlag, Berlin.

Buck, J. M., Adams, R. S., Cone, J., Conkle, M. T., Libby, W. J., Eden, C. J. and Knight, M. J. 1970. California Tree Seed Zones. USDA Forest Service.

Campbell, R. K. 1984. Procedures for determining the biological limits of breeding zones in the Pacific Northwest, pp. 24—33. In: Proc. Progeny testing, servicewide genetics workshop, Charleston, SC, December 5—9, 1983. USDA Forest Service, Timber Management.

Campbell, R. K. 1986. Mapped genetic variation of Douglas-fir to guide seed transfer in southwest Oregon. Silvae Genet. 35(2—3): 85—96.

Campbell, R. K. and Sugano, A. J. 1987. Seed zones and breeding zones for sugar pine in southwestern Oregon. USDA Forest Serv., Res. Pap. PNW-RP-383, 18 pp.

Campbell, R. K., Pawuk, W. A. and Harris, A. S. 1989. Microgeographic genetic variation of Sitka spruce in southeastern Alaska. Can. J. For. Res. 19: 1004—1013.

Cavalli-Sforza, L. L. and Edwards, A. W. F. 1967. Phylogenetic analysis: Models and estimation procedures. Evolution 21: 550—570.

Conkle, M. T. 1974. Enzyme poymorphism in forest trees, pp. 95—105. In: Proc. third North Amer. for. biol. workshop, Colorado State University, 9—12 Sept., 1974. College of Forestry and Natural Resources, Colorado State University, Fort Collins, CO.

Conkle, M. T. and Westfall, R. D. 1984. Evaluating breeding zones for ponderosa pine in California, pp. 89—98. In: Progeny testing, proc. servicewide genetics workshop, Charleston, SC, December 5—9, 1983. USDA Forest Service, Timber Management.

Conkle, M. T. and Westfall, R. D. 1987. Douglas-fir isozyme test results. USFS, Pac. SW For. & Range Expt. Stn., Unpublished report to the USFS R-5 Regional Tree Improvement Program, 5 pp.

Conkle, M. T. and Westfall, R. D. 1988. White fir report. USFS, Pac. SW For. & Range Expt. Stn., Unpublished report to the USFS R-5 Regional Tree Improvement Program, 13 pp.

Dickinson, T., Knowles, P. and Parker, W. H. 1988. Data set congruence in northern Ontario tamarack (*Larix laricina*, Pinacea). Syst. Bot. 13(3): 442—455.

Dorman, K. W. 1976. The Genetics and Breeding of Southern Pines. USDA, Agric. Handbook 471, 407 pp.

Draper, N. R. and Smith, H. 1966. Applied Regression Analysis. John Wiley & Sons, Inc., New York, 407 pp.

Dunn, G. and Everitt, B. S. 1982. An Introduction to Mathematical Taxonomy. Cambridge University Press, Cambridge.

Efron, B. 1979. Bootstrap methods: another look at the jacknife. Ann. Stat. 7: 1—26.

El-Kassaby, Y. A. 1990. Genetic Variation within and among conifer populations: review and evaluation of methods, pp. 59—74. In: Hattemer, H. H., Fineschi, S., Cannata, F. and Malvoti, M. E. (Eds) Biochemical Markers in the Population Genetics of Forest Trees. SPB Academic Publishing, The Hague.

Ellstrand, N. C. 1992. Gene flow among seed plant populations. This issue (pp. 241—256).

Epperson, B. K. 1992. Spatial structure of genetic variation within populations of forest trees.This issue (pp. 257—278).

Epperson, B. K. 1990. Spatial patterns of genetic variation within plant populations, pp. 229—253. In: Brown, A. H. D., Clegg, M. T., Kahler, A. L. and Weir, B. S. (Eds) Plant Population Genetics, Breeding, and Genetic Resources. Sinauer Associates Inc., Sunderland, MA.

Falkenhagen, E. R. 1985. Isozyme studies in provenance research of forest trees. Theor. Appl. Genet. 69(4): 335—347.

Feret, P. P. and Bergmann, F. 1976. Gel electrophoresis of proteins and enzymes, pp. 49—77. In: Miksche, J. P. (Ed) Modern Methods in Forest Genetics. Springer-Verlag, New York.

Gittins, R. 1968. Trend-surface analysis of ecological data. J. Ecology 56(3): 845—869.

Gittins, R. 1985. Canonical Analysis. A Review with Applications in Ecology. Springer-Verlag, New York.

Griffin, J. R. and Critchfield, W. B. 1982. The distribution of forest trees in California. USDA For. Serv., Research Pap. PSW-82, 118 pp.

Guries, R. P. 1984. Genetic variation and population differentiation in forest trees, pp. 119—131. In: Proc. eighth N. Amer. for. biol. workshop, Utah State Univ., Logan, UT, Jul. 30—Aug. 1, 1984. Dept. For Res., Utah State Univ., Logan, UT.

Guries, R. P. and Ledig, F. T. 1981. Genetic structure of populations and differentiation in forest trees, pp. 42—47. In: Proc. symposium on isozymes of North American forest trees and forest insects, Berkeley, CA, July 27, 1979. USDA For. Serv. PSW For. & Range Expt. Stn., Gen. Tech. Rep. PSW-48.

Hamrick, J. L. and Godt, M. J. W. 1990. Allozyme Diversity in plant species, pp. 43—63. In: Brown, A. H. D., Clegg, M. T., Kahler, A. L. and Weir, B. S. (Eds) Plant Population Genetics, Breeding, and Genetic Resources. Sinauer Associates Inc., Sunderland, MA.

Kempthorne, O. 1969. An Introduction to Genetic Statistics. Iowa State Univ. Press, Ames, IA.

Kinloch, B. B., Westfall, R. D. and Forrest, G. I. 1986. Caledonian Scots pine: origins and genetic structure. New Phytol. 104: 703—729.

Kitzmiller, J. H. 1976. Tree Improvement Master Plan for the California Region. USDA Forest Service, San Francisco, 123 pp.

Kitzmiller, J. H. 1990. Genetic variation and adaptability of Douglas-fir in northwestern California. In: Proc. Joint Mtg. W. For. Genet. Assoc. & IUFRO Working Parties S2.02—05. 06, 12, & 14, Douglas-fir, Contorta Pine, Sitka Spruce, & Abies Breeding & Genetic Resources, Olympia, WA, Aug. 20—21, 1990.

Knowles, P. 1985. Comparison of isozyme variation among natural stands and plantations: jack pine and black spruce. Can. J. For. Res. 15: 902—908.

Ledig, F. T. 1988. The conservation of diversity in forest trees. BioSci. 38(7): 471—479.

Lee, P. J. 1969. The theory and application of canonical trend surfaces. J. Geology 77(3): 303—318.

Loveless, M. D. 1992. Isozyme variation in tropical trees: patterns of genetic organization. This issue (pp. 67—94).

Lundkvist, K. 1979. Allozyme frequency distributions in four Swedish populations of Norway spruce (*Picea abies* K.). I. Estimations of genetic variation within and among populations, genetic linkage and a mating system parameter. Hereditas 90: 127—143.

Manly, B. F. J. 1986. Multivariate Statistical Methods. A Primer. Chapman and Hall, New York, 159 pp.

Merkle, S. A., Adams, W. T. and Campbell, R. K. 1988. Multivariate analysis of allozyme variation patterns in coastal Douglas-fir from southwest Oregon. Can. J. For. Res. 18: 181—187.

Millar, C. I., Strauss, S. H., Conkle, M. T. and Westfall, R. D. 1988. Allozyme differentiation and biosystematics in the California closed-cone pines (Subsection *Oocarpae* Little & Critchfield, Genus *Pinus*). Syst. Bot. 13(3): 351—370.

Millar, C. I. and Westfall, R. D. 1992. Allozyme markers in forest genetic conservation. This issue (pp. 347—371).

Mitton, J. B. 1992. The dynamic mating systems of conifers. This issue (pp. 197—216).

Mitton, J. B., Linhart, Y. B., Hamrick, J. L. and Beckman, J. S. 1977. Observations on the genetic structure and mating system of ponderosa pine in the Colorado Front Range. Theor. Appl. Genet. 51: 5—13.

Moran, G. F. and Adams, W. T. 1989. Microgeographical patterns of allozyme differentiation in Douglas-fir from southwest Oregon. For. Sci. 35(1): 3—15.

Morrison, D. F. 1990. Multivariate Statistical Methods. McGraw-Hill, New York.

Muona, O. 1990. Population genetics in forest tree improvement, pp. 282—298. In: Brown, A. H. D., Clegg, M. T., Kahler, A. L. and Weir, B. S. (Eds) Plant Population Genetics, Breeding, and Genetic Resources. Sinauer Associates Inc., Sunderland, MA.

Namkoong, G. 1969. Nonoptimality of local races, pp. 149—153. In: Proc. tenth south. for. tree improv. conf. Texas For. Serv., Texas A & M Univ. Press, College Sta., TX.

Namkoong, G. 1985. Considerations of respect for future forest populations, pp. 117—121. In: Proc. N. Amer. for. commission tree improv, Study Group. The state of the art in forest genetics and breeding. Int. Meeting, Satellite to the IX World Forestry Congress, Mexico City, Jul. 5, 1985.

Namkoong, G. 1990. Forest genetics: 2050. The quick and the dead. In: Proc. Joint Mtg. W. For Genet. Assoc. & IUFRO Working Parties S2.02-05. 06, 12, & 14, Douglas-fir, Contorta Pine, Sitka Spruce, & Abies Breeding & Genetic Resources, Olympia, WA, Aug. 20—21, 1990.

Namkoong, G. and Kang, H. 1990. Quantitative genetics of forest trees, pp. 139—188. In: Janick, J. (Ed) Plant Breeding Reviews, Vol 4. Timber Press, Portland, OR.

Namkoong, G., Kang, H. C. and Brouard, J. S. 1988. Tree Breeding: Principles and Strategies. Springer-Verlag, New York.

Neale, D. B. and Adams, W. T. 1985. Allozyme and mating-system variation in balsam fir (*Abies balsamea*) across a continuous elevational transect. Can. J. Bot. 63(2): 2448—2453.

Neale, D. B., Devey, M. E., Jermstad, K. D., Ahuja, M. R., Alosi, M.C. and Marshall, K. A. 1992. Use of DNA markers in forest tree improvement research. This issue (pp. 391—407).

Nei, M. 1978. Estimation of average heterozygosity and genetic distance from a small number of individuals. Genetics 89: 583—590.

Rehfeldt, G. E. 1986. Adaptive variation in *Pinus ponderosa* from Intermountain regions. I. Snake and Salmon River Basins. For. Sci. 32(1): 79—92.

Rehfeldt, G. E. 1990a. Gene resource management: Using models of genetic variation in silviculture, pp. 31—44. In: Proc. genetics/silviculture workshop, Wenatchee, WA. Aug 27—31, 1990, USDA Forest Service.

Rehfeldt, G. E. 1990b. Genetic differentiation among populations of *Pinus ponderosa* from the upper Colorado River Basin. Bot. Gaz. 151(1): 125—137.

SAS Institute Inc. 1985. SAS® User's Guide: Statistics. SAS Institute Inc., Cary, NC, 956 pp.

Smouse, P. E. and Williams, R. C. 1982. Multivariate analysis of HLA-disease associations, Biometrics 38(3): 757—768.

Smouse, P. E., Spielman, R. S. and Park, M. H. 1982. Multiple-locus allocation of individuals to groups as a function of the genetic variation within and differences among human populations. Am. Nat. 119(4): 445—463.

Spielman, R. S. and Smouse, P. E. 1976. Multivariate classification of human populations. I. Allocation of Yanomama Indians to villages. Am. J. Hum. Genet. 28(4): 317—331.

Swofford, D. L. 1981. On the utility of the distance Wagner procedure, pp. 25—43. In: Funk, V. A. and Brooks, D. R. (Eds) Advances in Cladistics. Allen Press, Lawrence, KS.

Thorpe, R. S. 1985. Clines: Character number and the multivariate analysis of simple patterns of geographic variation. Biol. J. Linn. Soc. 26(3): 201—214.

Thorpe, R. S. 1987. Complex clines: the predictivity of complicated patterns of geographic variation portrayed by multivariate analysis. Biol. J. Linn. Soc. 31: 75—88.

Tigerstedt, P. M. A. 1973. Studies on isozyme variation in marginal and central populations of *Picea abies*. Hereditas 75: 47—60.

Wartenberg, D. 1985. Canonical trend surface analysis: A method for describing geographic patterns. Syst. Zool. 34(3): 259—279.

Weir, B. S. 1990. Genetic Data Analysis. Sinauer Associates, Inc., Sunderland, MA.

Westfall, R. D. 1991. Developing seed transfer zones. In: Fins, L. and Friedman, S. T. (Eds) Manual of Quantitative Forest Genetics (in press).

Williams, B. K. and Titus, K. 1988. Assessment of sampling stability in ecological applications of discriminant analysis. Ecology 69(4): 1275—1285.

Yeh, F. C., Cheliak, W. M., Dancik, B. P., Illingworth, K. Trust, D. C. and Pryhitka, B. A. 1985. Population differentiation in lodgepole pine, *Pinus contorta* spp. *latifolia*: a discriminant analysis of allozyme variation. Can. J. Genet. Cytol. 27: 210—218.

New Forests **6**: 311–328, 1992.

Review paper

The use of electrophoretic markers in seed orchard research

N. C. WHEELER and K. S. JECH

Weyerhaeuser Forestry Research Center, 505 North Pearl Street Centralia, Washington 98531, USA

Received 8 October 1990; accepted 10 April 1991

Key words: allozyme, isozymes, mating systems, paternity analyses, pollen contamination, supplemental mass pollination

Application. Gene flow is a critical determinant of population genetic structure, playing an important role in forest tree improvement, conservation genetics, and the potential for the escape of engineered genes. A review of gene flow among plant populations yields the following generalization: although gene flow varies substantially among species, populations, seasons, and even individual plants, at physical isolation distances of hundreds to thousands of meters, gene flow levels are frequently sufficient to counteract genetic drift and moderate levels of directional selection. Gene flow will be hard to predict a priori. Therefore, any project or management plan depending on a given level of gene flow or reproductive isolation should include a genetically-based method to measure the actual gene flow among the populations involved.

Abstract. Biochemical markers are ideally suited for addressing questions concerning genetic or parental identity. For production seed orchards, such questions are common as a consequence of the uncertainty of paternity in non-controlled cross situations. During the last decade, electrophoretic procedures have been used extensively to investigate issues such as:

- clonal identification,
- pollen contamination levels,
- mating systems,
- supplemental mass pollination (SMP) verification,
- patterns of gene flow, and
- levels of pollen competition.

A number of patterns have emerged from this work. Most notable, contamination rates are disturbingly high, clonal phenology and pollen application methodology influence SMP success, and phenology and proximity are important in determining paternal success. Results are discussed with reference to orchard management prescriptions.

This paper was distibuted within Weyerhaeuser Company as an internal research report.

Introduction

The use of segregating electrophoretic variants (allozymes) for genetic studies in forest trees has continuously increased since the technology was introduced to the profession in the early 1970s (Conkle 1971, 1972; Feret and Bergmann 1976). During the last ten years, a gradual shift in research emphasis from population genetic studies in natural stands, to applied tree improvement issues has occurred and is expected to continue in the future (Rudin 1986). In particular, studies in seed orchards have proliferated in recent years. Previous reviews of the use of allozymes in tree breeding have identified most of the topics discussed in this paper (Rudin and Lindgren 1977; Adams, 1981a, 1983; Rudin 1986; Cheliak and others 1987b). However, substantive additions to the seed orchard literature and changes in research emphasis warrant a renewed look at progress in understanding the dynamics of reproductive biology and orchard management techniques.

We have chosen to treat seed orchard electrophoretic studies under the following general headings: Pedigree Verification, Supplemental Mass Pollination Verification (determination of success), Selection, and Orchard Genetic Efficiency. The latter of these categories is further divided into the areas of Contamination, Mating Systems, and Differential Reproductive Success. Individual study results will be summarized where relevant, however, this paper is not intended to serve as a complete bibliographic reference (see Paule 1990).

Pedigree verification

A prerequisite to most seed orchard allozyme studies is the genotyping of orchard pedigrees, typically clones, at selected enzyme systems. Such genotyping may be perceived as being indirectly useful. However, the ability to unambiguously differentiate between pedigrees, using single or multilocus genotypes, has direct utility as well, most notably for checking genetic integrity of orchard stock.

Early studies demonstrated that most conifers possess enough genetic variability to fingerprint virtually every pedigree in an orchard (Hunter 1977; Conkle and Adams 1977; Adams and Joly 1980). Subsequent surveys to confirm clonal identification have generally revealed that mislabeling of orchard ramets is relatively common. Adams (1983), using allozyme analyses of both seed and needle tissue, found that 2—13% of the ramets assayed in a Douglas-fir (*Pseudotsuga menziesii*) orchard were mislabeled. Similarly, Harju and Muona (1989) found 7—10% mislabeled ramets in two. Scots pine (*Pinus sylvestris*) orchards. In a survey of two loblolly pine (*Pinus taeda*) orchards, we identified mislabeled ramets in about 10% of the clones evaluated (Wheeler and Jech, unpub. data). One instance was particularly noteworthy. We surveyed 20 ramets of a highly valued loblolly pine clone located in two separate orchards (different organizations) and found four distinct multilocus genotypes. Which clone of the four was used for the original progeny testing is yet to be determined, but doing so should be relatively easy. A similar case occurred in a Douglas-fir orchard, but with an interesting twist. Genetic test results from two separate trials gave conflicting breeding values for one of the orchard clones. A survey of all the ramets of that clone revealed 3 of 11 were mislabeled, probably at time of grafting. Breeding for the two tests took place on separate ramets, one of which was mislabeled.

Given what appears to be a ubiquitous problem in seed orchards, we recommend that orchard managers consider pedigree validation as a routine function at time of orchard establishment, or at least prior to orchard production and *definitely* before breeding! The costs of mislabeling can be significant and will compound as improvement programs move through generations.

Another proposed use of allozymes is for identification and certification of bulk orchard or individual family seedlots (Adams 1983; Cheliak and others 1987b). This is particularly of concern in certain European countries where genetic improvement and deployment issues are legislated (Müller-Starck 1986, 1987). While it is relatively easy to identify individual progeny arrays from a given cross or open-pollinated family (Adams and others 1988) or bulked clone cultivars (Bergmann 1987), identifying the origin of multi-family seedlots from entire orchards poses difficulties. Problems are compounded as the number of pedigrees in an orchard increases and the contribution of those genotypes varies (i.e., different number of ramets, differential fecundity, annual variation, etc.). Adams (1983) suggested that the most effective means of identifying orchard seedlots was by using arrays of alleles that occur in low frequency. The expectation is that every orchard, or orchard block, will possess its own unique set of rare alleles and their presence in progeny arrays would be diagnostic. Muller-Starck (1986, 1987) has used a combination of analytical procedures, including measures of genetic distance and population differentiation, for discriminating between provenances and seed orchards, based on gene frequency data. He reports reasonable success. However, given the relatively modest differences that exist in gene frequencies among natural and artificial populations of most conifers, the discretionary powers of most analytical methods cited here must be considered, generally, as inadequate.

Supplemental mass pollination (SMP) verification (success)

Operational SMP in wind-pollinated conifer seed orchards has gradually increased since Woessner and Franklin (1973) touted its value for increasing gain and filled seed yields. Although a number of studies of SMP efficacy have been conducted on orchard-sized trees (reviewed by Bridgwater and others 1990), most have relied on filled seed counts, cone survival, or colored pollen as indicators of SMP success. Such measures are not directly equated with fertilization success of supplementally applied pollen, and may be quite misleading. Direct determination of the paternal contribution to fertilized seed is required to measure SMP success with certainty. Electrophoretic procedures are ideally suited to the purpose (Wheeler and Jech 1985).

The manner in which allozymes have been used to quantify SMP success has varied. Joly and Adams (1983) predicted the proportion of hybrid seed produced by SMP of 6 pitch pine (*Pinus rigida*) clones with loblolly pine clonal *pollen mixes* by using a single-locus marker common to loblolly pine but rarely found in pitch pine. By knowing the frequency of the marker in the pollen mix and the embryo gene pool, they were able to estimate the frequency of successful SMP to be approximately 11%.

Ideally, the pollen used in SMP studies should be unambiguously marked, either by a single locus, homozygous for an allele occurring rarely in the orchard studied, or by a multilocus marker unique to the pollen donor. Estimates of success, in such cases, can be obtained directly from the paternal gene frequencies of the fertilized embryos, although slight correction factors may be required if the markers used are not unique in the orchard (Yazdani and others 1986; Wheeler and Jech 1985). Markers found in the heterozygous condition can be used effectively by simply multiplying by two, the frequency of the marker found in embryos. This approach was used in an operational SMP trial in loblolly pine where success varied from 48—69% (Blush 1987). Use of heterozygotic markers assumes that pollen gametes carrying alternative alleles contribute equally to the seed pool; this assumption may not always be met (Adams and Joly 1980).

We found both single-locus and multilocus markers to be particularly effective in evaluating factors that influence operational SMP success including selection of optimal pollination devices, identifying optimal timing and number of applications, extent of pollen storage and dilution effects, clonal effects, etc. (Wheeler and Jech 1985, 1988, unpub. data). In our studies, reproductive phenology of the receptor clone had the greatest influence on SMP success. SMP success on clones that flowered early, before the majority of pollen within the orchard released, averaged 78%

(Table 1), while success rates for intermediate phenology clones were significantly less, with a range of 25 to 60% (Wheeler and Jech 1988). Success was determined by simply counting seed embryos that carried the marker genotypes.

Selection

Over the past ten years, considerable attention has been directed at evaluating the opportunity of using allozymes for indirect selection in forest trees. The most common relationships sought have been between single- or multilocus genotypes, or measures of heterozygosity and expressions of growth, fecundity and resistance to biotic or abiotic agents (Ledig and others 1983; Mitton and Grant 1984; Linhart and Mitton 1985; Bush and others 1987; Strauss and Libby 1987; Ducharev and others 1987; Geburek and others 1987; see Scholz and others 1989). Most of these studies have been performed on experimental populations, both natural and artificial, rather than populations originating in or from seed orchards. A few of the exceptions are noted below.

Müller-Starck and others (1983) and Muller-Starch and Ziehe (1984) investigated the effect of segregation at an LAP locus on reproductive selection in Scots pine. They concluded the locus had a selective function

Table 1. Supplemental mass pollination (SMP) success determined electrophoretically for seven early flowering Douglas-fir clones (age 20, whole-crown pollination; Wheeler and Jech 1988). Filled seed per cone (*FS/C*) are also noted, for treated and open-pollinated control trees

	Treatment			
	SMP		Control	
Parent	% Success[1]	*FS/C*	% Success	*FS/C*
1	73	63.7	0	52.9
2	85	35.7	0	11.0
3	74	51.4	0	40.9
4	66	43.9	0	16.5
5	93	48.1	7	17.3
6	60	13.4	0	5.2
7	92	50.6	0	27.1
Mean	78	43.8	1	24.4

[1] Success results based on a sample of 60 seed per condition per tree.

in sexual reproduction, and that selection acted asymmetrically in the female and male components of each clone.

Bongarten and others (1985) predicted the levels of isozyme heterozygosity in 178 full-sib crosses of Douglas-fir that were created from genotyped seed orchard parents and which were evaluated in replicated genetic tests. They found virtually no relationship between growth performance or stability of progenies and general levels of heterozygosity although a very modest negative correlation between frequency of rare alleles and family growth was observed. In the same study, fecundity of parents was contrasted directly with parental heterozygosity and allozyme genotypes. No relationships were found (Bongarten, unpubl. data). Jech and Wheeler (unpubl. data) attempted to further evaluate the relationship between rare alleles and poor growth and fecundity performance in their Douglas-fir orchards. The study was hampered by a disproportionate loss of clones with rare alleles due to roguing based on progeny test data. Seed crops of 24 clones possessing 28 rare alleles at 16 loci were screened in germination tests. At least 2 loci, IDH and UGP, demonstrated strong segregation distortion in speed and completeness of germination (Table 2). Rare alleles were under-represented in seed that germinated at normal speed, but over-represented in non-germinating seed. Distortion was significant only when the rare allele was contributed by the female parent. An extended study of progeny from crosses made amongst the same clones, and growing in genetic tests, showed that certain alleles (i.e., IDH-9) were virtually unrepresented at age eight in the field (Table 2). Altokhov and others (1983) have observed similarly strong selection against rare alleles in Scots pine. This type of information could be used to select against the inclusion of certain clones in production seed orchards, although in reality, the utility of such an approach is probably limited. Still, segregation distortion appears to be relatively common (Adams and Joly 1980; Altokhov and others 1986; Cheliak and Pitel 1984; Neale and others 1984; Strauss and Conkle 1986), and should be more thoroughly investigated.

The most direct application of isozymes for selection in seed orchards is probably that of Copes (1978). He used peroxidase and esterase loci activity differences (gene expression) to predict, with 90—100% accuracy, whether Douglas-fir graft unions would be compatible or not. Chung (1984) has used allozymes to determine whether seedlings produced in seed orchards containing pitch pine and loblolly pine were hybrids.

While use of allozymes shows promise for indirect or direct selection in seed orchards, at present there are few, if any, tree improvement programs actively doing so. The greatest promise for allozyme selection will probably arise from linkage associations between isozyme loci and Mendelian traits such as disease, insect or pollution resistance.

Table 2. Segregation distortions for rare alleles at two loci, IDH and UGP, in controlled-crosses of Douglas-fir

Cross (Female × Male)	Germination Speed	*N*	Locus (alleles)			P^1
			(2)	IDH	(9)	
78 × 55	Fast	124	81.4		18.6	**
	Slow	254	67.7		32.3	**
	Non-germ	100	28.0		72.0	**
55 × 78	Fast	72	45.8		54.2	ns
	Slow	125	62.4		37.6	**
78 × 54	Fast	100	78.0		22.0	**
	Slow	20	60.0		40.0	ns
	Non-germ	10	30.0		70.0	ns
54 × 78	Fast	60	46.7		53.3	ns
	Slow	15	33.3		66.6	*
	Non-germ	75	44.0		56.0	ns
			(1)	IDH	(9)	
68 × 94	Fast	25	72.0		28.0	*
	Slow	25	52.0		48.0	ns
	Non-germ	120	45.0		55.0	ns
			(1)	UGP	(4)	
68 × 94	Fast	25	76.0		24.0	**
	Slow	25	80.0		20.0	**
	Non-germ	120	29.2		70.8	**
68 × 94 (8-year-old test trees)	Progeny vegetative bud sample	41	(1)	IDH	(9)	
			97.6		2.4	**
			(1)	UGP	(4)	
			73.2		26.8	**

[1] ** = Proportions followed by ** are statistically significant at 0.01 level;
* = 0.05 level; and
ns = non-significant, based on Chi-square analysis with 1:1 expectation.
Clones 68 and 78 are heterozygous for the rare IDH 9 allele; clone 68 is heterozygous for the rare UGP 4 allele. The segregation of IDH alleles for these 2 clones only are shown.

Orchard genetic efficiency

The extent to which the genetic potential of wind-pollinated seed orchards is realized has been referred to as an orchard's "efficiency" (Woessner and

Franklin 1973), or more recently, "genetic efficiency" (Adams and Joly 1980; El-Kassaby and others 1984). To obtain full genetic efficiency in an orchard, several conditions or assumptions must be met (Eriksson and others 1973; Woessner and Franklin 1973):

(1) pollen contamination from outside orchard sources is negligible,
(2) all clones (families) contribute equally to the gene pool (this assumes synchrony of flowering, similar fecundity, and uniform pollen distribution),
(3) self-fertilization or related matings are negligible, and
(4) all possible crosses are equally compatible.

In short, if orchard seed is to reflect its theoretical expectation of genetic gain, the orchard itself must represent a nearly perfect, closed, panmictic population. Electrophoretic studies have demonstrated that most of these assumptions are violated, to some extent or another, in conifer seed orchards. Each of the points above are briefly reviewed, in the general order of their importance to deviation from panmixis.

Contamination

The influx of unimproved pollen sources into seed orchards has long been a concern to orchard managers but until recently, the extent to which contamination occurred remained unknown. The introduction of electrophoretic techniques provided a definitive procedure for estimating pollen contamination by using genetic markers to evaluate paternal contribution to the orchard seed pool. A number of analytical procedures have been developed to use paternity information for estimating contamination. Depending on the procedure used, the genetic information required to estimate contamination is usually some combination of orchard parental, seed pool and/or outside population gene and genotype frequencies.

The simplest approach is the "unique marker" method which is dependent on the presence of a unique genetic marker (usually a single locus) in populations outside the seed orchard, but not in the orchard itself. By knowing the frequency of the unique allele in the natural population and in the orchard seed pool, contamination estimates can be made (Friedman and Adams 1981). The use of this procedure is constrained by the need for unique alleles in background pollen pools, their typically low frequency of occurrence, and the possibility they are under-represented in the seed pool (see the section on "Selection" above).

In situations where gene frequencies in orchards and surrounding stands are quite distinct, multi-locus procedures based strictly on gene frequency differences among orchard parents and orchard seed pools can be used to estimate migration of contaminant pollen into the orchard (El-Kassaby and Ritland 1986). However, these conditions are rarely met in conifer species and furthermore, assumptions regarding effective orchard paternal gene frequencies need to be made.

Paternity exclusion techniques have proven very effective in determining contamination rates (reviewed in Lowe and Wheeler 1990). With these procedures, the potential pollen gamete genotypes that can be produced by orchard parents are determined on the basis of clonal genotypes. These are checked against the paternal genotypes of outcrossed seed from the orchard crop. Seed which could not have been sired by orchard pollen are listed as unambiguous contaminants. The fewer the number of clones (pedigrees) in the orchard and the greater the number of polymorphic loci used, the greater the ability to identify contaminants. In reality, all contaminants cannot be identified because of the overlap in genotype frequencies between orchard and outside populations. Thus, straight exclusion techniques give only a minimum estimation of contamination. Actual contamination rates are probably much higher (Harju and Muona 1989). Smith and Adams (1983) developed a procedure that theoretically accounted for the shortcomings of the "paternity exclusion" procedure by estimating a *probability of detection* of contaminant genotypes. This probability is based on information regarding gene and genotype frequencies in surrounding stands and the orchard itself. This analytical procedure is probably the most applicable of those available today. However, a new Expectation-Maximization (EM) procedure that simultaneously estimates outcrossing and contamination holds promise (Yeh, pers. comm.).

The overwhelming evidence of contamination studies performed to date is that migration rates of unimproved pollen into seed orchards is distressingly high (Table 3). Rates ranging from 21 to 75% in pine and 1 to 89% in Douglas-fir have been observed. Clearly, contamination represents the single greatest deterrent to meeting genetic expectation in conifer seed orchards. The routine use of electrophoretic procedures and/or pollen trapping techniques to characterize contamination dynamics in new and old orchards is encouraged (Lowe and Wheeler 1990).

Mating systems

Perhaps more attention has been given to this category of study than any other yet discussed. Investigations have focused on three general areas of interest: genetic representation in orchard seed pool, rates of selfing and outcrossing, and patterns of mating within the orchard. Because of the large number of studies in this area, we have chosen to cite a representa-

Table 3. Estimates of pollen contamination in conifer seed orchards based on electrophoretic data

Species	Est. procedure	% Contamination	Author
Pinus taeda	Marker Group	28	Friedman and Adams 1981
	Pater. Exclusion	36	Friedman and Adams 1985
Pinus	Pater. Exclusion (min)	21—36	El-Kassaby and others 1989
sylvestris	Pater. Exclusion	17—39	Harju and Muona 1989
	Pater. Exclusion (min)	38	Nagasaka and Szmidt 1985
Pseudotsuga	Gene Frequency	0.2	El-Kassaby and Ritland 1986
menziesii	Pater. Exclusion	29—52	Smith and Adams 1983
	Pater. Exclusion	4—25	Wheeler and Jech 1985
	Pater. Exclusion	44—89	Fast and others 1986

Analytical procedures are noted.
Min refers to a minimum estimate only.

tive few, and encourage interested readers to pursue the literature on their own.

Geneticists have long been concerned that restricted population sizes inherent to seed orchards could result in significant declines in genetic variability, relative to natural populations. At least with respect to genetic variation at the allozyme level, this concern does *not* generally appear to be supported. That is, seed orchards retain nearly as much genetic diversity as natural populations, even when population sizes are as low as 25—30 clones (Adams 1981b; Szmidt 1987; Muona and Harju 1989). Such comparisons are generally made using estimates of observed and expected heterozygosity, and frequencies of rare alleles. In fact, in at least one instance, observed heterozygosity of a seed crop from a subset of clones growing in an orchard was actually greater than it was for the seed crop from the ortets of those clones growing in natural stands (Rudin and others 1986). To be sure, orchard restriction in population size can result in loss of rare alleles (Muona and Harju 1989; Wheeler and Jech, unpubl. data), but this seems to have little effect on overall genic diversity.

Another concern of forest geneticists is that the orchard seed crop does not represent the ideal panmictic orchard parental mix. Empirical and model based studies based on measures of fecundity and phenological synchrony among clones suggest these concerns are well-founded. Electrophoretic studies have shown that gene frequencies in parental and progeny gene pools are typically quite similar but in some cases, non-random, sub-population structure within the orchard crop may occur due to those factors noted above (Muller-Starck and others 1983; El-Kassaby and others 1988; Barrett and others 1987).

The greatest emphasis in studies of mating systems in forest trees has been on estimation of selfing (s) and outcrossing (t) rates (reviewed in Adams and Birkes 1989). Intense interest in this topic has led to considerable refinement in analytical procedures to deal with electrophoretic data (Muller-Starck 1976; Shaw and Allard 1982; Cheliak and others 1983; Neale and Adams 1985; Ritland and El-Kassaby 1985). Most recent papers have reported estimates based on both single- and multi-locus estimation procedures.

With the notable exception of a black spruce (*Picea mariana*) orchard in Canada (Barrett and others 1987; $t = 0.84$), outcrossing rates in conifer seed orchards appear to be consistently greater, on average, than 0.90 (range = 0.84—0.99, reviewed in Adams and Birkes 1989). Where comparisons have been made, outcrossing rates appear to be higher in orchards than in natural populations of the same species (Shaw and Allard 1982; Rudin and others 1986; Muona and Harju 1989) or even the same clones (Szmidt 1987). This is counter to early speculation that inbreeding in orchards would be greater than natural stands because of the large number of ramets of individual clones in close proximity within the orchard. It is now typically felt that inbreeding is greater in natural populations because of family structure (Shaw and Allard 1982).

For most species studied, outcrossing estimates among individual clones are quite variable, ranging from 0.5 to 1.0 (Shaw and Allard 1982; Omi and Adams 1986; Moran and others 1980; Erickson 1987). This information is of considerable practical value; orchard managers can tailor harvest regimes or supplemental mass pollination activities based on estimates of self-fertility. Mating parameter studies have also been used to evaluate orchard management techniques such as top pruning (Omi and Adams 1986), bloom delay (El-Kassaby and others 1986), and flower stimulation (Wheeler and Jech unpubl. data), and effects of biological variables such as crown position (Shen and others 1981; El-Kassaby and others 1986; Adams and Birkes 1989) and reproductive phenology (El-Kassaby and others 1988) on mating systems.

Study of out-cross fertilization patterns within seed orchards has been the most recent focus of mating system studies. As with previous topics of interest, there has been a considerable investment in development of analytical procedures to better evaluate existing electrophoretic data (reviewed in Adams and Birkes 1989; Wheeler and others 1990). Briefly, analytical procedures can be assigned to one of the following approaches:

1. rare marker,
2. paternity analysis, or
3. model.

The first approach relies on the tracking of pollen distributed from a single clone possessing a rare allele (Müller-Starck 1976; Erickson 1987). The procedure is somewhat cumbersome and is constrained by the need for the presence of rare alleles. The second approach uses multilocus markers to assign paternity to offspring of individual mother trees (Neale 1984; Devlin and others 1988). This approach is considerably more flexible and efficient in terms of information gained per unit of research effort. The power of the technique is enhanced as the number of variable loci increases and the number of clones in an orchard decreases. In reality, it is generally impossible to unambiguously assign all seed to a given parent and because paternity must often be inferred, estimation biases may occur. The model approach estimates mating parameters directly from a probability model rather than indirectly from inferred parental pairs (Adams and Birkes 1989), thus eliminating the estimation bias associated with paternity analyses. Schoen and Stewart (1986) applied a mating model to a white spruce (*Picea glauca*) seed orchard and determined that average paternal contribution of the clones studied varied from near 0 to over 14% across several female receptors. Adams and Birkes (1989) subsequently described a variation on the model approach based on the concept of neighborhoods. This model simultaneously calculates a parameter estimate for s, m (pollen migration or contamination) and d (a distance function, between paternal and maternal parents).

In general, studies of pollen distribution into and within seed orchards reveal that gene flow is under the influence of several, often interacting, environmental and biological factors (Wheeler and others 1990). Two recurring although seemingly paradoxical conclusions regarding gene flow in orchards are

1. likelihood of gene flow between two individuals is greatest when they are in close proximity and phenological synchrony, and
2. long distance pollen flight often accounts for the majority of successful fertilizations for a given female.

Differential reproductive success

The last assumption pertaining to orchard genetic efficiency is that all matings are equally compatible. We refer to this factor as differential reproductive success (DRS) and distinguish it from differential parental contributions which is more a function of fecundity and phenology. DRS is of practical significance to the tree breeder because progeny from polycross mating designs may not represent all pollen parents equally, thereby biasing estimates of genetic parameters and breeding values

(Moran and Griffin 1985). In addition to the implied effects to orchard efficiency, DRS is of practical significance to the orchard manager because it could influence successful implementation of supplemental mass pollination (Apsit and others 1989).

In conifers, two approaches to the study of DRS have been used. The first employed controlled crosses using polymixes of three to nine pollen parents and electrophoretic procedures were used to assign paternity to each control-crossed seed (Moran and Griffin 1985; Cheliak and others 1987a; Wiselogel and van Buijtenen 1988). With the exception of the latter of these studies, results of these investigations suggest that certain pollen parents were more successful at fertilizing embryos than were others, although their results were somewhat compromised by the inability to unambiguously identify paternity for every seed.

A second group of experiments was based on series of 2 pollen parent mixes crossed to arrays of females and replicated over years (O'Malley and Wheeler 1986; Apsit and others 1989; Nakamura and Wheeler 1990). These studies concluded that:

- differences in male reproductive success, based on proportion of ovules fertilized under competitive conditions, is genetically determined and repeatable over years,
- selection among pollen parents occurs post-zygotically, and
- male reproductive success is a function of female genotype (i.e., there is strong male-female complementarity; Table 4).

El-Kassaby and Ritland (unpubl. data) note that studies of pollen mixes of variable proportions in Douglas-fir suggest a presence of frequency-dependent male reproductive success in polycrosses.

Table 4. Proportion of seed sired by tester pollen parents when pollen was mixed 1:1, by weight, with a marker pollen parent homozygous for an IDH allele

Paternal parent		Proportion of seed sired by tester pollen for seed parent				
Marker	Tester	25	26	27	28	29
2	1	0.44	0.39	0.42	0.47	0.40
2	2	0.41	0.45	0.48	0.50	0.51
2	3	0.60	0.65	0.66	0.60	0.65
2	4	0.47	0.48	0.62	0.70	0.54

All tester and female parents were homozygous for an alternative allele.

[1] Excerpted from Apsit, Nakamura and Wheeler 1989, TAG 77: 681–684.

Future directions

With few exceptions, electrophoretic markers are better suited to seed orchard studies requiring knowledge of paternity than to any other area of forestry research. Their use in this context should expand and be considered routine orchard management practice for quality control for aggressive tree improvement programs. In particular, markers should be used for:

1. Contamination — For most programs there is no greater threat to genetic gain erosion than through pollen contamination. A thorough understanding of the process and means of reducing it is necessary.
2. SMP — This highly regarded, but still little used, technology can only develop with a good understanding of process efficiency obtained through paternity analyses.
3. Mating patterns — The least well-understood factor in wind-pollinated orchards is the dynamics of pollen movement within an orchard. Studies such as those defined by Adams and Birkes (1989) are critical to extending our knowledge base, and should be given priority in the near future.
4. Selection — Efforts to identify single or multilocus genotypes associated with resistance to biotic or abiotic factors must continue.

It should be noted here that some of the applications mentioned in this manuscript could be equally-well or better addressed using new and developing molecular genetic technologies (Bernatsky and Tanksley 1989), although constraints of labor and material costs may make such applications infrequent. Use of molecular markers in seed orchard research was treated independently elsewhere in this proceedings.

References

Adams, W. T. 1981a. Applying isozyme analyses in tree-breeding programs. In: M. T. Conkle (Ed), Proc. Symp. Isozymes North Am. Forest Trees and Forest Insects, Berkeley, CA, USDA Forest Serv. Gen. Tech. Rep. PSW-48, pp. 60—64.

——. 1981b. Population genetics and gene conservation in Pacific Northwest conifers. In: Scudder, G. G. E. and Reveal, J. L. (Eds) Evolution Today, Proc. Sec. Int. Congr. System. and Evol. Biol., Hunt Institute for Botanical Documentation, Pittsburg, pp. 401—415.

——. 1983. Application of isozymes in tree breeding. In: Tanskley S. D. and Orton, T. J. (Eds) Isozymes in Plant Genetics and Breeding. Elsevier Science Publishers, Amsterdam, Part A, pp. 381—400.

Adams, W. T. and Birkes, D. S. 1989. Estimating mating patterns in forest tree populations.

In: Proc. International Workshop on plant biology, biochemical markers in population genetics of forest trees. Inst. for Agroforestry of the National Res. Council of Italy (CNR). Porano-Orvieto, Italy, Oct. 1988.

Adams, W. T. and Joly, R. J. 1980. Allozyme studies in loblolly pine seed orchards: clonal variation and frequency of progeny due to self-fertilization. Silvae Genetica 29(1): 1—4.

Adams, W. T., Neale, D. B. and Loopstra, C. A. 1988. Verifying control crosses in conifer tree improvement programs. Silvae Genetica 37(3—4): 147—152.

Altukov, Yv. P., Dukharev, V. A. and Zhirotovskii, L. A. 1983. Selection against rare electrophoretic protein variants and the rate of spontaneous mutability in populations. Genetika 19(2): 264—276.

Altukov, Yv. P., Krutovskii, K. V., Gafarou, N. I., Dukharev, V. A., and Morozov, G. P. 1986. Allozyme variability in a natural population of Norway Spruce (*Picea abies* (L.) Karst). I. Polymorphism systems and mechanisms of their genetic control. Genetika 22(8): 2135—2151.

Apsit, V. J., Nakamura, R. R. and Wheeler, N. C. 1989. Differential male reproductive success in Douglas-fir. Theor. Appl. Genet. 77: 681—684.

Barrett, J. D., Knowles, P. H. and Cheliak, W. M. 1987. The mating system in a black spruce clonal seed orchard. Can. J. For. Res. 17(5): 379—382.

Bergmann, F. 1987. Characterization of multiclonal aspen cultivars using isozyme electrophoresis. For. Ecol. and Manage. 22(1—2): 167—172.

Bernatsky, R. and Tanksley, S. D. 1989. Restriction fragments as molecular markers for germ-plasm analysis and utilisation. In: Brown A. D. H., Marshall D. R., Frankel O. H. and Williams J. T. (Eds) The Use of Plant Genetic Resources. Cambridge University Press, Cambridge, 353—362.

Blush, T. 1987. An operational trial of supplemental mass pollination in a loblolly pine seed orchard. In: Proc. 20th South. For. Tree Improve. Conf., Charleston, SC.

Bridgwater, F., Blush, T., and Wheeler, N. C. 1990. Supplemental mass pollination. In: Proc. Pollen management workshop, Southern Res. Inf. Exchange Group Meeting, Macon, Georgia, July 18—19.

Bongarten, B. C., Wheeler, N. C. and Jech, K. S. 1985. Isozyme heterozygosity as a selection criterion for yield improvement in Douglas-fir. In: New ways in forest genetics. Proc. 19th Can. Tree Improv. Assoc., Quebec City, Quebec, Canada, pp. 121—127.

Bush, R. M., Smouse, P. E. and Ledig, F. T. 1987. The fitness consequences of mutiple-locus heterozygosity: the relationship between heterozygosity and growth rate in pitch pine (*Pinus ridida* Mill.). Evolution 41: 787—798.

Cheliak, W. M., Morgan, K. , Strobeck, C., Yeh, F. C. H. and Dancik, B. P. 1983. Estimation of mating system parameters in plant populations using the EM algorithm. Theor. Appli. Genet. 65: 157—161.

Cheliak, W. M. and Pitel, J. A. 1984. Genetic control of allozyme variants in mature tissues of white spruce trees. J. of Heredity 75: 34—40.

Cheliak, W. M., Skroppa, T. and Pitel, J. A. 1987a. Genetics of the polycross. I. Experimental results from Norway spruce. Theor. Appli. Genet. 73: 321—329.

Cheliak, W. M, Yeh, F. C. H. and Pitel, J. A. 1987b. Use of electrophoresis in tree improvement programs. For. Chron. 63: 89—96.

Chung, M. S. 1984. Allozyme variation of *Pinus rigida* Mill. in an F1-hybrid seed orchard and estimation of the proportion of F1-hybrid seeds by allozyme analysis. J. Korean For. Society 66: 109—117.

Conkle, M. T. 1971. Inheritance of alcohol dehydrogenase and leucine aminopeptidase isozymes in knobcone pine. For. Science 17(2): 190—194.

Conkle, M. T. 1972. Analyzing genetic diversity in conifers — isozyme resolution by starch gel electrophoresis. USDA For. Serv. Res. Note, PSW-264 pp. 1—5.

Conkle, M. T. and Adams, W. T. 1977. Use of isozyme techniques in forest genetic research. In: Proc. 14th South. For. Tree Improve. Conf. Gainesville, FL. pp. 219—226.

Copes, D. L. 1978. Isoenzyme activities differ in compatible and incompatible Douglas-fir graft unions. For. Science 24: 297—303.

Devlin, B., Roeder, K. and Ellstrand, N. C. 1988. Fractional paternity assignment: theoretical development and comparison to other methods. Theor. Appli. Genet. 76: 369—380.

Ducharev, V. A., Romanovskij, H.G. and Rjabokon, S.M. 1987. [Heterozygosity and seed production in Scots pine. Lesovedenije 22(2): 87—89.

El-Kassaby, Y. A., Fashler, A. M. K. and Sziklai, O. 1984. Reproductive phenology and its impact on genetically improved seed production in a Douglas-fir seed orchard. Silvae Genet. 33: 120—125.

El-Kassaby, Y. A., Parkinsson, J. and Devitt, W. J. B. 1986. The effect of crown segment on the mating system in a Douglas-fir (*Pseudotsuga menziesii* (Mirb.) Franco) seed orchard. Silvae Genetica 35: 149—155.

El-Kassaby, Y. A. and Ritland, K. 1986. Low levels of pollen contamination in a Douglas-fir seed orchard as detected by allozyme markers. Silvae Genetica 35: (5—6)224—229.

El-Kassaby, Y. A., Ritland, K., Fashler, A. M. K. and Devitt, W. J. B. 1988. The role of reproductive phenology upon the mating system of a Douglas-fir seed orchard. Silvae Genetica 37(2): 76—82.

El-Kassaby, Y. A., Rudin, D., and Yazdani, R. 1989. Levels of outcrossing and contamination in two *Pinus sylvestris* L. seed orchards in northern Sweden. Scand. J. For. Res. 4: 41—49.

Erickson, V. J. 1987. The influence of distance and floral phenology on pollen gene flow and mating system patterns in a coastal Douglas-fir seed orchard. M.S. Thesis, Oregon State Univ., Corvallis, OR, USA.

Eriksson, G., Jonsson, A. and Lindgren, D. 1973. Flowering in a clone trial of *Picea abies* Karst. Studia Forestalia Suecica Nr 110, Royal College of Forestry, Stockholm.

Fast, W., Dancik, B. P. and Bower, R. C. 1986. Mating system and pollen contamination in a Douglas-fir clone bank. Can. J. For. Res. 16: 1314—1319.

Feret, P. P. and Bergmann, F. 1976. Gel electrophoresis of proteins and enzymes, pp. 49—77. In: Miksche J. P. (Ed) Modern methods in forest genetics. Springer Verlag, Hamburg.

Friedman, S. T. and Adams, W. T. 1981. Genetic efficiency in loblolly pine seed orchards, pp. 213—224. In: Proc. 16th South Forest Tree Improve. conf., Blacksburg, VA.

Friedman, S. T. and Adams, W. T. 1985. Estimation of gene flow into two seed orchards of loblolly pine (*Pinus taeda* L.). Theor. Appli. Genet. 69: 609—615.

Geburek, T., Scholz, F., Knabe, W. and Vornweg, A. 1987. Genetic studies by isozyme gene loci on tolerance and sensitivity in an air polluted *Pinus sylvestris* field trial. Silvae Genetica 36(2): 49—53.

Harju, A. and Muona, O. 1989. Background pollination in *Pinus sylvestris* L. seed orchards. Scand. J. For. Res. 4(4): 513—520.

Hunter, S. C. 1977. An electrophoretic analysis of isoenzyme variation in a Piedmont loblolly pine seed orchard. M.S. Thesis, North Carolina State Univ., Raleigh, 48 pp.

Joly, R. J. and Adams, W. T. 1983. Allozyme analysis of Pitch X Loblolly pine hybrids produced by supplemental mass pollination. For. Science 29(2): 423—432.

Ledig, F. T., Guries, R. P. and Bonefeld, B. A. 1983. The relation of growth to heterozygosity in pitch pine. Evolution 37(6): 1227—1238.

Linhart, Y. B. and Mitton, J. B. 1985. Relationships among reproduction, growth rates, and protein heterozygosity in ponderosa pine. Amer. J. Bot. 72(2): 181—184.

Lowe, W. and Wheeler, N. C. 1990. Pollen contamination in seed orchards. In: Proc. Pollen management workshop, Southern Res. Inf. Exchange Group Meetings, Macon, Georgia.

Mitton, J. B. and Grant, M. C. 1984. Associations among protein heterozygosity, growth rate, and developmental homeostasis. Ann. Rev. of Ecol. Syst. 15: 479—499.

Moran, G. F. and Griffin, A. R. 1985. Non-random contribution of pollen in polycrosses of *Pinus radiata* D. Don. Silvae Genetica 34: 117—121.

Moran, G. F., Bell, J. C. and Matheson, A. C. 1980. The genetic structure and levels of inbreeding in a *Pinus radiata* D. Don seed orchard. Silvae Genetica 29: 190—193.

Müller-Starck, G. 1976. A simple method of estimating rates of self-fertilization by analyzing isozymes in tree seeds. Silvae Genetica 25: 15—17.

——. 1986. Genetic means of verifying observance of the law. II. Genetic characterization of orchard produced seeds. In: Muhs, H. J. (Ed) Biochemical genetics and legislation of forest reproductive material. Mitteilungen der Bundesforschungsanstalt fur Forst- und Holzwirtschaft 154: 67—74.

——. 1987. Genetic differentiation among seed samples from provenances of *Pinus sylvestris* L. Silvae Genetica 36(5—6): 232—238.

Müller-Starck, G. and Ziehe, M. 1984. Reproductive systems in conifer seed orchards. 3. Female and male fitnesses of individual clones realized in seeds of *Pinus sylvestris* L. Theor. Appl. Genet. 69: 173—177.

Müller-Starck, G., Ziehe, M. and Hattemer, H. H. 1983. Reproductive systems in conifer seed orchards. 2. Reproductive selection monitored at an LAP gene locus in *Pinus sylvestris* L. Theor. Appli. Genet. 65: 309—316.

Muona, O. and Harju, A. 1989. Effective population sizes, genetic variability, and mating system in natural stands and seed orchards of *Pinus sylvestris*. Silvae Genetica 38(5—6): 221—228.

Nagasaka, K. and Szmidt, A. E. 1985. Multilocus analysis of external pollen contamination of a Scots pine (*Pinus sylvestris* L.) seed orchard. In: Gregorius, H. R. (Ed) Population genetics in Forestry. Lecture notes in Biomathematics 60: 134—138.

Nakamura, R. and Wheeler, N. C. 1990. Pollen competition and paternal success in Douglas-fir. Evolution (in press).

Neale, D. B. 1984. Population genetic structure of the Douglas-fir shelterwood regeneration system in southwest Oregon. Ph.D. Thesis, Oregon State Univ., Corvallis, OR, USA.

Neale, D. B. and Adams, W. T. 1985. The mating system in natural and shelterwood stands of Douglas-fir. Theor. Appli. Genet. 71: 201—207.

Neale, D. B., Weber, J. C. and Adams, W. T. 1984. Inheritance of needle tissue isozymes in Douglas-fir. Can J. Gen. Cyto. 26: 459—468.

O'Malley, D. and Wheeler, N. C. 1986. Differential pollination success of two-pollen parent controlled crosses in Douglas-fir. In: Proc. IUFRO Joint Meeting of working parties on Breeding Theory, Progeny Testing and Seed Orchards, Williamsburg, Virginia (Abstract).

Omi, S. K. and Adams, W. T. 1986. Variation in seed set and proportions of outcrossed progeny with clones, crown position, and top pruning in a Douglas-fir seed orchard. Can. J. For. Res. 16(3): 502—507.

Paule, L. 1990. Bibliography: Isozymes and forest trees (1968—1989). Sveriges Lantbruks-universitet, Institutionen for skoglig genetik och vaxtfysiologi Rapport —9.

Ritland, K. and El-Kassaby, Y. A. 1985. The nature of inbreeding in a seed orchard of Douglas fir as shown by an efficient multi-locus model. Theor. Appl. Genet. 71: 375—384.

Rudin, D. 1986. Developmental trends in the field of biochemical genetics of forest trees. In: Proceedings 18th IUFRO World Congress, Division 2, Vol. 2, pp. 577—588.

Rudin, D. and Lindgren, D. 1977. Isozyme studies in seed orchards. Studia Forestalia Suecica 139: 1—23.

Rudin, D., Muona, O. and Yazdani, R. 1986. Comparison of the mating system of *Pinus sylvestris* in natural stands and seed orchards. Hereditas 104: 15—19.

Schoen, D. J. and Stewart, S. C. 1986. Variation in male reproductive investment and male reproductive success in white spruce. Evolution 40(6): 1109—1120.

Scholz, F., Gregorius, H. R. and Rudin, D. (Eds) 1989. Genetic Effects of Air Pollutants in Forest Tree Populations. Springer Verlag, Berlin, Heidelberg.

Shaw, D. V. and Allard, R. W. 1982. Estimation of outcrossing rate in Douglas-fir using isozyme markers. Theor. Appl. Genet. 62: 113—120.

Shen, X. H., Rudin, D. and Lindgren, D. 1981. Study of pollination patterns in a Scots pine seed orchard by means of isozyme analysis. Silvae Genetica 30(1): 7—15.

Smith, D. B. and Adams, W. T. 1983. Measuring of pollen contamination in clonal seed orchards with the aid of genetic markers. In: Proc. 17th South. For. Tree Improv. Conf., 64—73.

Strauss, S. H. and Conkle, M. T. 1986. Segregation, linkage and diversity of allozymes in Knobcone pine. Theor. Appl. Genet. 72: 483—493.

Strauss, S. H. and Libby, W. J. 1987. Allozyme heterosis in radiata pine is poorly explained by overdominance. American Naturalist 130(6): 879—890.

Szmidt, A. E. 1987. Genetic composition of seed orchard crops. For. Ecol. Manag. 19: 227—232.

Wheeler, N. C. and Jech, K. S. 1985. Estimating supplemental mass pollination (SMP) success electrophoretically. In: 19th Can. Tree Improv. Conf. pp.111—120.

Wheeler, N. C. and Jech, K. S. 1988. Supplemental mass pollination (SMP) in Douglas-fir seed orchards: Biological efficiency and economic evaluation. Weyer. Co. For. Res. Tech. Rep. # 050—3210/25. 31 pp.

Wheeler, N. C., Adams, W. T. and Hamrick, J. L. 1990. Pollen distribution in wind-pollinated seed orchards. In: Proc. Pollen Management Workshop, Southern Res. Inf. Exchange Group Meetings, Macon, Georgia.

Wiselogel, A. E. and van Buijtenen, J. P. 1988. Probability of equal mating in polymix pollinations by loblolly pine (*Pinus taeda* L.). Silvae Genetica 27: 129—134.

Woessner, R. H. and Franklin, E. C. 1973. Continued reliance on wind-pollinated southern pine seed orchards, is it reasonable? pp. 64—73. In: Proc. 12th South. For. Tree Improv. Conf.

Yazdani, R., Hadders, G. and Szmidt, A. E. 1986. Supplemental mass pollination in a seed orchard of *Pinus sylvestris* L. investigated by isozyme analyses. Scand. J. For. Res. 1(3): 309—315.

New Forests **6**: 329—345, 1992.

Review paper

Effect of forest management on gene pools

OUTI SAVOLAINEN and KATRI KÄRKKÄINEN
Department of Genetics, University of Oulu, SF-90570 Oulu, Finland

Received 25 October; accepted 13 May 1991

Key words: allozymes, forest management, genetic markers, natural regeneration, seed orchards

Application. Genetic consequences of various modes of forest regeneration are reviewed. Forest regeneration plans should take into consideration both genetic diversity and inbreeding levels. Many forest regeneration practices may influence adaptive quantitative variation in ways that are not reflected by allozyme variation.

Abstract. The influence of current forest management practices on gene pools is discussed. These forest management methods range from natural regeneration to replacing local species with exotics. Genetic changes caused by drift, alterations in the mating system, and directional selection are considered. A baseline is provided by information on genetic variation and demography of natural populations. While biochemical markers are useful for monitoring broad changes in genetic diversity or levels of inbreeding, changes of adaptive characters are likely to go undetected because of the low correlation in level and pattern of variation between biochemical and adaptive characters.

Introduction

Forest management has influenced gene pools of forest trees for long periods. Ancient people of the Middle East used the Cedar of Lebanon (*Cedrus libani* Loud.) for ship building and construction. They cut selectively, picking out the trees with straight trunks. As a result of millennia of dysgenic selection, they left behind only small areas of cedar forests with crooked stems and long branches (Beekeman 1949; Makkonen 1967). In Sweden and Finland, early harvesters of Scots pine (*Pinus sylvestris* L.) cut the tallest trees close to waterways for several centuries. The negative influence of these cutting habits are supposedly still seen along the most heavily used streams in Sweden (Lindquist 1946), but no scientific data are available. In Central Europe, replacing of native species and mixing of seed sources during the centuries has been so extensive that hardly any "natural" forests are left.

At present many tree species are domesticated, and more species are entering tree breeding and forest management programs. We should be well aware of the genetic changes that are caused by forestry practices (e.g., Ledig 1986). Some populations should be left in their natural state for ecosystem conservation and for future plant breeding needs. Tree breeding, considered as part of forest management (see Daniels 1984), by definition influences the genetic composition of populations. Some forest management practices may also result in unintentional changes in gene pools. Are we managing genetic resources better than the ancient Phoenicians did with their cedars? How is modern forestry practice influencing the genetic composition of populations?

The natural state of forest tree populations, their genetic structure and demography, forms a point of reference for managed populations. The natural populations of all forest trees share many important traits. Most trees have high outcrossing rates. Temperate wind-pollinated conifers have average outcrossing rates of 0.9. Some insect pollinated trees have lower outcrossing rates, as for example many eucalypts with an average of 0.7 (see Muona 1989, Table 1 for review).

The outcrossing rates were measured from mature seeds. However, in the adult stage the genotypic proportions in most cases have been found to be in Hardy-Weinberg proportions. Seed production is high, and only a very low proportion of the seed survives to the adult stage. A *Pinus sylvestris* stand may start from a seed cohort of one to two million seeds per hectare, whereas there are less than a thousand adult trees per hectare in a mature stand (Koski and Tallqvist 1978). Some demographic data on *Pinus sylvestris* are given in Table 1. During the early life stages there is heavy mortality. Much of the mortality is random, but intense selection may also take place. For example, most trees display heavy inbreeding depression (see Sedgley and Griffin 1989, p. 208). Inbreds are eliminated in early development, among much random mortality, for example in *Pinus sylvestris* (Yazdani et al. 1985). In this species, as in many others, there is extensive pollen flow (Koski 1970). Due to this pollen migration, and due to segregation, many individual seedlings will not be adapted to the local climate. Part of the selection probably eliminates these individuals. In experimental plantings with seedlings from local seed of *Pinus sylvestris*, Eriksson et al. (1980) observed 20—40% mortality in the first 20 years.

Natural populations may have genetic spatial structure, which would be eliminated by forest regeneration by sowing or planting. However, pollen and seed dispersal are so efficient in most conifers that family structure seems to be weak or nonexistent (e.g., Epperson and Allard 1989). Bird-dispersed pines may occur in clusters of relatives (Furnier et al. 1987), and animal pollinated species may have some family structure (Perry and Knowles 1991). Moran et al. (1989a) suggested that spatial genetic stucturing could explain the higher level of inbreeding in natural stands than in a seed orchard found in *Eucalyptus regnans* Mueller. Microenvironmental selection could also create spatial sturucture, but this is likely reflected in adaptive quantitative characters, usually not in allozymes (e.g., Campbell 1979).

Natural selection in the wild is expected to maximize total fitness, which consists of a viability and a fertility component. Forestry practices shift this goal: instead of maximizing seed yield, the aim is usually to maximize vegetative growth. In trying to achieve the goal, forestry practices interfere with various life stages. Selective cutting, either picking the best trees or leaving the best trees to reproduce, may differ from the effects of natural viability selection. Natural regeneration changes densities and may alter mating systems of reproducing populations. In artificial regeneration, seeds of a different origin may be used. Production of genetically improved material brings about many desired and maybe some other changes in the genetic composition of the new population. Thus forestry influences both the viability and fertility components of selection.

The intensity of forest management varies from natural regeneration to intensive industrial forestry. Forestry can cause many kinds of genetic changes. We shall consider whether forestry practices cause directional changes, whether drift due to small population sizes is important in causing decreases in diversity, and whether management practices cause alterations in the mating system that in turn influence the level of inbreeding. Forest management could also break microgeographic patterns of variation, where they exist. The aim is to evaluate critically what changes genetic markers, allozymes or restriction fragment length polymorphisms, will be able to detect, and where other kinds of methods must be employed. We start with the most natural forestry practices and then consider modern intensive industrial forestry.

Natural regeneration

Selective cutting

The ancient way to regenerate forest was to harvest the best trees and leave the rest to reproduce. The effects of this kind of management depend on the intensity of selection on parental trees (Zobel and Talbert 1984). Even if heritabilities for height growth are fairly low in the natural populations, long lasting intensive phenotypic selection may bring about genetic change, as in *Cedrus libani*.

In shelterwood and seed tree regeneration, a relatively few large trees are left to reproduce. Will this kind of stands maintain sufficient genetic variability, or does selection of big trees result in directional change in genetic composition? Table 2 shows that the level of genetic variability is little altered in shelterwood and seed tree stands (Neale 1985; Yazdani et al. 1985; Cheliak et al. 1988). Nor was evidence of directional changes in frequencies of genetic markers found. Thus, selection for size has not influenced allozyme frequencies systematically, and population sizes are not small enough for drift to have a measureable influence in one generation. From allozyme evidence, shelterwoods or seed tree stands seem to be a very natural way to regenerate forests. Forest fires have similar effects: after fires the biggest trees are most likely to remain alive and contribute progeny to the next generation (Wyant et al. 1986).

Effects of the density of reproducing trees

In seed tree and shelterwood stands the density is lower than in forests normally. If only a few trees are reproducing, the probability of inbreeding increases. The level of inbreeding can rise because of higher levels of selfing or matings between relatives.

Farris and Mitton (1984) found some evidence of higher levels of selfing in low density stands of *Pinus ponderosa* Laws. On the other hand, Neale (1985), Neale and Adams (1985a), and Yazdani et al. (1985) did not find any additional selfing in low density stands of *Pseudotsuga menziesii* (Mirb.) Franco or *Pinus sylvestris*. In *Pseudotsuga menziesii* (Neale and Adams 1985a), the differences between densities were large, but no sign of increased selfing was found (see Table 3). In animal pollinated plants, the density of populations could be more important, as Ellstrand et al. (1978) found in *Helianthus annuus*.

The level of inbreeding is also affected by the size of the trees. Both very small conifer trees (Neale and Adams 1985b) and very big ones (Shea 1987) can have a higher level of selfing than average size trees. In small trees this may be due to weak separation of male and female strobili (Neale and Adams 1985b), and in big trees due to their high production of pollen (Shea 1987). However, increased selfing is not a problem in natural stands because of the early selection against inbreds. Loehle and Namkoong (1987) have pointed out that selection of the biggest trees could also mean that seed yields are lowered, if there is a negative correlation between seed production and growth, as was found by Schmidtling

(1981). However, as long as a sufficient number of trees are left, such selection would be advantageous for future timber production.

Role of natural regeneration

Natural regeneration allows only minor improvement because of low heritabilities. However, this kind of regeneration will probably be used in many areas for a long time to come. In some areas artificial regeneration is too expensive, in others natural regeneration is the method that works best. Environmental considerations may also lead to this kind of regeneration (Mahlein and Hemstrom 1988). Of the total area of *Pinus sylvestris* regenerated in Finland, almost half is natural (Yearbook of Forest Statistics 1987).

Artificial regeneration: bulk seed

Origin of the seed source

In large clearcuts, where no shelterwood is left, regeneration must be artificial. Since seed orchards of many species are young, and their seed production insufficient, bulk seed lots from natural forests are commonly used (Zobel and Talbert 1984). Foresters have often used seed lots from outside the local area in order to improve yields, e.g., in northern Finland when regenerating *Pinus sylvestris*. When the biology of the species is not known sufficiently well, such transfers have resulted in failure of regeneration due to heavy mortality. Many wind pollinated conifers have large continuous populations, but they are highly differentiated in many characters. For example, *Pinus sylvestris* has steep clinal variation in growth characteristics. In common garden experiments with Finnish material (lat. 60—70 °N), Mikola (1982) found a two month range in bud-setting in the autumn. The correlation of median budset date with latitude of origin was −0.968. Similar latitudinal and altitudinal differentiation has been found in *Pseudotsuga menziesii* (Rehfeldt 1989) and *Pinus contorta* Loud. (Rehfeldt 1988). In transfer experiments such latitudinal and altitudinal differentiation is observed clearly. Even short range transfers of seedlings either to the north or to higher altitudes result in increased mortality in *Pinus sylvestris* (Eriksson et al. 1980). Provenance differentiation may also exist for other characters, e.g., disease resistance in *Pseudotsuga menziesii* (McDermott and Robinson 1989). Much of the quantitative variation is presumably adaptive in present environments, however, geographic variation may also be generated by historical patterns of migration (Lagercrantz and Ryman 1990).

In contrast to studies of adaptive traits, studies of biochemical markers have shown a near lack of differentiation between populations, with very low G_{ST} values, usually less than 10 % (Hamrick 1983; Muona 1989). A cline of acid phosphatase allozymes in *Picea abies* (L.) Karst is one of the rare exceptions (Bergmann 1978), but Tigerstedt (1973) found no differences between a northern and a southern population in the same species at the same locus. Some indication of edaphic differentiation was given by data of Mitton et al. (1989) in *Picea engelmannii* (Parry) Engelm. and Millar (1989) in *Pinus muricata* D. Don.

Common garden experiments show strong differentiation with respect to latitude or of origin in timing of growth or survival in *Pinus sylvestris* and several other species. Allozyme variability shows very low levels of differentiation. This poses a limit on the usefulness of allozymes in controlling seed origin.

Direct seeding

Regeneration by sowing is used for *Pinus sylvestris* in Finland on about 10% of the yearly regeneration (Yearbook of Forest Statistics 1987). The seed used for this purpose is normally bulk seed, from any natural stands or sometimes from selected seed collection stands. The number of seeds per hectare will be initially quite high, e.g., 50 000 (Table 1). This is, however, far lower than would obtain in a natural population, but does allow for heavy mortality in the early life stages. If the seed source was initially chosen correctly, the genotypic composition of the adult stand may resemble that of local stands. As more genetically improved seed becomes available, the use of sowing will probably decline.

Seedlings from bulk seed

We will here discuss general questions dealing with regenerating with seedlings, specifics of seed orchard seed will be dealt with later. Seedlings are produced in nurseries, under close to optimal conditions. These are then planted at a low density in regeneration areas (Table 1). Selection intensities in nurseries may be low, or the selection may differ from that in natural stands (see Kitzmiller 1990). The mortality in natural populations brings about genetic changes, e.g., elimination of inbreds. Some mortality is probably also due to lack of climatic adaptation in the seedlings (due to segregation and pollen migration). If such selection does not operate in nurseries, the result may be poor growth or heavy mortality in the field.

Table 1. Demography of natural and artificial regeneration of *Pinus sylvestris* in Finland

Life stage	Natural stands	Sowing	Planting
Seeds	2 000000	50000	
Seedlings (3 yr.)	50000		2500
Adult trees	<1000	<1000	<1000

(*n* of individuals/hectare).

This may result in stands that are too scarce. Such growing conditions will lead to decreased quality of the timber (see e.g., Barbour and Kellogg 1990).

A possible way to study selection in nurseries using allozymes is to monitor the elimination of inbreds, or to follow allelic frequency changes. We compared genetic changes that took place in a natural stand of *Pinus sylvestris* during the first years to those in a nursery (Muona et al. 1988), but found only small differences at marker loci. It would be important to monitor the fate of inbreds in nurseries in species where the level of selfing is considerable and inbreds in natural stands are eliminated in the early years, as e.g., *Eucalyptus regnans* (Moran et al. 1989a).

Regeneration with seed orchard seed

Seedlings from seed orchard seeds are the most extensively used genetically improved material for regeneration of coniferous forests. Wheeler (1992, this issue, pp. 311–328) has reviewed the use of allozymes in monitoring seed orchard function. Our perspective will deal with those genetic changes that are really a byproduct of the intended genetic improvement.

Genetic variability and inbreeding

Seed orchards should provide genetically variable seeds. Most of the effort in measuring genetic variability in any organism has centered on expected heterozygosity. Measured in this way, drift depletes genetic variability very slowly, at the rate of $1/2N_e$ per generation (where N_e is the effective population size). It is evident that seed orchards with even moderate numbers of clones or genotypes will lose very little of the expected heterozygosity in one generation. The regenerated forests are not going to continue with this small population size. Instead, they will be regenerated with seed from orchards again. Additive genetic variability behaves as expected heterozygosity, and is likely to be retained to a high degree. However, some of the important genetic variability may be in the form of rare alleles, e.g., disease resistance. Seed orchards typically contain such low numbers of clones that even fairly common alleles are lost (see Hattemer et al. 1982). Empirical results on variability in natural stands and seed orchards crops confirm that expected heterozygosities are similar in natural and managed populations (Table 2).

Table 2. Expected heterozygosities (H_e) at isozyme loci in natural and managed forests

Species	Natural	Managed	Stage	Reference
Pinus sylvestris	0.28	0.27 (SO)	seeds	Muona and Harju 1989
Pinus sylvestris		0.35 (SW)	seeds	Yazdani et al. 1985
Pseudotsuga	0.21	0.23 (SO)	seeds	Shaw and Allard 1982
menziesii	0.22	0.23 (SW)	adults	Neale 1985

SO — seed orchard, SW — shelterwood or seed tree stand.

The effective population size is usually much lower than the true number of clones because of variation in progeny number. In the orchards studied by Muona and Harju (1989), the effective sizes were about half the number of clones. Even so, little variability is lost in one generation. Even if variability was lost, it would be partly compensated for by background pollination, which is extensive and difficult to avoid in the Finnish orchards studied. The number of alleles in the seed crop was clearly higher than in the orchard parental clones, due to background pollination (Harju and Muona 1989). These studies suggest that hardly any changes of variability occur in seed orchards at allozyme loci, though other genetic changes may occur (see below).

Table 3 shows that in general seed orchards have no more inbreeding than natural stands. In *Eucalyptus regnans* less inbreeding was found in an orchard (Moran et al. 1989a).

Directional genetic changes due to fertility selection?

In the ideal seed orchard no fertility selection occurs. However, clones do not contribute equally to the offspring generation. There is extensive variation in female flowering and cone crops between the clones of e.g., *Pinus sylvestris* (e.g., Jonsson et al. 1976; Muona and Harju 1989), *Picea abies* (Eriksson et al. 1973), *Pseudotsuga menziesii* (El-Kassaby et al.

Table 3. Multilocus estimates of outcrossing (t_m) in natural and managed forests

Species	Natural	Managed	Reference
Pinus sylvestris	0.94	0.98 (SO)	Muona and Harju 1989
Picea mariana	0.70	0.84 (SO)	Barrett et al. 1987
Pseudotsuga menziesii	0.98	0.95 (SW)	Neale and Adams 1985
Eucalyptus regnans	0.74	0.91 (SO)	Moran et al. 1989a

SO — seed orchard, SW — shelterwood or seed tree stand.

1989) and *Pinus kesiya* Royle ex Gordon (Sirikul and Luukkanen 1987) (see Muona 1989 for further references). The coefficients of variation of cone production ranged from 0.5 to 1 in *Pinus sylvestris*. Likewise, male fertility varies between clones in *Pinus sylvestris* and in other species (Muona and Harju 1989). The variation in male strobilus production is also reflected in the actual participation in fertilizations of the different clones, as shown by paternity analysis for *Picea glauca* (Moench) Voss by Schoen and Stewart (1986) and *Pinus sylvestris* by Muona and Kärkkäinen (1989). However, phenological effects may also be very important in determining mating success (Erickson and Adams 1989). In addition, male fertilities may vary after pollination (Schoen and Cheliak 1987). Total fertility variation thus is an important selective factor. This will probably not influence the frequencies of biochemical markers, which are likely to vary independently of the flowering variables. However, flowering variation may be negatively correlated with growth (e.g., Loehle and Namkoong 1987). Surprisingly, few data are available on such correlations (Sedgley and Griffin 1989). Skroppa and Tutturen (1985) found no correlation between flowering and growth in young *Picea abies*, but Schmidtling (1981) demonstrated a negative correlation in *Pinus taeda* L. Nikkanen and Velling (1987) found a negative correlation between vegetative growth and flowering in clones of *Pinus sylvestris*. Loehle and Namkoong (1987) also suggested that variation in sexual reproduction may be negatively correlated with resources allocated to defenses against pathogens.

Even if there are some potential problems with seed orchards, genetic gains are achieved even in first generation orchards. Matheson et al. (1986) showed that in *Pinus radiata* Don trees from first generation seed orchards had considerably better volume growth (15—30%) than control seed lots, due to genetic improvement.

Correlations between levels of variability

The use of biochemical markers in monitoring variability within populations is based on the idea that they reflect the genome in general. How true is this assumption? There are examples of widespread species that have extensive allozyme and morphological and physiological variability, as for example *Pinus sylvestris* (Mikola 1982; Szmidt and Muona 1985; Muona and Harju 1989). *Picea omorika* (Pančić) Purkyne is a species with low morphological variability (Langner 1959), but it has been shown to have extensive genetic variability at marker loci (Kuittinen et al. 1991). One of the important trees in the reforestation of tropical areas is *Acacia mangium* Willd. This species is known to have hardly any allozyme variability, with expected heterozygosity 0.02 (Moran et al. 1989b). However, the populations contain much genetic variability for growth and physiological characters (Atipanumpai 1989). Finally, *Pinus torreyana* Parry ex Carr. has little variability in both kinds of traits (Ledig and Conkle 1983). Despite these examples of lack of concordance between levels of variability, allozymes are obviously useful when variability is monitored over time in a population or species, even if bottlenecks may have somewhat differing influences on variation at single loci and in quantitative characters (Lande 1980).

Clonal forestry

Clonal forestry is currently used for many broad-leaved species, e.g., eucalypts and poplars. Among conifers, *Cryptomeria japonica* (L. F.) D. Don, *Pinus radiata* and *Picea abies* are species where the technology is available (Zobel and Talbert 1984). As techniques are developed, the share of clonal forestry will increase. Modern biotechnology can be most rapidly integrated into tree breeding with the aid of vegetative propagation (Cheliak and Rogers 1990). In theory clonal forestry allows complete control over the genotypic composition of the plantation. The risks associated with a genetically uniform plantation with a low number of clones have been discussed intensively for years (see e.g., Benzer et al. 1988 for references). Many technical problems, for example with respect to genetic variation between genotypes in the ease of cloning, remain to be solved. As discussed above in connection with flowering variation, genetic correlations could lead to concomitant selection on other traits. The clones to be used in plantations are thoroughly tested for their production and adaptive characters. Marker genes have a minor role in monitoring genetic changes due to these practices. In species where regeneration takes place by seeds, the markers allow observation of the genetic effects of the reproductive cycle. Since this cycle is missing in clonal propagation, the genetic composition of the population can be monitored directly. Marker genes are used as technical aids, for example in identifying clones.

Exotic species

Exotic tree species have been used extensively for centuries to increase yields (Zobel et al. 1987). In Europe, conifer forests have to a large degree replaced native deciduous forests. The species that are used as exotics are often ecologically and genetically well known, as *Pinus contorta*, *Pinus radiata*, many eucalypts or acacias (see e.g., Turnbull 1987). One of the reasons for growing exotics rather than local species is that their biology and silvicultural requirements are well known (Evans 1987; Booth et al. 1988). Similar genetic considerations apply to these species as have been discussed above. Genetic markers are helpful in assessing how much of the variation in native populations has been included with the introductions. For older introductions, the initial sources may be unknown. Moran and Bell (1987) were able to infer the main sources of Australian *Pinus radiata* by comparing the original and introduced populations.

When exotics are planted, native forests are being replaced. In Europe this occurred so long ago that it is hardly remembered. At present, the disappearance of tropical rain forests is a major concern. Currently, every minute, 20 ha of forest are being converted or destroyed, and less than one hectare for every 10 hectares is reforested (Burley 1988). Much of the disappearance of forests is not due to "forest management," but to diverse social and economic pressures (FAO 1985). Plantation forestry may sometimes reduce pressure on natural tropical forests, because the yields can be four to ten times higher than in natural forests (FAO 1985; Gladstone and Ledig 1990). The protection of genetic diversity in tropical forests should be of interest to foresters. The genetic resources of tropical species are needed for future breeding, and for ecosystem stability (Ledig 1986).

Conclusions

Foresters interfere massively with many aspects of the trees' life cycle in tree breeding and silviculture. Forestry practices range from selective cutting and leaving of seed trees to replacing local species with exotics. It is important to be aware of the kinds of genetic changes that are to be expected along with these practices. There are both short term and long term considerations. In the short term, forestry practices should result in productive stands that are able to tolerate the environmental variation for the duration of the rotation. Natural regeneration and regeneration with bulk or orchard seed should mostly satisfy the requirements for genetically variable non-inbred seed. Foresters are also concerned with the quality of the wood produced (e.g., Barbour and Kellogg 1990). If plantations start with a low density, and heavy mortality occurs, the resulting stand will be quite sparse. Even with good genetic material, the result may be of poorer quality wood due to excessive branching. In using clonal forestry, the level of genetic variability may decrease considerably from that of natural populations. However, in this case foresters opt to take a calculated risk. The desired level of diversity will depend on the species, environment, and rotation age.

The longer term concern is the maintenance of reservoirs of genetic variability. This is needed for current breeding programs, and more importantly, for future needs. Important changes in the environmental conditions occur with climates being changed by the greenhouse effect and environmental pollution. Also, the requirements set by industry for wood production may change. As more and more forest land comes under intensive management practices, do any forests remain in a completely natural condition? Even if populations are maintained in a natural condition, isolation is not possible, and pollen from commercial stands will form part of the next generation's gene pool. Forest geneticists are exploring the need for differently managed populations for breeding and conservation (e.g., Namkoong 1989).

Current public debate on forestry practices deals mainly with two topics. In temperate areas, the debate revolves around choices between intensive industrial forestry and a more "natural" kind of management, often called New Forestry (Hasenkamp 1978; Mahlein and Hemstrom 1988; Franklin 1988). Genetics does not enter into the discussion often, except at the level of maintaining species diversity (e.g., Mahlein and Hemstrom 1988). When diversity within tree species is considered, most alternative modes of regeneration will likely maintain sufficient diversity.

The other intensely discussed topic is tropical forestry. Genetic resources are frequently mentioned as a reason for conserving native rainforests (e.g., Burley 1988). Again, the genetic issues primarily concern species-level diversity. The extinctions of species due to modification of environment are taking place so rapidly that the question of loss of genetic variability within species is moot: all individuals may have disappeared before drift has a chance to operate. The many ecological causes of extinction often operate before lack of genetic variation becomes an issue (Lande 1988).

What role do genetic markers have in following these changes? So far, the research has mainly used allozymes, but in the near future more numerous DNA markers will become available (Neale 1992, this issue, pp. 391—407). Allozymes have proved useful in measuring mating systems and inbreeding levels. So far, there have been few reports of a useful correlation between levels and patterns of variability of allozymes and adaptive morphological and physiological characters. This limits the usefulness of allozyme markers in monitoring the important genetic changes. Random DNA-level markers are not likely to fare much better. However, by saturating chromosomes, eventually we should be able to use DNA-markers to identify important quantitative trait loci (QTLs). If sufficiently tight associations between markers and QTLs are found, then maybe DNA markers can provide a short cut to monitoring to what happens with adaptive characters, at least in specific families or populations. Strauss and Tsai (1988) demonstrated that there is some clinal variation in the number of rDNA repeat units in *Pseudotsuga menziesii.* Methods are being improved for detecting variation at single copy regions (Neale 1992, this issue, pp. 391—407). The main use of this variation will certainly be to aid directly in tree breeding, but when such variation is detected, it can also be used for monitoring genetic changes in cultivated and conservation populations.

Acknowledgments

We thank V. Koski, J. Lumme, T. G. Harmsen, T. Prout and P. M. A. Tigerstedt for discussions, encouragement, and comments on the manuscript. Our research has been financially supported by the National Research Council for Agriculture and Forestry and the Foundation for Natural Resources in Finland.

References

Atipanumpai, L. 1989. *Acacia mangium*: Studies on the genetic variation in ecological and physiological characteristics of a fast growing plantation tree species. Acta Forestalia Fennica 206: 1—92.

Barbour, R. J. and Kellogg, R. M. 1990. Forest management and end-product quality: a Canadian perspective. Can. J. For. Res. 20: 405—415.

Barrett, J. W., Knowles, P. and Cheliak, W. M. 1987. The mating system in a black spruce clonal seed orchard. Can. J. For. Res. 17: 379—382.

Beekeman, W. B. 1949. Hout in alle tijden. I. Prae-historie en Historie tot en met de Iraniers. Kluwer, Deventer/Batavia/ Bandoeng/Semarang/Soerabaia, 714 pp.

Benzer, B. G., Foster, G. S., Hellberg, A. R. and Podzorski, A. C. 1988. Genotype x environment interaction in Norway spruce involving three levels of genetic control: seed source, clone mixture and clone. Can. J. For. Res. 18: 1172—1181.

Bergmann, F. 1978. The allelic distribution at an acid phosphatase locus in Norway spruce (*Picea abies*) along similar climatic gradients. Theor. Appl. Genet. 52: 57—64.

Booth, T. H., Nix, H. A., Hutchinson, M. F. and Jovanovic, T. 1988. Niche analysis and tree species introduction. For. Ecol. Manage. 23: 47—59.

Burley, F. W. 1988. The tropical forestry action plan: recent progress and new initiatives. pp. 403—408. In: Wilson, E. O. (Ed) Biodiversity. National Academy Press, Washington, DC, 521 pp.

Campbell, R. K. 1979. Genecology of Douglas-fir in a watershed in the Oregon Cascades. Ecology 60: 1036—1050.

Cheliak, W. M., Murray, G. and Pitel, J. A. 1988. Genetic effects of phenotypic selection in white spruce. For. Ecol. Manage. 24: 139—149.

Cheliak, W. M. and Rogers, D. L. 1990. Integrating biotechnology into tree improvement programs. Can. J. For. Res. 20: 452—463.

Daniels, J. D. 1984. Role of tree improvement in intensive forest management. For. Ecol. Manage. 8: 161—195.

El-Kassaby, Y. A., Fashler, A. M. K. and Crown, M. 1989. Variation in fruitfulness in a Douglas-fir seed orchard and its effect on crop-management decisions. Silvae Genet. 38: 113—121.

Ellstrand, N. C., Torres, A. M. and Levin, D. A. 1978. Density and the rate of apparent outcrossing in *Helianthus annuus* (Asteraceae). Syst. Bot. 3: 403—407.

Epperson, B. K. and Allard, R. W. 1989. Spatial autocorrelation analysis of the distribution of genotypes within populations of lodgepole pine. Genetics 121: 369—377.

Erickson, V. J. and Adams, W. T. 1989. Mating success in a coastal Douglas-fir seed orchard as affected by distance and floral phenology. Can. J. For. Res. 19: 1248—1255.

Eriksson, G., Andersson, S., Eiche, V., Ifver, J. and Persson, A. 1980. Severity index and transfer effects on survival and volume production of *Pinus sylvestris* in northern Sweden. Studia Forestalia Suecica 156: 1—32.

Eriksson, G., Jonsson, A. and Lindgren, D. 1973. Flowering in a clone trial of *Picea abies.* Studia Forestalia Suecica 110: 1—44.

Evans, J. 1987. Site and species selection — changing perspectives. For. Ecol. Manage. 21: 299—310.

FAO (Food and Agriculture Organization) 1985. Tropical Forestry Action Plan. Food and Agriculture Organization of the United Nations, Committee on Forest Development in the Tropics, Rome, 159 pp.

Farris, M. A. and Mitton, J. B. 1984. Population density, outcrossing rate, and heterozygote superiority in ponderosa pine. Evolution 38: 1151—1154.

Franklin, J. F. 1988. Structural and functional diversity in temperate forests. pp. 166—175. In: Wilson, E. O. (Ed) Biodiversity 5. National Academy Press, Washington, DC.

Furnier, G. R., Knowles, P., Clyde, M. A. and Dancik, B. P. 1987. Effects of avian seed dispersal on the genetic structure of whitebark pine populations. Evolution 41: 607—612.

Gladstone, W. T. and Ledig, F. T. 1990. Reducing pressure on natural forests through high yield forestry. For. Ecol. Manage. 35: 69—78.

Hamrick, J. L. 1983. The distribution of genetic variation within and among natural populations. pp. 335—348. In: Scoenewald-Cox, C. M., Chambers, S. M., MacBryde, B. and Thomas, W. L. (Eds) Genetics and conservation. Benjamin Cummings, Menlo Park.

Harju, A. and Muona, O. 1989. Background pollination in *Pinus sylvestris* seed orchards. Scand. J. For. Res. 4: 513—520.

Hasenkamp, J. G. 1978. Betriebswirtschaftliche Auswirkungen der naturgemässen Waldwirtschaft. Der Forst- und Holzwirt 33: 278—280.

Hattemer, H., Gregorius, H.-R., Ziehe, N. and Müller-Starck, G. 1982. Klonenanzahl forstlichen Samenplantagen und Genetische Vielfalt. Allg. Forst U. J. Ztg 153: 183—191.

Jonsson, A., Ekberg, I. and Eriksson, G. 1976. Flowering in a seed orchard of *Pinus sylvestris* L. Studia Forestalia Suecica 135: 1—38.

Kitzmiller, J. H. 1990. Managing genetic diversity in a tree improvement program. For. Ecol. Manage. 35: 131—149.

Koski, V. 1970. A study of pollen dispersal as a mechanism of gene flow. Communicationes Instituti Forestalis Fenniae 75(3): 1—30.

Koski, V. and Tallqvist, R. 1978. Tuloksia monivuotisista kukinnan ja siemensadon määrän mittauksista metsäpuilla. Folia Forestalia 364: 1—60.

Kuittinen, H., Muona, O., Kärkkäinen, K. and Borzan, Z. 1991. Serbian spruce, a narrow endemic, contains much genetic variation. Can. J. For. Res. 21: 363—367.

Lagercrantz, U. and Ryman, N. 1990. Genetic structure of Norway spruce (*Picea abies*): concordance of morphological and allozymic variation. Evolution 44: 38—53.

Lande, R. 1980. Genetic variation and phenotypic evolution during allopatric speciation. Am. Nat. 116: 463—479.

——. 1988. Genetics and demography in biological conservation. Science 241: 1455—1460.

Langner, W. 1959. Selbstfertilität und Inzucht bei *Picea omorica* (Pancic) Purkyne. Silvae Genet. 8: 84—93.

Ledig, F. T. 1986. Conservation strategies for forest gene resources. For. Ecol. Manage. 14: 77—90.

Ledig, F. T. and Conkle, M. T. 1983. Gene diversity and genetic stucture in a narrow endemic, torrey pine (*Pinus torreyana* Parry ex Carr.). Evolution 37: 70—85.

Lindquist, B. 1946. Den skogliga rasforskningen och praktiken. Svenska Skogsvårdsföreningens förlag. Stockholm, 176 pp.

Loehle, C. and Namkoong, G. 1987. Constraints on tree breeding: growth tradeoffs, growth strategies, and defense investments. Forest Sci. 33: 1089—1097.

Mahlein, D. and Hemstrom, M. 1988. Ecosystem management, a new forestry. Forest Watch 8(8): 8—10.

Makkonen, O. 1967. Ancient forestry. An historical study. Part I. Facts and information on trees. Acta Forestalia Fennica 82.3: 1—84.

Matheson, A. C., Eldridge, K. G., Brown, A. G. and Spencer, D. J. 1986. Wood volume gains from first-generation radiata pine seed orchards. CSIRO Div. For. Res. User Ser no. 4, 13 pp.

McDermott, J. M. and Robinson, R. A. 1989. Provenance variation for disease resistance in *Pseudotsuga menziesii* to the Swiss needle-cast pathogen, *Phaeocrytopus gaeumannii*. Can. J. For. Res. 19: 244—246.

Mikola, J. 1982. Bud-set phenology as an indicator of climatic adaptation of Scots pine in Finland. Silva Fennica 16: 178—184.

Millar, C. 1989. Allozyme variation of bishop pine associated with pygmy-forest soils in northern California. Can. J. For. Res. 19: 870—879.

Mitton, J. B., Stutz, H. P., Schuster, W. S. and Shea, K. L. 1989. Genotypic differentiation at PGM in Engelmann spruce from wet and dry sites. Silvae Genet. 38: 217—221.

Moran, G. F. and Bell, J. C. 1987. The origin and genetic diversity of *Pinus radiata* in Australia. Theor. Appl. Genet. 73: 616—622.

Moran, G. F., Bell, J. C. and Griffin, A. R. 1989a. Reduction in levels of inbreeding in a seed orchard of *Eucalyptus regnans* F. Muell. compared with natural populations. Silvae Genet. 38: 32—36.

Moran, G. F., Muona, O. and Bell, J. C. 1989b. *Acacia mangium*: a tropical forest tree of the coastal lowlands with low genetic diversity. Evolution 43: 231—235.

Muona, O. 1989. Population genetics in forest tree improvement. pp. 282—298. In: Brown, A. H. D., Clegg, M. T., Kahler, A. L. and Weir, B. S. (Eds) Plant population genetics, breeding and genetic resources. Sinauer, Sunderland, Mass.

Muona, O. and Harju, A. 1989. Effective population sizes, genetic variability, and mating system in natural stands and seed orchards of *Pinus sylvestris*. Silvae Genet. 38: 221—228.

Muona, O., Harju, A. and Kärkkäinen, K. 1988. Genetic comparison of natural and nursery grown seedlings of *Pinus sylvestris* using allozymes. Scand. J. For. Res. 3: 37—46.

Muona, O. and Kärkkäinen, K. 1989. Reproductive effort and reproductive success in *Pinus sylvestris*. Abstracts of 2nd European Conference on Evolutionary Biology, Rome, Italy, 1989.

Namkoong, G. 1989. Population genetics and the dynamics of conservation. pp. 161—181. In: Knutson, L. and Stoner A. K. (Eds) Biotic diversity and germplasm preservation, global imperatives. Kluwer Academic Publishers, Dordrecht.

Neale, D. B. 1985. Genetic implications of shelterwood regeneration of Douglas-fir in Southwest Oregon. Forest Science 31: 995—1005.

——. 1992. Use of DNA markers in forest tree improvement research. This issue (pp. 391—407).

Neale, D. B. and Adams, W. T. 1985a. The mating system in natural and shelterwood stands of Douglas-fir. Theor. Appl. Genet. 71: 201—207.

——. 1985b. Allozyme and mating-system variation in balsam fir (*Abies balsamea*) across a continuous elevational transect. Can. J. Bot. 2448—2453.

Nikkanen, T. and Velling, P. 1987. Correlations between flowering and some characteristics of grafts of *Pinus sylvestris*. For. Ecol. Manage. 19: 35—40.

Perry, D. J. and Knowles, P. 1991. Spatial genetic structure within three sugar maple (*Acer saccharum* Marsh.) stands. Heredity 66: 137—142.

Rehfeldt, G. E. 1988. Ecological genetics of *Pinus contorta* from the Rocky Mountains (USA): a synthesis. Silvae Genet. 37: 131—135.

——. 1989. Ecological adaptations in Douglas-fir (*Pseudotsuga menziesii* var. glauca): a synthesis. For. Ecol. Manage. 28: 203—215.

Schmidtling, R. C. 1981. The inheritance of precocity and its relationship with growth in loblolly pines. Silvae Genet. 30: 188—192.

Schoen, D. J. and Cheliak, W. M. 1987. Genetics of the polycross. 2. Male fertility variation in Norway spruce, *Picea abies* (L.) Karst. Theor. Appl. Genet. 74: 554—559.

Schoen, D. J. and Stewart, S. C. 1986. Variation in male reproductive investment and male reproductive success in white spruce. Evolution 40: 1109—1120.

Sedgley, M. and Griffin, A. R. 1989. Sexual reproduction of tree crops. Academic Press, London, 378 pp.

Shaw, D. V. and Allard, R. W. 1982. Isozyme heterozygosity in adult and open-pollinated embryo samples of Douglas-fir. Silva Fennica 16: 115—121.

Shea, K. 1987. Effects of population structure and cone production on outcrossing rates in Engelmann spruce and subalpine fir. Evolution 41: 124—136.

Sirikul, W. and Luukkanen, O. 1987. Promotion of flowering using exogenous plant hormones in tropical pines. For. Ecol. Manage. 19: 155—161.

Skroppa T. and Tutturen, R. 1985. Flowering in Norway spruce seed orchards. Silvae Genet. 34: 90—95.

Strauss, S. H. and Tsai, C.-H. 1988. Ribosomal gene number variability in Douglas-fir. J. Heredity 79: 453—458.

Szmidt, A. E. and Muona, O. 1985. Genetic effects of Scots pine (*Pinus sylvestris* L.) domestication. pp. 241—252. In: Gregorius H.-R. (Ed) Population genetics in Forestry. Lecture notes in biomathematics 60. Springer Verlag, Berlin.

Tigerstedt, P. M. A. 1973. Studies on isozyme variation in marginal and central populations of *Picea abies*. Hereditas 75: 47—60.

Turnbull, J. W. 1987. (Ed) Australian Acacias in developing Countries. ACIAR Proceeding no. 16. 196 pp.

Wheeler, N. 1992. The use of electrophoretic markers in seed orchard research. This issue (pp. 311—328).

Wyant, F. G., Omi, P. N. and Laven, R. D. 1986. Fire induced mortality in Colorado ponderosa pine/Douglas-fir stand. Forest Sci. 32: 49—60.

Yazdani, R., Muona, O., Rudin, D. and Szmidt, A. E. 1985. Genetic structure of a *Pinus sylvestris* seed-tree stand and naturally regenerated understory. Forest Sci. 31: 430—436.

Yearbook of forest statistics 1987. Folia Forestalia 715: 1—245.

Zobel, B. J. and Talbert, J. T. 1984. Applied forest tree improvement. John Wiley and Sons. New York. 505 pp.

Zobel, B. J., van Wyk, G. and Stahl, P. 1987. Growing exotic forests. John Wiley, New York. 508 pp.

New Forests **6**: 347–371, 1992.

Review paper

Allozyme markers in forest genetic conservation

C. I. MILLAR and R. D. WESTFALL

Center for Conservation of Genetic Diversity, Institute of Forest Genetics, Pacific Southwest Forest and Range Experiment Station, USDA Forest Service, Box 245, Berkeley, CA 94701, USA

Received 8 October 1990; accepted 2 April 1991

Key words: genetic diversity, isozymes, rare and endangered species, in-situ conservation, ex-situ conservation

Application. Allozyme markers are useful in three main aspects of forest genetic conservation: assessing the amount and distribution of genetic variation; determining sampling plans for in-situ and ex-situ conservation; and monitoring changes in genetic diversity. Since allozymes represent only one type of polymorphism, they are best used in conjunction with other traits when information about adaptive variation is needed.

Abstract. Genetic diversity is important in tree-breeding, in managing rare and endangered tree species, and in maintaining healthy populations of widespread native tree species. Allozymes are useful in determining genetic relationships among species, where they can be used to assess affiliations of rare taxa and predict relative endangerment among species. Because allozymes sometimes yield different information about genetic variation within species than revealed by other traits, when estimates of total or adaptive genetic variation are important, allozymes are best used in conjunction with other traits. Allozymes are useful for measuring direct allelic diversity when designing ex-situ and in-situ conservation strategies. We demonstrate an application of canonical trend-surface analysis for determining locations of in-situ genetic conservation areas. Allozymes also serve as useful markers in monitoring the effects of forest management and other environmental changes on genetic diversity.

Introduction

Forest genetic conservation has its origins in the crop gene-resources movement (Frankel 1977). Initial concern focused primarily on economically important tree species, and on the importance of conserving germplasm for future benefits to tree breeding. Because commercial tree species, unlike agronomic crops, had long rotation times and were planted in wildland settings, diversity was important for adaptability as well as increased productivity. Within such a context, the goals and techniques of genetic conservation became interwoven with long-term tree-improvement strategies (e.g., FAO 1975; Kitzmiller 1976, 1990; Zobel 1977; Namkoong 1984).

Increasing concern for environmental degradation has broadened the scope of genetic conservation to include non-commercial species, especially rare or endangered taxa, for their intrinsic values (Namkoong 1986; Ledig 1986a, 1988a, b). Forest geneticists view as serious losses the species extinctions and systematic impoverishment of populations that are occurring not just in tropical ecosystems, but in temperate forests as well. Trees are also increasingly valued for their role in ecosystem functioning. As structurally large organisms, trees create the physical and biological habitats on which many other plants and animals depend for their survival.

Allozymes in the service of genetic conservation

For many of the same reasons that allozyme markers have been useful in other aspects of forest genetics, they find applications in a wide range of genetic conservation situations. Research and application in forest genetic conservation can be divided into three steps: Assessing, conserving, and monitoring genetic diversity. Genetic conservation borrows heavily from the cumulative research of forest genetics, using information about genetic architecture, mating systems, gene flow, etc. as input for many conservation analyses. It is because so much basic knowledge exists about population genetic parameters in trees that we have been able to progress in conservation analyses beyond what has been possible in lesser known plants.

Assessing genetic diversity

Knowledge of baseline genetic diversity among and within species is needed for nearly all conservation analyses and applications. This information, whether measured or predicted, is essential for deciding what and how to conserve, for assessing whether genetic changes are occurring, and for determining whether genetic parameters are even relevant in a conservation plan. Assessing baseline diversity goes beyond the inventory or description of taxa and genetic variation. In most cases it includes an analysis of pattern in the data and attempts to interpret the pattern in light of evolutionary history and ecological context.

Systematic studies

Knowledge of species boundaries and genetic relationships among species are critical to conservation. Both by legal mandates and general consensus among conservation biologists, genetic resources of an individual species are usually given higher priority for protection against extinction than populations within a species. Although in the temperate zone species extinction in historical times has been infrequent in tree species, 233 woody plant species have been identified by the Center for Plant Conservation as being of "significant conservation concern" in the United States (Falk 1990; Ledig 1991). For example, American chestnut (*Castanea dentata*) has been all but eliminated by an introduced blight early in this century (Burnham 1988). American elm (*Ulmus americana*) and white pines (*Pinus* sect. *Strobus*) (Kinloch 1972) also have been devastated by introduced pathogens. The ranges of Port-Orford-cedar (*Chamaecyparis lawsoniana*) (Zobel et al. 1985) and Florida torreya (*Torreya taxifolia*) (Godfrey and Kurz 1962) are being fragmented by diseases. Ozone and other pollutants are narrowing the gene pools of pines around urban areas, such as the Los Angeles basin (Miller 1973). Genetic diversity in Fraser fir (*Abies fraseri*) and red spruce (*Picea rubens*) in the Appalachains is being lost as a result of combined onslaughts by insects, climate change, heavy metal pollution and acid deposition (Hamburg and Cogbill 1988). In tropical forests, the rate of species extinction is high (Wilson 1988).

Allozymes are useful for evaluating species boundaries because they provide more-or-less standard yardsticks that make comparisons among taxa meaningful (Strauss 1992, this issue, pp. 125—158). They are also less subject to problems of homoplasy (that is, similar traits occurring in distinct taxa that are not due to common ancestry) than traits traditionally used to evaluate taxonomy. Because endangered taxa at the species level are eligible for protection under endangered species legislation, taxonomic determinations carry legal implications. Although listing under plant protection legislation involves an evaluation of many factors affecting endangerment, one important consideration is the degree of genetic distinction of the taxon. Allozymes can provide a quantitative estimate of distinction, thereby helping to determine conservation priorities (Millar and Libby 1991). In some cases, allozymes may indicate that a taxon traditionally accepted as a rare species (and maybe already listed) is actually well within the range of variation of a more widespread species. In other cases, allozymes may draw attention to distinct taxa that hadn't been recognized on other grounds. Because of inconsistent correlations of allozyme with other traits (Hamrick and Godt 1989), conservation priorities probably should not be determined legalistically on allozymes alone, although they are important when combined with other data.

The cypresses (*Cupressus*) of southwest United States and Baja California provide an example. As many as 14 species (Wolf 1948) and as few as 7 (Little 1970) have been recognized, all of which occur as mosaics of disjunct populations associated with sterile soils. Santa Cruz cypress (*C. abramsiana*) is federally and state-listed as endangered. We analyzed allozyme variation in 62 populations at 25 loci of all taxa in Baja California and the United States, and found patterns of relationships that differed in some cases from those suggested by the published taxonomies.[1] For example, there was no evidence that Santa Cruz cypress is a distinct taxon. Its five populations were well within the range of variation of the Sargent cypress (*C. sargentii*) complex. The Santa Cruz cypress populations had lower genetic distance to populations of other taxa than to each other. The five populations of Tecate cypress (*C. forbesii*) in California, which were petitioned for state listing, were also within the range of another taxon, the Arizona cypress complex (*C. arizonica, C. glabra*). A sixth population of Tecate cypress, however, on Guadalupe Island, Mexico, was distinct from this group. Cuyamaca cypress (*C. stephensonii*), known from one small and declining population in the United States, but not currently being considered for listing, was a distinct taxon allozymically. In almost all taxa, the populations with the smallest sizes were more divergent.

In another example, Prober et al. (1990a) analyzed allozyme variation of the 12 "green ash" species of *Eucalyptus*, which include several rare species of conservation interest, and identified several relationships that had been either poorly resolved or intractable to morphological analysis. A major difference was the strong allozyme evidence for monophyly of members of the "dendromorpha" group, a relationship not indicated by morphology. Based primarily on analysis of allozyme distance and genetic diversity, the rarity in *E. paliformis* and *E. rupicola* was shown to be old, whereas rarity in *E. burgessiana* was mostly likely due to recent divergence. The allozyme analyses, together with morphological data, provided important information for setting conservation priorities and managing these rare species.

In some cases, distinct populations given species status are identified by allozyme markers as hybrid populations. For instance, in the Black Hills of South Dakota and Wyoming, distinct oak populations were originally classified as *Quercus mandanensis.* Allozyme analysis suggested instead that these populations resulted from hybridization between closely related

Q. macrocarpa and *Q. gambelii* with extensive backcrossing to *Q. macrocarpa* (Schnabel and Hamrick 1990). Thus, by clarifying the close relationship of these populations to non-endangered species, conservation efforts might be redirected to distinct endangered species.

Diversity within species

The literature on genetic architecture in natural populations of forest trees in rich (Conkle; Dancik; Müller-Starck; Moran; Loveless; Smouse; Hanover, this issue), and the envy of conservation biologists who work with lesser known taxa. Although most of these studies were not carried out intentionally for conservation purposes (and will not be reviewed here), like studies on systematics, they provide critical baseline data for conservation. For a species that has been identified for conservation efforts, information about genetic diversity within species provides the raw data for determining conservation guidelines.

Genetic conservation has been the major motivation for assessing diversity within rare or threatened tree species. Torrey pine (*Pinus torreyana*), endemic to two small and widely disjunct populations in California, had declined to about 500 trees by 1921 (Dusek 1985). No allozyme variation was found within either of the populations at any of 59 isozyme loci, although there were fixed differences at two loci between the populations (Ledig and Conkle 1983). The lack of genetic variability may have contributed to Torrey pine's decline; several growth and reproductive traits characteristic of the species seem to be maladaptive (e.g., seeds are retained in cones even when open; seeds germinate in cones; seed maturity takes six months longer than usual for related pines, Ledig 1986a). Active protection of both populations has increased the census to about 10,000 trees (Ledig and Conkle 1983). Should the numbers decline again, for example due to pest or pathogen epidemic or to climate changes, a small amount of genetic diversity could be reintroduced by crossing trees from the two populations.

Not all rare tree species are depauperate of genetic diversity. Monterey cypress (*Cupressus macrocarpa*) exists in two small California coastal populations with a total population about 10,000 (Conkle 1987). Monterey cypress resembles Torrey pine in its narrow endemism, habitat and ecology, and evolutionary history. Unlike Torrey pine, however, Monterey cypress maintains high heterozygosity within populations (0.18) with geographically structured variation both within and between the two nearby populations (5 km apart) (Conkle 1987). Although parts of the native populations are protected, expanding urban development on the Monterey peninsula creates a need for mitigation, and artificial regeneration is commonly undertaken. The genetic data indicate that propagules should come from very local samples and should include many trees in each sample.

The general criticism that allozymes are neutral markers and do not adequately estimate adaptive (e.g., growth and form) variation (e.g., Hamrick and Godt 1989; El-Kassaby 1990) is especially relevant to conservation. Measures based on different loci and different traits often yield discordant results about geographic patterns (Smouse, this issue, pp. 179—196; Hanover, this issue, pp. 159—178; McDonald 1983). The lessons learned in forestry about seed transfer and the importance of local adaptation make it important that adaptive variation not be underestimated in conservation. Multivariate analyses of allozyme data seem to correlate with patterns based on growth and form better than single-locus methods (Conkle and Westfall 1984; Guries 1984).

Where datasets are available on different kinds of traits, one possibility is to use all genetic information together to assess patterns of variation important for conservation, as has been done in numerical taxonomy (Sneath and Sokal 1973). If different traits give discordant patterns of variation, the species is divided into the smallest reasonable units suggested by cumulative analysis of individual patterns. An example using bishop pine (*Pinus muricata* is given in Fig. 1. This approach weights traits equally (as has been suggested in crop germplasm sampling, Marshall 1989), accepting the conservative position that we cannot know what genes are or will be adaptive in future environments. This approach, of course, depends on the number of traits available, which, if large, could result in over-partitioning the taxon. An alternative approach would be to use the trait or markers that give the most sensitive and random sample of the genome.

Correlates of genetic variation

Given the importance of baseline genetic diversity for conservation and the difficulty of obtaining such data, there has been an urgency in the conservation field to find correlates of genetic variation. The underlying hope is that some readily observable traits or attributes of species might allow indirect estimation of genetic architecture, obviating the need for direct genetic studies.

The most comprehensive efforts have been made over the past 15 years by Hamrick and colleagues (Hamrick and Godt 1989; Hamrick, this issue, pp. 95—124), who have analyzed correlations of life-history traits of plants with allozyme statistics from hundreds of studies. As the number of species included in the analyses increased (449 species in Hamrick and

Allozymes	Growth & Form	Cone Shape	Mono-terpenes	Stomata	Conservation Units
1	1	1	1	1	1
2	1	1	1	1	2
2	1	1	2	1,2	3,4
3	2	2	2	2	5
4	3	2	2	2	6
5	4	2	3	2	7
5	4	2,3	3	2	7,8
6	5	3	4	2	9
7	6	2,3	4	2	10
8	5	2,3	3	2	11

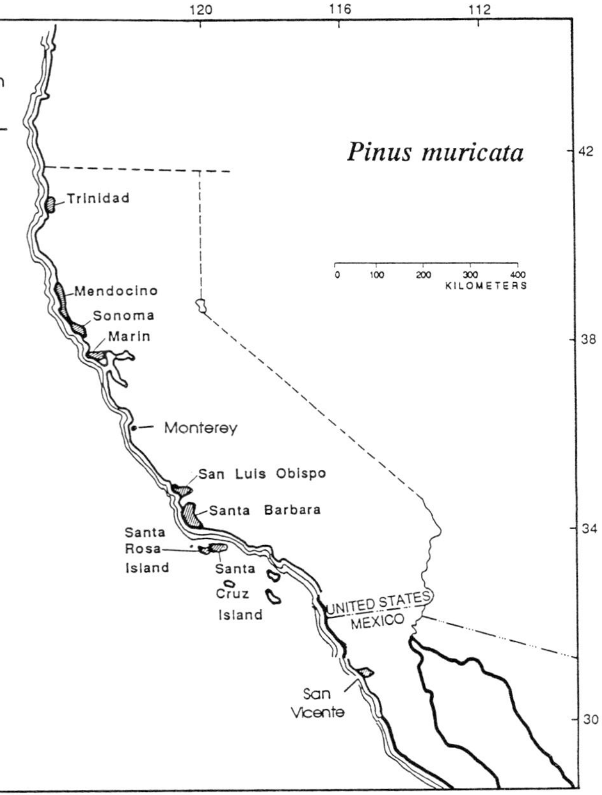

Fig. 1. Patterns of genetic variation in bishop pine (*Pinus muricata*) based on various traits, and conservation units derived from this information. Similarities among populations in each trait are indicated by identical numbers among populations; genetically dissimilar stands within a population are indicated by two numbers together. Despite quantitative variation in some traits, the groupings represent distinct clusters in all traits. Conservation units with different numbers should be treated as discrete groups in planning in-situ or ex-situ conservation. Data from Millar and Critchfield 1986; Millar et al. 1988; Millar 1989; unpub. data.

Godt 1989), certain traits appeared increasingly correlated with genetic diversity. Of special significance to conservation is the observation that geographic range accounts for the largest proportion of variability in genetic diversity at the species level. Widespread species contain significantly more allozyme diversity than narrowly distributed species. Narrowly distributed species, however, partition variation within species in the same way as more widespread species do. Partitioning of variation within species is correlated most highly with plant breeding system. Although these models provide generalizations, only about 30—50% of the variation among species is accounted for by the model (Hamrick and Godt 1989). Thus, direct genetic analyses of individual species are still necessary when genetic structure must be reliably ascertained.

Other correlations may exist at regional levels. For example, expected allozyme heterozygosities in conifers of western North America increase clinally with decreasing latitude (Ledig 1987, 1988b). This may result primarily from effects of the Pleistocene on conifer distributions. Northern populations were extirpated or severely challenged during glacial periods, whereas more southerly populations persisted, retaining high diversity. Following the return to warmer interglacial periods, northern regions were recolonized, with concomitant drift effects eroding genetic variability in the newly founded populations. For conservation, the most diverse conifer populations in western North America are southern ones; northern populations, although lower in total diversity, may contain unique alleles.

Conserving genetic diversity

Assessment of genetic diversity is only the prelude to genetic conservation. Whether conservation acts reactively to correct critical situations involving extremely endangered taxa or proactively to maintain diversity of currently stable species, decisions must be made about how and what germplasm to conserve. Traditionally germplasm has been conserved either in-situ, where organisms interact in natural or restored populations in wild habitat and condition, or ex-situ, where germplasm is stored in isolation from the natural habitat. Comprehensive programs that integrate both methods are most effective practically (Falk 1990; Millar and Libby 1991). In all cases, sampling decisions must be made about what genes to sample and from what locations.

Population viability analysis — Sampling in rare and endangered species

Soule (1987) has argued that the question "how much is enough" is the quintessential issue in conservation biology. For rare and endangered species, population viability analysis (PVA) attempts to define minimum conditions for a population or species to maintain itself over an agreed-on time with an agreed-on degree of certitude (reviewed in Soule 1987). PVA is an analysis of risk because it acknowledges the impossibility of guaranteeing absolute survival of a group indefinitely and the need for society to decide on an acceptable risk. The time scale usually used is on the order of centuries and the probability usually over 95%. Many factors influence viability of populations, and these are categorized as genetic, demographic, environmental, and catastrophic (Shaffer 1981). Theoretical constructs have been developed that independently model genetic and demographic effects on viability, and significant advances in considering both aspects have been made. Nevertheless, we are still far from being able to model or measure all the main factors and their complex interactions.

Attempts at empirical PVA have been limited almost exclusively to animals, especially birds and large vertebrates, although application of theoretical PVA to plants (Menges 1990a) and empirical assessment of plants (Menges 1991) is beginning. In forest trees, although comprehensive analyses have not been made, some of the genetic components of population viability have been evaluated. In general, inbreeding has been viewed as the principal genetic threat to short-term population survival and genetic drift as the principal threat to long-term survival (Frankel and Soule 1981). General estimation of these parameters for trees is reviewed in Epperson, Mitton, Ellstrand, Smouse, and Hanover in this issue. Specific aspects relevant to conservation are considered here.

The relationship between allozyme heterozygosity and fitness is especially important to conservation. In trees, allozyme heterozygosity has been shown to be positively correlated with fitness (usually measured as growth) in many but not all cases (Mitton and Grant 1984; Ledig 1986b). Much discussion has centered on the source of fitness, whether due to heterosis or release from inbreeding depression. Regardless of the nature of the effect, outbreeding trees have been shown repeatedly to carry high genetic loads and suffer inbreeding depression (reviewed in Ledig 1986b). Species that are low in allozyme diversity, especially those that are naturally rare or those that have undergone bottlenecks in their geologic past generally suffer lower inbreeding depression than species that are high in diversity and have maintained large populations sizes for long times. In species that survive bottlenecks and/or persist as rare populations, most seriously deleterious alleles seem to have been purged, although mildly maladaptive genes may be fixed (Ledig 1986b). By contrast, species whose distributions have been reduced by human actions may be at greatest risk of accelerated decline in population viability and numbers due to rapid exposure of recessive deleterious alleles in homozygous condition. This effect has been demonstrated repeatedly in animal breeding programs when formerly large and outbred populations are reduced to small effective population sizes (Lande and Barrowclough 1987).

In an effort to assess relative viability and endangerment of two rare *Eucalyptus* species, *E. paliformis* and *E. parvifolia*, Prober et al. (1990b) measured several ecological as well as genetic parameters. In addition to allozyme structure, they estimated numbers of individuals (population size), number of breeding individuals (trees bearing fruit, an estimate of effective population size), stand densities and age structure, fire history, and gene flow among the populations. Considering these factors together, they were able to show that *E. paliformis*, although lower in genetic diversity than *E. parvifolia*, had high and stable effective population sizes, indicative of an old rare species. By contrast, *E. parvifolia*, which had high genetic diversity indicative of a formerly widespread species, had only recently undergone reduction in size. They estimated that it had low effective population sizes, was under threat of swamping through hybridization from several widespread *Eucalyptus* species due to its being relegated to marginal sites for the species, and therefore was at greater risk of extinction.

When populations of rare species cannot be protected in-situ, their germplasm may be removed and stored ex-situ. Brown and Briggs (1990) use the neutral allele model to determine the minimum number of individuals to sample for a rare species. They argue that rare alleles are probably insignificant to species survival and that the effort involved in obtaining them jeopardizes protection of other species. They show that allelic content of a sample is proportional to the logarithm of population size and sample size, a relationship independent of mating structure and population structure. This relationship has been shown to hold for the proportion of alleles with frequency greater than 5% surviving a bottleneck (Sirkkomaa 1983). From this "law of diminishing returns", Brown and Briggs argue that the first ten individuals randomly sampled from a population are as important genetically as the next ninety. Although small samples may preserve alleles, these guidelines seem to ignore the effect of small samples on genotypic compositions, the degree of inbreeding depression in the species, and subsequent survival probabilities.

Sampling in widespread species

For widespread species where imminent extinction is unlikely, the objective for conservation extends beyond sustaining minimum conditions for survival to promoting healthy populations with sufficient genetic diversity to adapt to environmental stresses. Thus, sampling strategies depend on how much diversity is necessary to maintain forest health. This issue is nearly as intractable as that of minimum viable population size, although theoretical aspects have been considered (reviewed in Brown and Moran 1981).

1. *Sampling for ex-situ conservation*

Since the amount of time and money available constrains sampling, the topic of sampling efficiency has received considerable attention in the crop literature. Sampling to capture unique or locally rare alleles or to replicate natural allele frequencies in a sample may not be worth the effort necessary to get them (Marshall 1989). Efficient sampling captures the maximal amount of useful variation within practical limits (Allard 1970). The number of alleles per locus, usually measured by allozyme statistics, has been considered the best measure of genetic diversity (Marshall and Brown 1975; Asins and Carbonell 1987). Efficient sampling would occur when at least one copy of each important allele in a population was preserved. Marshall and Brown (1981) define priority of alleles in a population by subdividing alleles into four classes, based on frequency within the population (> 5% and < 5%) and distribution among populations (> 10% and < 10%). Highest priority in sampling should be given to the class of common local alleles. These alleles are apparently maintained in frequencies higher than the mutation rate and may be important elements of geographic diversity. Common widespread alleles are likely to be contained in all sampling strategies and rare local alleles are difficult to obtain. Modifications to this approach and applications to more sophisticated approaches when more information is available have been developed (in Marshall 1989).

Several allozyme datasets of trees have been divided into these classes. For six conifers, percent of alleles in the locally common class ranged from 0—17% (Adams 1981; Brown and Moran 1971). In 20 California conifers, this range was 0—32% (Table 1). The size of this class was unrelated to geographic distribution of the species (rare versus widespread; disjunct versus continuous populations), taxonomic affiliation, or ecological status. With a few exceptions, there was a common pattern for the classes of common widespread and rare local alleles to dominate, together accounting for over 80% on average, whereas proportions in the two other classes varied. The distribution patterns of alleles in fifteen *Eucalyptus* species (Prober et al. 1990b) were similar to the conifers, although in the eucalypts, the general pattern had fewer common widespread alleles and more common local alleles. This may be due to insect pollination in the eucalypts versus wind pollination in conifers. Efficient sampling strategies

Table 1. Percents of alleles with four types of distributions among populations of 20 California confiers

Species	No. pops. sampled	No. alleles	Common Wide-spread	Common Local	Rare Wide-spread	Rare Local	Source
Pseudotsuga menziesii	62	217	44	32	5	19	(1)
Pinus ponderosa	60	199	54	28	1	18	(1)
Cupressus macnabiana	10	93	48	16	15	20	(1)
Pinus muricata	7	86	57	14	7	22	(2)
Cupressus sargentii	9	83	53	10	8	29	(1)
C. goveniana	2	48	65	10	2	23	(1)
C. forbesii	5	77	53	10	6	30	(1)
C. bakeri	6	67	64	9	1	25	(1)
C. abramsiana	4	55	65	7	5	22	(1)
Chamaecyparis lawsoniana	9	105	58	6	15	22	(3)
Cupressus nevadensis	4	64	59	5	14	22	(1)
C. macrocarpa	2	44	80	5	9	7	(4)
C. pygmaea	5	53	67	4	11	17	(1)
Pinus radiata	5	91	52	4	21	12	(2)
P. torreyana	2	60	97	3	0	0	(5)
P. attenuata	4	63	54	3	16	27	(2)
P. jeffreyi	14	52	78	0	19	4	(6)
P. washoensis	3	50	52	0	20	28	(7)
Calocedrus decurrens	3	122	33	0	43	24	(8)
Sequoiadendron giganteum	34	15	100	0	0	0	(9)

Common alleles are defined as having frequencies over 0.05 and widespread alleles are defined as being distributed in more than 10% of the populations. (1) Millar, unpubl. data; (2) Millar et al. 1988; (3) Millar and Marshall 1991; (4) Conkle 1987; (5) Ledig and Conkle 1983; (6) Furnier 1984; (7) Niebling and Conkle 1990; (8) Harry 1984; (9) Fins and Libby 1982.

for species having larger classes of locally common alleles would be to concentrate efforts on increasing geographic coverage and reducing the number of samples at a location.

The approach of Marshall and Brown (1975) has been criticized for its apparent emphasis on allozyme loci. It has been argued that sampling strategies for conservation should not be designed to capture allozyme variants per se since they are not directly useful to breeders, nor do they necessarily correlate with the quantitative variation that is useful (in Marshall 1989). These objections are met by using datasets from as many kinds of traits as are available, weighting each equally, considering the classes of alleles separately for each dataset, and sampling according to the most conservative set of guidelines.

Sampling for maintenance of a germplasm collection involves many of the same considerations as the initial sampling from natural populations. This issue is of increasing concern in crop gene-bank management, as the size of collections has increased to the point where many are impractically large and impossible to maintain (Frankel and Soule 1981; Marshall and Brown 1981; Marshall 1989). Ways of identifying unnecessary redundancy for the purpose of purging gene banks have been discussed (Roos 1988; Peeters and Martinelli 1989). Crossa (1989) reviews sampling design for regenerating collections with alleles of known frequencies, including effects of sampling neutral alleles.

2. *Sampling for in-situ conservation*

The concept of conserving genetic resources of trees in special genetic reserves or genetic resource management units (GRMUs) is not new (Hagman 1973; Krugman 1984; Ledig 1986a, 1988a, b; Millar and Libby 1991), but the idea has been little developed and even less applied. Although most natural areas (National Parks, Wilderness Areas, etc.) serve as *de facto* gene reserves, in practice their size and location are determined almost exclusively on non-genetic, primarily esthetic and ecological, grounds. If these natural areas are allowed to serve as the primary gene-pool reserves for forest trees, then we must accept that in-situ genetic coverage will be a hodge-podge of protection, and the job of systematically conserving genetic resources must be done in some other way, most likely through ex-situ preservation.

Far preferable is a systematic approach to in-situ genetic conservation, in which genetic architecture is first analyzed and locations of GRMUs are then determined from the genetic analyses. Established natural areas could serve as GRMUs when they coincide with the locations and specifications determined by genetic analyses. In other cases, GRMUs would be newly designated following the genetic analysis. GRMUs, unlike other natural areas, are not exclusive of many land uses, including timber harvest. A forested area designated as a GRMU must only be managed to meet minimum genetic specifications (Krugman 1984; Ledig 1988a; Millar and Libby 1991). If a GRMU cannot be established where recommended, then it represents a known gap that can be compensated in other ways.

One approach to estimating genetically adequate sizes of GRMUs for widespread species calls for conserving populations that are large enough to maintain current levels of genetic diversity. Since there are many measures of diversity (Brown 1978; Brown and Weir 1983; Asins and Carbonell 1987), which often yield different relative patterns, deciding on the appropriate one is important. One method for quantitative analysis derives from the equilibrium equation for neutral mutations (Crow and Kimura 1970; Brown and Moran 1981). Effective population sizes required to maintain expected allozyme heterozygosities in equilibrium for most forest trees are large. For instance, in ponderosa pine (*Pinus ponderosa*) and Douglas-fir (*Pseudotsuga menziesii*) populations, heterozygosities ranged from 0.082 to 0.267 (Millar and Libby 1991). When conservative mutation rates are used (10^{-7}) in the equilibrium formula, the effective population sizes to maintain these heterozygosities range from 223,300 to 910,600. At appropriate stand densities, these numbers would correspond to areas far larger than are currently protected or potentially can be protected.

If the analysis is calculated to conserve heterozygosity over a finite rather than equilibrium time, then the required population sizes are smaller. Measures other than expected heterozygosity, such as number of alleles or effective number of alleles (Marshall and Brown 1975), may be more valuable indicators of diversity, and may give different results. For example, for a 90% probability of conserving all alleles of frequency equal to or greater than 0.01 for ten generations, the effective population size would have to be 916 (from Gregorius 1980).

Several criticisms of this approach have been raised. Neutral mutations may not be the variants of concern to conservation. The rates of advantageous mutations are not known, but are probably much lower. Once exposed to selection, however, these alleles will be kept in populations of modest size. Furthermore, Lande and Barrowclough (1987) have shown that estimates of variability based on single-locus statistics such as allozymes will be much lower than estimates based on traits with quantitative inheritance even if neutral mutation rates prevail. This is because variance in quantitative traits is obtained by mutation in any of the genes controlling the traits, thus yielding a higher effective mutation rate for the trait than occurs for single allozyme loci. Another criticism questions the value of maintaining current levels of heterozygosity. Most populations are in non-equilibrium states (Namkoong 1986) and current levels of population heterozygosity may have little to do with population stability or robustness.

Guidelines for deciding how many and where GRMUs should be for trees have been general. In the absence of direct genetic data, seed zones representing regions of physiographic similarity could be adopted as the zones for genetic conservation, with GRMUs placed in each zone. To the extent that allozymes have improved the delimitation of seed zones, they also contribute to genetic conservation (Westfall and Conkle 1992, this issue, pp. 279—309).

The Washington State Department of Natural Resources established a network of gene pool reserves to conserve genetic resources of Douglas-fir (Wilson 1990). Reserves are designated in each 152 m elevation band within every seed zone in which the Department owns over 405 ha. Over 100 reserves have been established with average size of 10 ha. These areas are managed as strict reserves, with no timber harvest allowed.

In general, even if harvest were allowed on GRMUs, practical constraints will probably limit the ability to maintain as many or as large GRMUs for all commercial and widespread species as may be indicated by genetic analysis. We are developing an approach to designing systematic GRMU networks that incorporates pre-existing reserved areas where appropriate.[2] The process first identifies genetic targets, then analyzes gaps in protection. The gap analysis highlights areas that require greatest conservation attention.

We are studying the Eldorado National Forest (ENF) on the west slope of the central Sierra Nevada in California as an application of this approach.[3] Mixed conifer forests blanket the low and middle elevations, where most of the production forests occur (USFS 1989). Reserved areas of varying sizes and types are scattered throughout the Forest. Excellent genetic data on allozymes and quantitative traits are available for several of the commercial conifers that sample stands throughout the Forest. We focus on white fir allozymes here (from Conkel and Westfall 1988 unpublished report), but intend to incorporate datasets for different traits in a more comprehensive analysis of tree species on the ENF.

The first step was to describe the georgraphic patterns of allozyme variation in white fir throughout the ENF. A canonical trend-surface analysis (CTSA; Lee 1969) of white fir allozymes over the range of the species in the western Sierra Nevada (Conkle and Westfall 1988) provided the source of data for the EMF portion of white fir's range. CTSA provides scores for the individual trees in the analysis, with the scores measured in multi-locus standard deviation units. Trees were grouped according to their scores into one of four classes (A, B, C, D), with the classes defined as above (+) or below (−) the mean in each of three primary vectors (A was — for first vector, — for second vector, — for third vector; B was −, −, +; C was −, +, −; D was −, +, +, respectively). The complex variation of white fir on the ENF (Fig. 2) is better understood when compared to regional trends in the species throughout the western Sierra Nevada (Conkle and Westfall 1988). The bulk of the middle-elevation populations on the ENF are part of a broad class (denoted by As in Fig. 2) that extends north of the ENF. A higher elevation class, above 1675 m elevation, is limited mostly to the ENF (Bs in Fig. 2). The extreme southern part of the ENF contains complex white fir stands that are transitional to a broad class that extends south of the ENF. These

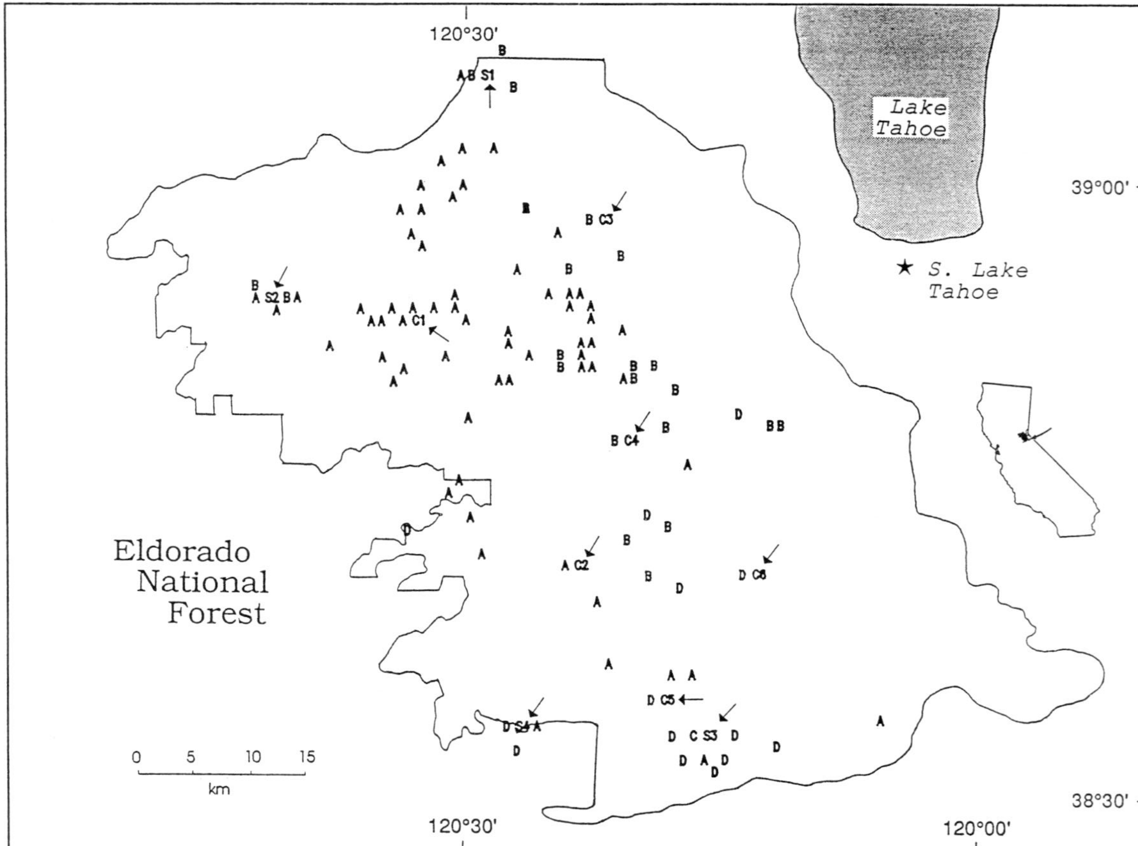

Fig. 2. Patterns of multi-locus allozyme variation in white fir on the Eldorado National Forest, California, as indicated by canonical-trend surface analysis. Letters indicate location of samples, and genetically similar samples are indicated with the same letter. Three conservation zones for white fir correspond to the A, B, and C/D pattern. Six core Genetic Resource Management Units (GRMUs) (C1—C6) and four supplemental GRMUs (S1—S4) are identified from the genetic data (arrows).

are the C and D classes (Fig. 2), which are subdivided by elevation. The transition between the A and D classes in the southern ENF represents the sharpest genetic shift in white fir on the ENF.

The data are further analyzed for genetic conservation purposes by progressive application of transfer-risk analysis (Campbell 1986; Westfall 1991). Based on the genetic data, three major zones are defined for white fir on the ENF that correspond to the A, B, and C/D classes described above. Zones here are defined by the risk that would be incurred if trees from one area were moved to another area (Westfall and Conkle, this issue, pp. 279—309). Risk measures the proportion of multi-locus genotypes that differs among locations, and is reflected by the changes in frequencies over space. Average transfer risk within zones is 5%; average risk among the extreme zones (i.e., A-D) is 22%. We consider two core GRMUs per zone, chosen to represent the average diversity within the zone, as the minimal goal (Millar and Libby 1991).

As a validation of the allozyme analysis, we computed transfer risks using growth, morphological, and phenological data from common-garden nursery (Hamrick 1976) and plantation (J. Jenkinson, U.S. Forest Service, unpubl.) trials of white fir. These tests contained central Sierra Nevada sources that spanned the elevation range of the allozyme dataset. Average transfer risk per 300 m elevation was 0.18 for the allozyme and 0.19 for the common-garden data. This close correspondence between independent traits confirms the values from the transfer risk analysis of allozyme data for white fir in the Sierra Nevada.

Locations of core GRMUs within zones were chosen such that the average transfer risk between them was greater than 6% (to capture some among-site diversity) and less than 12% (to avoid extremes of diversity within zones) (Table 2; Fig. 2). In addition to core GRMUs, supplemental GRMUs were nominated in areas that represent extremes in diversity, that is, where transfer risk within zones averaged greater than 10% to the core GRMUs (Fig. 2). In this way, six core GRMUs and four supplemental GRMUs were identified for the ENF.

The set of candidate GRMUs was then compared to a map of pre-existing areas such as Wilderness Areas, Spotted Owl Habitat Areas, etc., where management is compatible with GRMU status and where natural stands of white fir should exist. Pre-existing areas that coincide with candidate GRMUs were chosen first for GRMU status. Where candidate GRMUs are not included within pre-existing protected areas but protected areas exist nearby, the transfer risk of using nearby areas as GRMUs is assessed. On a genetic basis, these areas could acceptably serve GRMU status if their transfer risk is less than about 5—6%. Locations where risks between a candidate GRMU and its potential replacement site are greater

Table 2. Transfer risks among candidate Genetic Resource Management Units (GRMUs) identified for white fir on the Eldorado National Forest, California

GRMUs	Core						Supplemental		
	C1	C2	C3	C4	C5	C6	S1	S2	S3
Core									
C2	0.06								
C3	0.12	0.12							
C4	0.15	0.11	0.09						
C5	0.21	0.15	0.19	0.11					
C6	0.30	0.25	0.23	0.16	0.12				
Supplemental									
S1	0.11	0.14	0.09	0.13	0.22	0.26			
S2	0.11	0.07	0.09	0.06	0.12	0.19	0.11		
S3	0.16	0.10	0.17	0.09	0.05	0.17	0.18	0.09	
S4	0.19	0.13	0.20	0.15	0.11	0.21	0.22	0.12	0.10

Core GRMUs C1 and C2 are in the A zone; C3 and C4 in the B zone; and C5 and C6 in the D zone.

than 6% are identified as gaps in protection. These locations, especially any core sites, should receive attention for genetic conservation. Once replacement sites were identified as acceptable from the genetic model, they would be ground-checked for suitable species' habitat, and their long-term management status determined.

This process can be extended to multiple species or datasets based on different traits. Pre-existing areas that can act as acceptable GRMUs (i.e., within acceptable risk) for more than one species or dataset are given highest priority for GRMU status. Higher risks may be acceptable when attempting to designate GRMUs that represent many species.

Monitoring genetic diversity

Effective conservation requires an understanding not just of baseline genetic diversity but of changes in diversity. Conservationists are especially concerned with mitigating the effects of human actions on genetic diversity. Some actions have dispersed effects, such as introduction of exotic pathogens or insects, pollution of the atmosphere, and global climate change. Other actions are more localized, such as artificial reforestation and silvicultural practices.

Significant and consistent effects of atmospheric pollution on allozyme structure of trees have been reported (reviewed in Scholz et al. 1989). The effects include altered allele frequencies, decreases in expected heterozygosities, and decreases in multi-locus allozyme diversities. These effects were observed not just in allozyme structure of affected stands, but in comparisons of tolerant and sensitive clones subjected to specific polluting chemicals such as sulphur dioxide and ozone. A possible explanation for the relationship in these data between genetic diversity and air pollution is that certain genotypes (especially inbreds) are more susceptible to multiple stresses (such as air pollution, insect and disease, competition), and are most likely to succumb to cumulative effects (Ledig 1991).

In contrast to the relatively large and consistent effects on allozyme diversity related to atmospheric pollution, studies that monitored effects of forest management on allozyme diversity have not shown consistent results (Muona, this issue). Allozyme diversity was 33% higher on average in natural populations of Scots pine (*Pinus sylvestris*) and Monterey pine (*P. radiata*) (in Adams 1981), whereas in two studies of Douglas-fir, diversity was similar in natural populations and seed orchards (Adams 1981). In another study of Douglas-fir, allozyme diversity in seedlots from seed trees and commercial seedzone collections was about the same as diversity in seed orchards (Adams 1981).

Small or insignificant effects on allozyme diversity have been reported due to silvicultural manipulations in commercial conifers. With minor exceptions, there were no significant differences in allele diversity among samples from four life-cycle stages in Douglas-fir stands in Oregon managed under a shelterwood system (Neale 1985). This was true despite large reductions in the number of parent trees in the shelterwood. Excess homozygotes generated in seeds from natural regeneration in a Scots pine (*Pinus sylvestria*) shelterwood were eliminated by natural selection after 10—20 years in the field (Muona, this issue, pp. 329—345).

Allozymes are especially useful in detective aspects of conservation, where their application is similar to identifying clones and evaluating contamination in seed orchards (e.g., Friedman and Adams 1985; Rajora 1988). Catalina mahogany (*Cercocarpus traskiae*) is an extremely endangered tree species of Santa Catalina Island, California, with only seven remaining individuals. The species has declined in recent years due to habitat degradation and overgrazing from introduced ungulates, leading the Center for Plant Conservation to begin a rescue effort for the species (Reiseberg 1988). Electrophoretic analyses indicated that two of the Catalina mahoganies were hybrids with the common mahogany (*C. betuloides*) on the island. This knowledge enabled conservationists to vegetatively propagate the five pure Catalina mahoganies with the intent of re-introducing them into fenced habitat on the island.

We used allozymes to evaluate genetic contamination in the northernmost and smallest grove of giant sequoia (*Sequoiadendron giganteum*), where planting by local service clubs has resulted in establishment of young trees of unknown origin. One restoration plan for the grove recommended removing all trees of non-local origin. The possibility existed, however, that some of the young trees now in the grove might be natural regenerants. We genotyped mature and young trees and, by analyzing loci present in the young trees that were not present in the old-growth trees, determined that all but one young tree had to have been planted.[4]

Conclusions

Application of allozyme markers to forest genetic conservation is a natural extension of the uses to which allozymes have been applied in many areas of forest genetics. Allozymes are routinely used to estimate the amount and distribution of genetic diversity in natural populations and provide important baseline data to conservation. Although this application should continue, there is increasing need to focus on questions of special relevance to conservation, especially relating to how much variation is enough to ensure viability of rare taxa and to ensure healthy and robust forests of widespread species. Increased emphasis on monitoring changes in diversity due to human actions is also needed.

In many of these applications, the isozyme method is valuable because it provides direct information on allelic diversity that can be compared across taxa. For estimating relationships among species and microevolutionary parameters within species, such as gene flow and mating systems, and in conservation detective work, allozymes are excellent markers. In cases where allozymes are used to represent other kinds of genes (e.g., adaptive), or where conservation decisions are made on the basis of allozyme diversity per se, they may be misleading. Certain statistical methods may allow allozyme diversity to be partitioned such that the portion of allozyme diversity that correlates with adaptive variation is revealed. Conversely, variation may be assessed in multiple traits, and the conservative pattern indicated by all data be accepted.

Finally, many of the applications needed in genetic conservation may be served better by combining isozyme methods with new techniques emerging from molecular biology. The large number of DNA markers that can be analyzed, their broad representation throughout the genome, and the different scales of polymorphism of these markers may provide critical information for conservation. While nuclear markers may provide the large numbers of DNA sequences useful for comparing extant populations, organelle markers may be especially useful for tracing pedigrees and phylogenies of threatened and endangered organisms. These techniques offer a great potential for extending the biochemical approaches that isozyme analysis affords to many new conservation situations.

Acknowledgments

We thank M. T. Conkle and J. Jenkinson for allowing us to use unpublished white fir data, D. L. Delany for contributions to unpublished work included here, and the associate editor and two anonymous reviewers for comments on the manuscript.

Notes

1. Millar, C. I. and Delany, D. L. (in prep.) Biosystematics of cypresses in Southwestern United States and Baja California. Systematic Botany.
2. Millar, C. I. and Westfall, R. D. (in prep.) A sampling strategy for in-situ genetic conservation. Conservation Biology.
3. This analysis is undertaken as an exercise of the approach, and nothing indicated here binds on ENF management and planning.
4. Millar, c. I. and Delany, D. L. (in prep.) Genetic contamination and natural regeneration in the Placer Grove, giant sequoia. Madrono.

References

Adams, W. T. 1981. Population genetics and gene conservation in Pacific Northwest conifers, pp. 401—415. In: Scudder, G. G. and Reveal, J. L. (Eds) Evolution Today, Proceed. Second Internatl. Congr. Syst. Evol. Bio.

Allard, R. W. 1970. Population structure and sampling methods, pp. 97—107. In: Frankel, O. H. and Bennett, E. (Eds) Genetic Resources in Plants — Their Exploration and Conservation. Blackwell, Oxford.

Asins, M. J. and Carbonell, E. A. 1987. Concepts involved in measuring genetic variability and its importance in conservation of plant genetic resources. Evolutionary Trends in Plants 1(1): 51—62.

Brown, A. H. D. 1978. Isozymes, plant population genetic structure, and genetic conservation. Theor. Appl. Genet. 52: 145—157.

Brown, A. H. D. and Briggs, J. D. 1991. Sampling strategies for genetic variation in ex-situ collections of endangered plant species, pp. 99–122. In: Falk, D. and Holsinger, K. (Eds) Genetics and Conservation of Rare Plants. Oxford Univ. Press, NY.

Brown, A. H. D. and Moran, G. F. 1981. Isozymes and the genetic resources of forest trees, pp. 1—10. In: Conkle, M.T. (Ed.) Isozymes of North American Forest Trees and Forest Insects. USDA Forest Service, Gen. Tech. Rept. PSW-48.

Brown, A. H. D. and Weir, B. S. 1983. Measuring genetic diversity in plant populations, pp. 219—238. In: Isozymes in Plant Genetics and Breeding. Part A. Elsevier.

Burnham, C. R. 1988. The restoration of the American chestnut. Am. Sci. 76: 478—487.

Campbell, R. K. 1986. Mapped genetic variation of Douglas-fir to guide seed transfer in southwest Oregon. Silvae Genet. 35(2—3): 85—96.

Conkle, M. T. 1987. Electrophoretic analysis of variation in native Monetary cypress (*Cupressus macrocarpa*), pp. 249—256. In: Elias, T. S. (Ed) Conservation and Management of Rare and Endangered Plants. California Native Plant Society, Sacramento.

Conkle, M. T. 1992. Genetic diversity — seeing the forest through the trees. This issue (pp. 5—22).

Conkle, M. T. and Westfall, R. D. 1984. Evaluating breeding zones for ponderosa pine in California, pp. 89—98. In: Proceedings of the Service-wide Genetics Workshop, Charleston, SC, Dec. 5—9, 1983. USDA Forest Service.

Conkle, M. T. and Westfall, R. D. 1988. Allozyme variation of white fir in the Sierra Nevada of California. Unpublished report to the U.S.F.S. Regional Tree Improvement Program. Pacific Southwest Region.

Crossa, J. 1989. Methodologies for estimating the sample size required for genetic conservation of outbreeding crops. Theor. Appl. Genet. 77: 153—161.

Crow, J. F. and M. Kimura. 1970. An Introduction to Population Genetics Theory. Harper and Row, New York.

Dusek, K. H. 1985. Update on our rarest pine. Am. Forests: 26—29; 61—63.

El-Kassaby, Y. A. 1990. Genetic variation within and among conifer populations: Review and evaluation of methods, pp. 59—74. In: Hattemer, H. H., Fineschi, S., Cannata, F. and Malvolti, M. E. (Eds) Biochemical Markers in the Population Genetics of Forest Trees. APB Academic Publ. bv, The Hague.

Ellstrand, N. C. 1992. Gene flow among seed plant populations. This issue (pp. 241—256).

Epperson, B. K. 1992. Spatial structure of genetic variation within populations of forest trees. This issue (pp. 257—258).

Falk, D. A. 1990. Endangered forest resources in the U.S.: Integrated strategies for conservation of rare species and genetic diversity. For. Ecol. & Manage: 91–108.

FAO 1975. Methodology of conservation of forest genetic resources. FAO/UNEP, Rome, 127 pp.

Fins, L. and Libby, W. J. 1982. Population variation in *Sequoiadendron*: seed and seedling studies, vegetative propagation and isozyme variation. Silvae Genet. 31: 102—110.

Frankel, O. H. 1977. Philosophy and strategy of genetic conservation in plants. Proceed. Third World Consultation on Forest Tree Breeding 1: 6—11.

Frankel, O. H. and Soule, M. E. 1081. Conservation and Evolution. Cambridge Univ. Press, Cambridge.

Friedman, S. T. and Adams, W. T. 1985. Estimation of gene flow into two seed orchards of loblolly pine (*Pinus taeda*). Theor. Appl. Genet. 69: 609—615.

Furnier, G. 1984. Population Genetic Structure of Jeffrey Pine. Ph.D. dissertation. Oregon State University, Corvallis.

Godfrey, R. K. and Kurz, H. 1962. The Florida torreya destined for extinction. Science 136: 900—902.

Gregorius, H. R. 1980. The probability of losing an allele when diploid genotypes are sampled. Biometrics 36: 643—652.

Guries, R. P. 1984. Genetic variation and population differentiation in forest trees, pp. 119—131. In: Lanner, R. M. (Ed) Proceed. Eight North American Forest Biology Workshop. July 30—August 1, 1984, Logan, Utah.

Hagman, M. 1973. The Finnish standard stands for forestry research, pp. 67—78. In: Fowler, D. P. and Yeatman, C. W. (Eds) Proceed 13th Meeting of Committee on Forest Tree Breeding in Canada. Part 2, August 24—27, 1971. Prince George, British Columbia.

Hamburg, S. P. and Cogbill, C. V. 1988. Historical decline of red spruce populations and climatic warming. nature 331: 428—431.

Hamrick, J. L. 1976. Variation and selection in western montane species. II. Variation within and between populations of white fir on an elevational transect. Theor. Appl. Genet. 47(1): 27—34.

Hamrick, J. L., Godt, M. J. W. and Sherman-Broyles, S. L. 1992. Factors influencing levels of genetic diversity in woody plant species. This issue (pp. 95–124).

Hamrick, J. L. and Godt, M. J. 1989. Allozyme diversity in plant species, pp. 43—63. In: Brown, H. D., Clegg, M. T., Kahler, A. L., and Weir, B. S. (Eds) Plant Population Genetics, Breeding, and Genetic Resources. Sinauer Assoc., Sunderland, Mass.

Hanover, J. W. 1992. Applications of terpene analysis in forest genetics. This issue (pp. 159—178).

Harry, D. E. 1984. Genetic structure of incense-cedar (*Calocedrus decurrens*). University of California, Berkeley, Ph.D. dissertation, 163 pp.

Kitzmiller, J. H. 1976. Tree Improvement Master Plan for the California Region. USDA Forest Service, San Francisco.

Kitzmiller, J. H. 1990. Managing genetic diversity in a tree improvement program. For. Ecol. Manage. 13: 131—150.

Kinloch, B. B. 1972. Genetic variation in resistance to *Cronartium* and *Peridermium* rust in hard pines., pp. 445—462. In: Biology of rust resistance in forest trees. Proc. NATO-IUFRO Adv. Study Inst. Aug. 17—24, 1969. USDA Misc. Pub. 1221. Washington, D.C.

Krugman, S. L. 1984. Policies, strategies, and means for genetic conservation in forestry, pp. 71—78. In: Yeatman, C. W., Kafton, D. and Wilkes, G. (Eds) Plant genetic resources. A conservation imperative. Am. Assoc. Adv. Sci. Selected Symposium 87. Westview, Colorado.

Lande, R. and Barrowclough, G. 1987. Effective population size, genetic variation, and their use in population management, pp. 87—124. In: Soule, M. E. (Ed) Viable Populations for Conservation. Cambridge Univ. Press, Cambridge.

Ledig, F. T. 1986a. Conservation strategies for forest gene resources. For. Ecol. Manage. 14: 77—90.

——. 1986b. Heterozygosity, heterosis, and fitness in outbreeding plants, pp. 74—104. In: Soule, M. E. (Ed.) Conservation Biology: The Science of Scarcity and Diversity. Sinauer Assoc. Sunderland, Mass.

——. 1987. Genetic structure and the conservation of California's endemic and near-endemic conifers, pp. 587—594. In: Alias, T. S. (Ed) Conservation and Management of Rare and Endangered Plants. California Native Plant Society, Sacramento, California.

——. 1988a. The conservation of diversity in forest trees. Bioscience 38(7): 471—479.

——. 1988b. Conservation of genetic diversity: The road to La Trinidad. Leslie Schaffer Lectureship in Forest Science, Oct. 27, 1988. Vancouver, British Colombia.

——. 1992. Human impacts on genetic diversity in forest ecosystems. Oikos 63: 87–108.

Ledig, F. T. and Conkle, M. T. 1983. Gene diversity and genetic structure in a narrow endemic, Torrey pine (*Pinus torreyana*). Evolution 37: 79—85.

Lee, P. J. 1969. The theory and application of canonical trend surfaces. J. Geology 77(3): 303—318.

Little, E. L. 1970. Names of the new world cypresses. Phytologia 20(7): 429—445.

Loveless, M. D. 1992. Isozyme variation in tropical trees: patterns of genetic organization. This issue (pp. 67—94).

Marshall, D. R. 1989. Crop genetic resources: Current and emerging issues, pp. 267—388. In: Brown, A. H. D., Clegg, M. T., Kahler, A. L. and Weir, B. S. 1989. Plant Population Genetics, Breeding, and Genetic Resources. Sinauer Assoc. Sunderland, Mass.

Marshall, D. R. and Brown, A. H. D. 1975. Optimum sampling strategies in genetic conservation, pp. 53—80. In: Frankel, O. H. and Hawkes, J. G. (Eds) Crop Genetic Resources for Today and Tomorrow. Cambridge Univ. Press, Cambridge.

Marshall, D. R. and Brown, A. H. D. 1981. Wheat genetic resources, pp. 21—40. In: Evans, L. T. and Peacock, W. J. (Eds) Cambridge Univ. Press, Cambridge.

McDonald, J. F. 1983. The molecular basis of adaptation: A critical review of relevant ideas and observations. Annual Rev. Ecol. Syst. 14: 77—102.

Menges, E. S. 1990a. Population viability analysis for an endangered plant. Conservation Biology 4(1): 52—62.

Menges, E. S. 1991. The application of minimum viable population theory to planqts. In: Falk, D. and Holsinger, K. (Eds) Genetics and Conservation of Rare Plants. Oxford Univ. Press, Cary NC.

Millar, C. I. 1989. Allozyme variation of bishop pine associated with pygmy forest soils in northern California. Can. J. For. Res. 19: 870—879.

Millar, C. I. and Critchfield, W. B. 1986. Crossability and relationships of *Pinus muricata* (*Pinaceae*). Madrono 35(1): 39—53.

Millar, C. I. and Libby, W. J. 1991. Strategies for conservation of clinal, ecotypic, and disjunct population diversity in widespread species. In: Falk, D. and Holsinger, K. (Eds) Genetics and Conservation of Rare Plants. Oxford Univ. Press, Cary, NC.

Millar, C. I. and Marshall, K. A. 1991. Allozyme variation in Port-Oxford-cedar; Implications for genetic conservation. For. Sci. 37: 1060—1077.

Millar, C. I., Strauss, S. H., Conkle, M. T. and Westfall, R. D. 1988. Allozyme differentiation and biosystematics of the Californian closed-cone pines (*Pinus* subsect. *Oocarpae*). Syst. Bot. 13(3): 351—370.

Miller, P. L. 1973. Oxidant-induced community change in a mixed conifer forest, pp. 101—117. In: Naegle, J. A. (Ed) Air pollution damage to vegetation. Adv. Chem. Ser. 122.

Mitton, J. B. 1992. The dynamic mating systems of conifers. This issue (pp. 197—216).

Mitton, J. B. and Grant, M. C. 1984. Associations among protein heterozygosity, growth rate, and developmental homeostasis. Annual Rev. Ecol. Syst. 15: 479—499.

Moran, G. F. and Hopper, S. D. 1983. Genetic diversity and the insular population structure of the rare granite rock species. *Eucalyptus caesia*. Aust. J. Bio. 31: 161—172.

Moran, G. F. and Hopper, S. D. 1987. Conservation of the genetic resources of rare and widespread eucalypts in remnant vegetation, pp. 151—162. In: Saunders, D. A., Arnold, G. W., Burbidge, A. A. and Hopkins, A. J. M. (Eds) Nature Conservation: The Role of Remnants of Native Vegetation. Beatty, Australia.

Moran, G. F. 1992. Patterns of genetic diversity in Australian tree species. This issue (pp. 49—66).

Müller-Starck, G., Baradat, Ph. and Bergmann, F. 1992. Genetic variations within European tree species. This issue (pp. 23—47).

Namkoong, G. 1984. Strategies for gene conservation in forest tree breeding. In: Yeatman, C. W., Kafton, D. and Wilkes, G. (Eds) Plant genetic resources. A conservation imperative. Am. Assoc. Adv. Sci. Selected Symposium 87. Westview, Colorado.

Namkoong, G. 1986. Genetics and the forests of the future. Unasylva. 38: 2—18.

Neale, D. B. 1985. Genetic implications of shelterwood regeneration of Douglas-fir in southwest Oregon. Forest Sci. 31(4): 995—1005.

Niebling, C. R. and Conkle, M. T. 1990. Diversity of Washoe pine and comparisons with allozymes of ponderosa pine races. Can. J. For. Res. 20: 298—308.

Peeters, J. P. and Martinelli, J. A. 1989. Hierarchical cluster analysis as a tool to manage variation in germplasm collections. Theor. Appl. Genet. 78: 42—48.

Prober, S., Bell, J. C. and Moran, G. F. 1990a. A plylogenetic and allozyme approach to understanding rarity in three "green ash" eucalypts. Plant Syst. Evol. (in press).

Prober, S. M., Tompkins, C., Moran, G. F., and Bell, J. C. 1990b. The conservation genetics of *Eucalyptus paliformis* and *E. parvifolia*, two rare species from south-eastern Australia. Aust. J. Bot. 38: 79—95.

Rajora, O. P. 1988. Allozymes as aids for identification and differentiation of some *Populus maximowiczii* clonal varieties. Biochem. Syst. Ecol. 16: 635—640.

Reiseberg, L. H. 1988. Saving California's rarest tree. Center for Plant Conservation Newsletter 3(1): 1—8.

Roos, E. E. 1988, Genetic changes in a collection over time. HortScience 23(1): 86—90.

Shaffer, M. L. 1981. Minimum population sizes for species conservation. Bioscience 31: 131—134.

Schnabel, A. and Hamrick, J. L. 1990. Comparative analysis of population genetic structure in *Quercus macrocarpa* and *Q. gambelii*. Syst. Bot. 15(2): 240—251.

Scholz, F., Gregorius, H. R. and Rudin, D. 1989. Genetic Effects of Air Pollutants in Forest Tree Populations. Springer Verlag.

Sirkkomaa, S. 1983. Calculations on the decrease of genetic variation due to the founder effect. Hereditas 99: 11—20.

Smouse, P. E. and Bush, R. M. 1992. Evidence for the adaptive significance of allozymes in forest trees. This issue (pp. 179—196).

Sneath, P. H. A. and Sokai, R. R. 1973. Numerical Taxonomy. W. H. Freeman, San Francisco.

Soule, M. E. (Ed) 1987. Viable Populations for Conservation. Cambridge Univ. Press, Cambridge.

Strauss, S. H., Bousquet, J., Hipkins, V. D. and Hong, Y.-P. 1992. Biochemical and molecular genetic markets in biosystematic studies of forest trees. This issue (pp. 125–158).

USDA Forest Service. 1988. Eldorado National Forest Land and Resource Management Plan. Pacific Southwest Region.

Westfall, R. D. 1991. Developing seed transfer zones, pp. 313–398. In: Fins, L. and Friedman S. T. (Eds) Quantitative Forest Genetic Handbook Kluwer Academic Publishers, Dordrecht, The Netherlands.

Westfall, R. D. and Conckle, M. T. 1992. Allozyme markers in breeding zone designation. This issue (pp. 279—309).

Wilson, B. C. 1990. Gene pool reserves of Douglas-fir. For. Ecol. Manage. 35: 121—130.

Wilson, E. O. (Ed) 1988. Biodiversity. National Academy of Sciences, Washington, DC.

Wolf, C. B. 1948. The New World cypresses. Alisio. 1: 1—250.

Zobel, B. 1977. Gene conservation — as viewed by a forest tree breeder. For. Manage. 1: 399—344.

Zobel, D. B., L. F. Roth and Hawk, G. M. 1985. Ecology, pathology, and management of Port-Oxford-cedar (*Chamaecyparis lawsoniana*). USDA Forest Service, Gen. Tech. Rep. PNW—184.

New Forests **6**: 373—390, 1992.

Review paper

Nuclear, chloroplast, and mitochondrial DNA polymorphisms as biochemical markers in population genetic analyses of forest trees

DAVID B. WAGNER
Department of Forestry, University of Kentucky, Lexington, Kentucky 40546—0073, USA

Received 1 October 1990; accepted 15 July 1991

Key words: copy number, genome size, RFLP, PCR, DNA sequencing

Application. Study of DNA polymorphisms in forest trees has provided new insights into basic biology. Chloroplast, mitochondrial, and nuclear DNA polymorphisms are expected to play increasingly important roles for forest population geneticists, and further investigation of these polymorphisms may lead to applications in germplasm improvement and conservation programs.

Abstract. DNA analyses have been used only occasionally to investigate genetic polymorphisms in forest tree populations. Nonetheless, these analyses have already contributed to significant discoveries, such as paternal chloroplast and maternal mitochondrial DNA inheritance in Pinaceae. DNA polymorphisms will be increasingly exploited in the future by forest population geneticists, because available technology permits large sample sizes and yields excellent resolution. The utility of chloroplast, mitochondrial, and nuclear DNA polymorphisms is expected to be greatest when less expensive genetic markers are unavailable, insufficiently numerous, or ineffectively polymorphic. For example, DNA fingerprinting may permit the unambiguous elucidation of genetic relationships within and among populations of woody species.

Introduction

A major goal of population geneticists is to understand evolutionary mechanisms (Hartl and Clark 1989), but applications of population genetics transcend evolutionary inference. For example, germplasm conservation strategies have benefited from knowledge of population genetic principles (e.g., Brown 1978; Millar and Westfall 1992). Also, studies of population variation have led to discovery of important and useful genes, such as "quantitative trait loci" (QTLs), which may increase gains in genetic improvement programs (e.g., Paterson et al. 1991).

Regardless of specific goals or applications, empirical population genetics requires genetic variation (polymorphism). The most useful genetic polymorphisms for population genetic study vary qualitatively, permitting one-to-one mapping of phenotype onto genotype in large samples. Polymorphisms that provide this type of specific genotypic information can be considered "genetic markers." Population geneticists have traditionally employed unusual morphological phenotypes, proteins, or secondary compounds as genetic markers, but DNA analyses are now providing new markers (Leigh Brown 1989; Kreike et al. 1991; Szmidt and Wang 1991).

One type of DNA variation is copy number polymorphism, which is due to differences in the number of genomic occurrences of a particular DNA sequence (e.g., Strauss and Tsai 1988). DNA variation can also occur in base sequence, length, and arrangement. These latter types of variation are often detected as restriction fragment length polymorphisms (RFLPs), using enzymes (restriction endonucleases) that cut DNA molecules into restriction fragments of varying lengths. RFLP methodologies are technically straightforward and permit analyses of large numbers of samples (e.g., Govindaraju et al. 1989a).

DNA variation can also be studied by direct determination of DNA base sequences in a genomic region of interest. In principle, comparison of individuals is then possible at each base pair, which is a level of resolution not attainable with copy number variation or RFLPs (e.g., Kreitman 1983). Collection of sequence data from population samples now appears feasible because of the efficiency of the polymerase chain reaction (PCR) in DNA amplification (Saiki et al. 1988; Leigh Brown 1989).

In general, DNA markers are especially powerful for several reasons, including the following:

1) Potentially, a large number of polymorphisms (some of which are selectively neutral and some of which are under selection) can be identified in any taxon.
2) Widely differing levels of polymorphism can be studied.
3) DNA analyses allow investigation of not only coding, but also non-coding, variation.
4) Both Mendelian and non-Mendelian markers can be identified, because DNA resides in chloroplasts and mitochondria as well as in the nuclei.
5) It is usually possible to determine the mutational differences among DNA variants (e.g., point mutations, insertions/deletions, or rearrangements), which strengthens population analyses.

These advantages have been convincingly demonstrated on occasions when traditional genetic markers were deficient (Leigh Brown 1989; Hartl and Clark 1989; Strauss et al. 1991; Szmidt and Wang 1991). However,

DNA polymorphisms are not a panacea: other genetic markers are sometimes preferable because of their lower costs in personnel, reagents, laboratory facilities, and time.

DNA polymorphisms of forest tree populations have received only limited attention, in comparison with such variation in other organisms. Nevertheless, these new genetic markers have already permitted important insights into fundamental biological processes in forest trees. The purpose of this article is to describe the significant results and likely future roles of nuclear, chloroplast, and mitochondrial DNA analyses in population genetic studies of forest trees. This review will not address methodological choices or details of laboratory techniques, because these topics have been discussed previously (e.g., Gustafsson and Sitbon 1986; Strauss and Tsai 1988; Strauss et al. 1989; Neale and Williams 1991; Szmidt and Wang 1991).

Nuclear DNA polymorphisms

The first direct analyses of DNA variation in forest trees were reported more than twenty years ago and detected variance in DNA content (e.g., Burley 1965; Mergen and Thielges 1967; Miksche 1968, 1971; El-Lakany and Sziklai 1971; Dhir and Miksche 1974). Nuclear DNA presumably contributes to this variance, but organellar DNA may also play a role.

Copy number variation of specific nuclear sequences is related to variance in DNA content but is more precise and informative. For example, the observed 20-fold range in ribosomal (r) RNA gene copy number in radiata pine (*Pinus radiata* D. Don) is due to 10 or more nuclear loci (Cullis et al. 1988a, b). Ribosomal RNA gene concentration appears stable within individuals of Sitka spruce (*Picea sitchensis* (Bong.) Carr.) (Strauss and Howe 1990), but exhibits geographic trends in Douglas-fir (*Pseudotsuga menziesii* (Mirb.) Franco) (Strauss and Tsai 1988) and red spruce (*P. rubens* Sarg.) (Bobola et al. 1992b).

In addition to variation in copy number and genome size, nuclear RFLPs are available in forest trees. Scotch pine (*P. sylvestris* L.) (Sitbon and Gustafsson 1988), *P. radiata* (Cullis et al. 1988a), black spruce (*P. mariana* (Mill.) B.S.P.), *P. rubens* (Bobola et al. 1992a; Bobola et al. 1992b), poplars (*Populus* L.) (Smith and Sytsma 1990), and several oak (*Quercus* L.) species (Whittemore and Schaal 1991) provide examples of RFLPs in or near rRNA genes. Nuclear RFLP genotypes are variable in trembling aspen (*P. tremuloides* Michx.) and appear to distinguish this species from largetooth aspen (*P. grandidentata* Michx.) (G. R. Furnier, pers. comm.). RFLPs of highly-repeated sequences, possibly nuclear in origin, differentiate half-sib cell lines of yellow poplar (*Liriodendron tulipifera* L.) (Merkle et al. 1988).

Nuclear restriction fragments differ between two cottonwoods (*P. fremontii* Wats. and *P. angustifolia* James) and can be used to identify natural hybrids and hybrid derivatives. Joint analyses of RFLP and other data from these two species revealed that introgression is unidirectional and that susceptibility to aphid infestation is associated with repeated backcrossing to *P. angustifolia* (Keim et al. 1989a, b).

The greatest contribution of nuclear DNA markers to forest population genetics will likely be the large number of polymorphisms that can be studied. Nuclear genomes can be saturated with RFLPs (Paterson et al. 1991; Neale and Williams 1991), which would permit population genetic analyses of virtually any region of a forest tree genome. Random amplified polymorphic DNA (RAPD) markers, which are based on PCR amplification of DNA from arbitrary oligonucleotide primer sequences (Williams et al. 1990), may be more efficient and may ultimately become even more useful than RFLPs for the saturation of nuclear genomes with genetic markers. Saturated linkage maps, once available, can be used to resolve quantitative variation into Mendelian factors (Paterson et al. 1991). This would eventually allow direct study of allele frequencies of economically important traits. However, the cost of saturating a linkage map is high at present, and the benefits of such an investment are untested in forestry applications (e.g., Neale and Williams 1991).

In addition to the large number of polymorphisms that are identifiable by DNA analyses, each polymorphism is potentially highly variable. DNA fingerprinting now appears possible in woody taxa, including *Polyalthia glauca* (Hassk.) Mueller, orange (*Citrus sinensis* Osbeck), Torrey pine (*P. torreyana* C. Parry ex Carrière), *P. radiata*, eastern cottonwood (*P. deltoides* Marsh.), *P. tremuloides*, apple (*Malus* x *domestica* Borkh.), black cherry (*Prunus serotina* Ehrh.), blackberries and raspberries (*Rubus* L.), and box elder (*Acer negundo* L.) (Rogstad et al. 1988a, b; Ryskov et al. 1988; Nybom et al. 1989; Zimmerman et al. 1989; Nybom 1990; Nybom et al. 1990; Nybom and Schaal 1990a, b; Nybom and Rogstand 1990; Nybom and Hall 1991; Rogstad et al. 1991). DNA fingerprinting may eventually permit unambiguous elucidation of genetic relationships in forest tree populations.

Chloroplast (cp) DNA polymorphisms

The power of DNA analyses to identify polymorphisms extends beyond the nuclear genome to the organellar genomes. In fact, cpDNA variation in forest trees has received greater attention than nuclear DNA polymorphisms for several reasons. Chloroplast DNA is readily detected because of its occurrence in multiple copies per cell. Also, the cpDNA molecule is smaller and structurally simpler than nuclear DNAs, which allows straightforward molecular interpretations of polymorphisms (e.g., Palmer et al. 1988).

Inheritance tests provided some of the first reports of cpDNA variation in forest trees, because knowledge of the mode of inheritance is important for any new polymorphic marker. This knowledge is critical for chloroplast polymorphisms, which do not obey the rules of Mendelian inheritance. These polymorphisms are inherited maternally in most angiosperms (Sears 1980), but paternal contributions can be variable in taxa with biparental chloroplast inheritance (e.g., Chiu et al. 1988).

Chloroplast DNA inheritance appears predominantly paternal in conifers, including *P. menziesii* (Neale et al. 1986), larch (*Larix* Mill.) hybrids (Szmidt et al. 1987), pine (*Pinus* L.) hybrids (Wagner et al. 1987; Neale and Sederoff 1989; Wagner et al. 1992), spruce (*Picea* A. Dietr.) hybrids (Szmidt et al. 1988; Neale and Sederoff 1988; Stine et al. 1989; Stine and Keathley 1990), coast redwood (*Sequoia sempervirens* D. Don Endl.) (Neale et al. 1989), jack pine (*P. banksiana* Lamb.) (Wagner et al. 1989b), incense-cedar (*Calocedrus decurrens* [Torr.] Florin) (Neale et al. 1991), and arbor-vitae (*Thuja* L.) hybrids (A. E. Szmidt, pers. comm.). The coniferous mode of cpDNA inheritance contrasts sharply with the norm in angiosperms, but is consistent with the inheritance of a plastid mutant in Sugi (*Cryptomeria japonica* D. Don) (Ohba et al. 1971) and with ultrastructural evidence (Chesnoy 1987; Owens and Morris 1990; Owens and Morris 1991). Unfortunately, cpDNA inheritance data has not been reported for non-coniferous forest trees.

It would be premature to conclude that chloroplast inheritance is strictly paternal in conifers, because the available sample sizes are usually small and do not exclude the possibility of maternal leakage. In fact, progeny with non-paternal cpDNA genotypes were observed in *C. japonica* (Ohba et al. 1971), *P. menziesii* (Neale et al. 1986), *Larix* hybrids (Szmidt et al. 1987), *P. banksiana* (Wagner et al. 1989b), and *C. decurrens* (Neale et al. 1991). The causes of these non-paternal cpDNA genotypes are unresolved in most cases, but germplasm contamination is at least partially responsible in *P. banksiana* (Wagner et al. 1989b).

Most attempts to identify intra-specific cpDNA polymorphism in forest trees have succeeded, even with small sample sizes (e.g., Ali et al. 1991). In addition to the examples represented by inheritance tests (cited above), cpDNA polymorphisms are also known in lodgepole pine (*P. contorta*

Dougl.) (Wagner et al. 1987), *P. sitchensis* (Szmidt et al. 1988), white spruce (*Picea glauca* (Moench) Voss), (Szmidt et al. 1988; Stine et al. 1989), *P. densata* Mast. (Wang and Szmidt 1990), western white pine (*P. monticola* Dougl. ex D. Don) (White 1990b), Engelmann spruce (*P. engelmannii* Parry ex Engelm.) (Stine and Keathley 1990), eastern hemlock (*Tsuga canadensis* (L.) Carr.) (Wang 1990), *Liriodendron tulipifera* (Parks and Wendel 1990), white oak (*Q. alba* L.), bur oak (*Q. macrocarpa* Michx.), swamp chestnut oak (*Q. michauxii* Nutt.), post oak (*Q. stellata* Wang.) (Whittemore and Schaal 1991), sessile oak (*Q. petraea* (Matt.) Liebl.), pedunculate oak (*Q. robur* L.) (Kremer et al. 1991), loblolly pine (*P. taeda* L.), slash pine (*P. elliottii* Engelm.) (Wagner et al. 1992), Rocky Mountain juniper (*Juniperus scopulorum* Sarg.), eastern red cedar (*J. virginiana* L.) (S. G. Ernst, pers. comm.), Hartweg pine (*P. hartwegii* Lindl.) (J. A. Matos and B. A. Schaal, pers. comm.), and pinyon (*P. edulis* Engelm.) (P. Keim, pers. comm.).

Identification of cpDNA polymorphisms is encouraging, but the actual utility of these new genetic markers can only be evaluated in population surveys. Although such surveys are rare in forest trees, cpDNA polymorphisms are already clearly informative. For example, cpDNA variants provide markers of hybridization in *Picea* and *Pinus* (Szmidt et al. 1988; Sigurgeirsson et al. 1991; Wang and Szmidt 1990; Wagner et al. 1992). These markers provide evolutionary insights and also may reduce economic costs resulting from inappropriate deployment of cryptic hybrid seedlots (El-Kassaby et al. 1988, Wang and Szmidt 1990; Sutton et al. 1991a).

In contrast to the *Picea* and *Pinus* examples, cpDNA polymorphisms fail to distinguish closely related *Quercus* species in either North America or Europe, possibly because of inter-specific hybridization in *Quercus* (Whittemore and Schaal 1991; Kremer et al. 1991). This result indicates that cpDNA diversity within taxonomic groups has the potential to confound cpDNA phylogenies (Whittemore and Schaal 1991). Nonetheless, the *Quercus* cpDNA variability is geographically structured on both continents and may be informative for seed zone delineation.

The largest population survey of DNA variation in forest trees has examined cpDNA insertion/deletion polymorphisms in more than 1300 individuals from 152 populations of *P. banksiana* and *P. contorta*. Chloroplast DNA variability differentiates these two species and also resides within subspecies of *P. contorta*, within allopatric and sympatric populations, and even within chimeric individuals (Wagner et al. 1987; Govindaraju et al. 1988). Chloroplast DNA is also variable within individuals in *P. monticola* and *Picea*, and the hypothesis of occasional biparental inheritance has been advanced to account for these observations (White 1990b; Sutton et al. 1991a; but see Stine et al. 1989). This hypothesis is consistent with the reports of infrequent non-paternal cpDNA genotypes in coniferous progeny arrays (e.g., Wagner et al. 1989b). Although the significance of within-individual genetic variation has been debated at length (e.g., Painter 1966; Libby et al. 1969; Klekowski 1988), cpDNA markers may now facilitate further analyses of this type of variation.

One of the *P. banksiana* — *P. contorta* cpDNA polymorphisms has at least 27 distinct genotypes, 19 of which are associated with a region of natural hybridization (Govindaraju et al. 1989a), and mutation mapping experiments have revealed even greater levels of variation (Govindaraju et al. 1989b). Formally, cpDNA recombination is a possible cause of the unusual chloroplast genotypes that occur in sympatric populations of *P. banksiana* — *P. contorta* (Govindaraju et al. 1989a). These unusual genotypes are of special interest, because recombination between parental chloroplast genomes has not been observed in controlled matings of any land plant (e.g., Chiu and Sears 1985). If cpDNA recombination between parental genotypes is extremely rare, one would predict that recombinants might be detectable only in populations that have persisted for many generations. Thus, geographic regions of natural hybridization could permit detection and rigorous tests of cpDNA recombination, if an adequate number of parental markers can be identified in the chloroplast genome.

Several *P. banksiana* — *P. contorta* cpDNA genotypes are spatially patterned within sympatric populations (Wagner et al. 1991b). This pattern would ordinarily be unexpected for paternally-inherited factors in outcrossers (Epperson and Allard 1984; Cheliak et al. 1985; Epperson and Allard 1989), but effective gene flow between *P. banksiana* and *P. contorta* may be limited (Saylor and Smith 1966; Righter and Stockwell 1949). Inter-specific gene flow constraints could also explain the observed concordance within sympatric individuals of morphological taxonomic classification and cpDNA genotype (Wagner and Govindaraju 1988; D. B. Wagner and D. R. Govindaraju, unpubl. data). However, such constraints would not simultaneously explain the predominantly random association of allozymes and cpDNA genotypes in the same individuals (Wagner et al. 1989a; D. B. Wagner and B. P. Dancik, unpubl. data). Although several questions will remain unanswered until more data is available, it is clear that cpDNA variation is a significant component of a complex population genetic architecture in sympatric populations of *P. banksiana* — *P. contorta*.

The evolutionary conservatism of the chloroplast genome (Palmer et al. 1988; Clegg et al. 1991) appears to contradict the numerous examples of intra-specific cpDNA polymorphisms in forest trees. However, coniferous chloroplast genomes lack an inverted repeat that is typical of other plants (Strauss et al. 1988; Lidholm et al. 1988). This physical characteristic may contribute to the occurrence of DNA rearrangements in conifers (Ali et al. 1991). Moreover, cpDNA variability may be restricted to only a few mutation hotspots within a particular species (Ali et al. 1991), in which case there would be no contradiction.

Hotspots of chloroplast polymorphism are associated with the presence of cpDNA repeat units in several plants, including *P. menziesii* (Blasko et al. 1988; Strauss et al. 1988; Tsai and Strauss 1989; Ali et al. 1991). Theory predicts that within-individual organellar variation may be frequent when DNA repeats are responsible for polymorphism (Clark 1988). Therefore, the duplicated and repeated DNA sequences that occur in the chloroplast genomes of *P. banksiana—P. contorta* and *P. monticola* (Govindaraju et al. 1989b; White 1990a; Lidholm et al. 1991) are especially intriguing, because these taxa harbor cpDNA variation within individuals (Govindaraju et al. 1988; White 1990b).

Although most attempts to identify intra-specific cpDNA RFLPs in forest trees have succeeded, failure have been reported for European larch (*Larix decidua* Mill.), Japanese larch (*Larix leptolepis* Sieb. and Zucc.) (Szmidt et al. 1987), blue spruce (*P. pungens* Engelm.) (Stine et al. 1989), *P. taeda* (Neale and Sederoff 1989; Ali et al. 1991), Chinese pine (*P. tabulaeformis* Carr.), Yunnan pine (*P. yunnanensis* Franch.), Masson pine (*P. massoniana* Lamb.) (Wang and Szmidt 1990), *Liriodendron chinense* (Hemsl.) Sarg. (Parks and Wendel 1990), *P. deltoides*, black poplar (*P. nigra* L.) (Smith and Sytsma 1990), *Q. virginiana* var. *fusiformis* (Small) Sarg. (Whittemore and Schaal 1991), longleaf pine (*P. palustris* Mill.) (Wagner et al. 1992), and Michoacán pine (*P. michoacana* Martínez) (J. A. Matos and B. A. Schaal, pers. comm.). Within shortleaf pine (*P. echinata* Mill.), the only observed cpDNA variation is explainable by inter-specific hybridization (Wagner et al. 1991a). Despite their failure to detect intra-specific cpDNA polymorphism, most of the above investigations identified distinctions among closely-related species. For example, Szmidt et al. (1987) found three cpDNA variants in a total sample of only 11 individuals from species crosses of *Larix*.

Additional study of the above examples would be appropriate prior to concluding that cpDNA is monomorphic intra-specifically, because sample sizes were usually small in terms of numbers of trees and numbers of cpDNA base pairs. A priori, one would expect to screen a very large number of base pairs in the chloroplast genome before finding any polymorphism, because cpDNA evolves slowly (Palmer et al. 1988; Clegg et al. 1991).

The role of the long-lived, woody-perennial life form in producing and maintaining cpDNA polymorphisms is entirely unknown. Similarly, we understand little of the similarities and/or differences in gymnosperm vs. angiosperm cpDNA polymorphisms and in inter- vs. intra-specific cpDNA polymorphisms. Expanded investigation would be enlightening on these topics.

Mitochondrial (mt) DNA polymorphisms

Male sterility is one of the best known effects of plant mitochondrial polymorphisms (Pring and Lonsdale 1985). Consequently, current understanding of mitochondrial population genetics in plants arises primarily from analyses of male sterility polymorphisms (i.e., gynodioecy) (e.g., van Damme 1986).

Maternal inheritance of mitochondrial genomes predominates in angiosperms (Conde et al. 1979; Soliman et al. 1987), but conifers differ in this regard. Ultrastructural evidence indicates that mitochondria are contributed solely by the pollen donor in Cupressaceae and Taxodiaceae, while both paternal and maternal mitochondria are present at fertilization in Pinaceae and Taxaceae (Chesnoy 1987; Owens and Morris 1990, 1991). Mitochondrial restriction fragments are predominantly maternally inherited in *Pinus* species hybrids (Neale and Sederoff 1988; Wagner et al. 1991c), *P. taeda* (Neale and Sederoff 1989), *P. contorta* (Wagner et al. 1991c), and *Picea* species hybrids (Sutton et al. 1991b; M. Stine, pers. comm.). In contrast, mtDNA RELPs appear paternally inherited in *S. sempervirens* and *C. decurrens* (Neale et al. 1989, 1991). These RFLP results are consistent with the ultrastructural evidence, because they represent assays of "net" outcomes after reproduction and development. Mitochondrial inheritance data from non-coniferous trees are unavailable for comparison.

Despite the availability of mtDNA polymorphisms within at least six conifer species (see articles cited above), we know little about the genomic, taxonomic, or geographic distributions of mitochondrial variation. The molecular nature of mtDNA polymorphism has been inferred in only one forest tree species, *C. decurrens* (Neale et al. 1991), and the use of mitochondrial markers in population surveys is just beginning (e.g., Sutton et al. 1991b). However, mtDNA polymorphisms are under investigation in several laboratories (e.g., S. H. Strauss et al., pers. comm.; J. Dong, pers. comm.; T. Li, pers. comm.), and we can expect substantive new information in the near future.

Conclusions

Despite the limited resources directed toward population surveys of DNA polymorphisms in forest trees, DNA as a biochemical marker has advanced our understanding of biology and provides additional opportunities to enhance the quantity and quality of genetic markers. The new markers are expected to be of greatest value when other polymorphisms are inadequate, such as in studies of taxa that appear genetically depauperate. For example, although isoenzyme variation is unknown within *P. torreyana* populations (Ledig and Conkle 1983), DNA polymorphisms revealed two genotypes in a sample of only three individuals of this species (Rogstad et al. 1988b)!

Future searches for DNA polymorphisms are likely to benefit from a variety of technical advances, such as DNA amplification by PCR and the use of RAPD technology. The polymerase chain reaction has already been used successfully in forest trees (Bousquet et al. 1990) and could be combined with DNA sequencing to reveal polymorphism in genomic regions that appear monomorphic by less sophisticated approaches (Leigh Brown 1989). A practical consideration is that PCR permits production of virtually unlimited quantities of DNA from very small amounts of stable materials (e.g., seeds), thereby reducing or eliminating the need to collect and transport large quantities of comparatively unstable tissues (e.g., leaves).

The potentially large number and widely varying diversities of nuclear DNA polymorphisms will be useful for solving a variety of problems in forest genetics. Identification and analysis of QTLs (e.g., Neale and Williams 1991) may be one of the most significant applications of these polymorphisms. Nuclear DNA markers could also increase the resolution of parentage analyses (Adams 1991) and permit more thorough investigation of linkage disequilibrium (Bush and Smouse 1992). However, the future roles of nuclear DNA polymorphisms should themselves be considered critical topics for current research. The most appropriate roles will likely depend on important details regarding cost and distribution of polymorphism.

Recombinant DNA technology has provided the first markers suitable for organellar population genetics in trees. Organellar polymorphisms may now be identifiable in nearly any species of choice, and these markers promise much information. For example, we should expect gene flow studies to benefit from organellar markers, because current gene flow estimators often fail to distinguish gene flow by pollen vs. gene flow by seed in natural populations (Ellstrand 1992). In particular, the contrasting modes of cpDNA and mtDNA inheritance in Pinaceae may enhance simultaneous analyses of paternal and maternal lineages (Neale and Sederoff 1989).

Multilocus population genetics will benefit substantially from the availability of organellar markers. Cytonuclear disequilibria have been defined recently and can be more enlightening than the corresponding nuclear parameters (Asmussen et al. 1987; Schnabel and Asmussen 1989). Forest trees may be among the most informative of all organisms for empirical applications of the new cytonuclear theory, because they represent three inheritance patterns in a single life-form, i.e., mtDNA paternal, cpDNA paternal, nuclear loci Mendelian (as in *S. sempervirens* and *C. decurrens*); mtDNA maternal, cpDNA paternal, nuclear loci Mendelian (as in *Pinus* and *Picea*); and mtDNA maternal, cpDNA maternal, nuclear loci Mendelian (assuming that angiospermous forest trees behave as most other flowering plants).

Organellar polymorphisms should not be overlooked as a potential source of economically useful variation (McDaniel 1984; Pring and Lonsdale 1985; Börner and Sears 1986). Indeed, reciprocal effects occur in forest trees, although it is not yet known if these effects are due to cytoplasmic factors (e.g., Perry 1976; Wilcox 1983). Identification and detailed molecular investigation of organellar DNA polymorphisms, in concert with classical genetic experiments, may elucidate the causes of these reciprocal effects.

Clearly, DNA polymorphisms have enhanced the wide array of methodologies available to forest geneticists. It now appears possible to choose appropriate genetic markers based on questions at hand, rather than the reverse.

Acknowledgments

I am grateful to many teachers, colleagues and students for numerous discussions and for their generous willingness to share unpublished information. The manuscript was improved substantially by the enormously careful reading and suggestions of D. E. Harry, D. B. Neale and D. L. Copes. Support was provided by Kentucky Agricultural Experiment Station Research Funds (McIntire-Stennis Project KY00640), USDA Grants 85-FSTY-9-0149 and 90-37290-5681, USDA Forest Service Cooperative Agreements 19-88-032 and 19-88-033, and L'Institut National de la Recherche Agronomique. This is Journal Paper Number 90-8-40 from the Kentucky Agricultural Experiment Station.

References

Adams, W. T. 1992. Gene dispersal within forest tree populations. New Forests 6: 217—240.

Ali, I. F., Neale, D. B. and Marshall, K. A. 1991. Chloroplast DNA restriction fragment length polymorphism in *Sequoia sempervirens* D. Don Endl., *Pseudotsuga menziesii* (Mirb.) Franco, *Calocedrus decurrens* (Torr.), and *Pinus taeda* L. Theor. Appl. Genet. 81: 83—89.

Asmussen, M. A., Arnold, J. and Avise, J. C. 1987. Definition and properties of disequilibrium statistics for associations between nuclear and cytoplasmic genotypes. Genetics 115: 755—768.

Blasko, K., Kaplan, S. A., Higgins, K. G., Wolfson, R. and Sears, B. B. 1988. Variation in copy number of a 24-base pair tandem repeat in the chloroplast DNA of *Oenothera hookeri* strain Johansen. Curr. Genet. 14: 287—292.

Bobola, M. S., Eckert, R. T. and Klein, A. S. 1992a. Restriction fragment variation in the nuclear rDNA repeat unit within and between *Picea rubens* and *Picea mariana*. Can. J. For. Res. 22: 255—363.

Bobola, M. S., Smith, D. E. and Klein, A. S. 1992b. Five major nuclear ribosomal repeats represent a large and variable fraction of the genomic DNA of *Picea rubens* and *P. mariana*. Molec. Biol. Evol. 9: 125—137.

Börner, T. and Sears, B. B. 1986. Plastome mutants. Plant Molec. Biol. Rep. 4: 69—92.

Bousquet, J., Simon, L. and Lalonde, M. 1990. DNA amplification from vegetative and sexual tissues of trees using polymerase chain reaction. Can. J. For. Res. 20: 254—257.

Brown, A. H. D. 1978. Isozymes, plant population genetic structure and genetic conservation. Theor. Appl. Genet. 52: 145—157.

Burley, J. 1965. Karyotype analysis of Sitka spruce, *Picea sitchensis* (Bong.) Carr. Silvae Genet. 14: 127—132.

Bush, R. M. and Smouse, P. E. 1992. Evidence for the adaptive significance of allozymes in forest trees. This issue (pp. 179—196).

Cheliak, W. M., Dancik, B. P., Morgan, K., Yeh, F. C. H. and Strobeck, C. 1985. Temporal variation of the mating system in a natural population of jack pine. Genetics 109: 569—584.

Chesnoy, L. 1987. L'origine des organites du cytoplasme embryonnaire chez les Gymnospermes. Bull. Soc. Bot. Fr., Actual. Bot. 134: 51—56.

Chiu, W.-L. and Sears, B. B. 1985. Recombination between chloroplast DNAs does not occur in sexual crosses of *Oenothera*. Molec. Gen. Genet. 198: 525—528.

Chiu, W.-L., Stubbe, W. and Sears, B. B. 1988. Plastid inheritance in *Oenothera*: organelle genome modifies the extent of biparental plastid transmission. Curr. Genet. 13: 181—189.

Clark, A. G. 1988. Deterministic theory of heteroplasmy. Evolution 42: 621—626.

Clegg, M. T., Learn, G. H. and Golenberg, E. M. 1991. Molecular evolution of chloroplast DNA. pp. 135—149. In: Selander, R. K., Clark, A. G. and Whittam, T. S. (Eds) Evolution at the Molecular Level. Sinauer Assoc., Inc., Sunderland, MA.

Conde, M. F., Pring, D. R. and Levings, C. S. III, 1979. Maternal inheritance of organelle DNAs in *Zea mays — Zea perennis* reciprocal crosses. J. Hered. 70: 2—4.

Cullis, C. A., Creissen, G. P., Gorman, S. W. and Teasdale, R. D. 1988a. The 25S, 18S, and 5S ribosomal RNA genes from *Pinus radiata* D. Don. pp. 34—40. In: Cheliak, W. M. and Yapa, A. C. (Eds) Molecular Genetics of Forest Trees. Petawawa National Forestry Institute Inform. Rep. PI-X-80, Chalk River, Ontario, Canada.

Cullis, C. A. Creissen, G. P., Gorman, S. W. and Teasdale, R. D. 1988b. The 25S and 18S ribosomal DNAs from *Pinus radiata*. p. 463. In: Hanover, J. W. and Keathley, D. E. (Eds) Genetic Manipulation of Woody Plants. Plenum Press, New York, New York.

Dhir, N. K. and Miksche, J. P. 1974. Intraspecific variation of nuclear DNA content in *Pinus resinosa* Ait. Can. J. Genet. Cytol. 16: 77—83.

Dong. J. (pers. comm.) Department of Forestry, University of Kentucky, Lexington, KY.

El-Kassaby, Y. A., Sigurgeirsson, A. and Szmidt, A. E. 1988. The use of restriction analysis of chloroplast DNA in classifying hybrid spruce seedlots, pp. 67—88. In: Hällgren, J.-E. (Ed) Molecular genetics of forest trees: Proceedings of the Frans Kempe symposium in Umeå, June 14—16, 1988. Swedish University of Agricultural Sciences, Umeå, Sweden.

El-Lakany, M. H. and Sziklai, O. 1971. Intraspecific variation in nuclear characteristics of Douglas-fir. Adv. Front. Plant Sci. 28: 363—378.

Ellstrand, N. C. 1992. Gene flow among seed plant populations. This issue (pp. 241—256).

Epperson, B. K. and Allard, R. W. 1984. Allozyme analysis of the mating system in lodgepole pine populations. J. Hered. 75: 212—214.

Epperson, B. K. and Allard, R. W. 1989. Spatial autocorrelation analysis of the distribution of genotypes within populations of lodgepole pine. Genetics 121: 369—377.

Ernst, S. G. (pers. comm.) Department of Forestry, Fisheries and Wildlife, University of Nebraska, Lincoln, NE.

Furnier, G. R. (pers. comm.) Department of Forest Resources, University of Minnesota, St. Paul, MN.

Govindaraju, D. R., Dancik, B. P. and Wagner, D. B. 1989a. Novel chloroplast DNA polymorphism in a sympatric region of two pines. J. Evol. Biol. 2: 49—59.

Govindaraju, D. R., Wagner, D. B., Lidholm, J., Gustafsson, P. and Szmidt, A. E. 1989b. A chloroplast DNA probe identifies unexpectedly high levels of polymorphism in *Pinus banksiana* and *P. contorta*. South. forest tree improvement conf. 20: 147—152.

Govindaraju, D. R., Wagner, D. B., Smith, G. P. and Dancik, B. P. 1988. Chloroplast DNA variation within individual trees of a *Pinus banksiana — P. contorta* sympatric region. Can. J. For. Res. 18: 1347—1350.

Gustafsson, P. and Sitbon, F. 1986. Studies on genetic differences between pine species by modern DNA technology; the nuclear genome, pp. 253—268. In: Lindgren, D. (Ed) Provenances and forest tree breeding for high latitudes: Proceedings of the Frans Kempe symposium in Umeå, June 10—11, 1986. Swedish University of Agricultural Sciences, Umeå, Sweden.

Hartl, D. L. and Clark, A. G. 1989. Principles of Population Genetics. 2nd edition. Sinauer Associates, Inc., Sunderland, Massachusetts.

Keim, P. (pers. comm.) Department of Biology, Northern Arizona State University, Flagstaff, AZ.

Keim, P., Paige, K. N., Whitham, T. G. and Lark, K. G. 1989a. Genetic analysis of an interspecific hybrid swarm of Populus: occurrence of unidirectional introgression. Genetics 123: 557—565.

Keim, P., Whitham, T. G. and Lark, K. G. 1989b. The genetic structure of a *Populus* hybrid swarm and the correlation of insect susceptibility to specific genetic classes. Genetics 122: s30.

Klekowski, E. J., Jr. 1988. Mutation, Developmental Selection, and Plant Evolution. Columbia University Press, New York.

Kreike, J., Burg, K., Zechmeister, M., Haider, T. and Glössl, J. 1991. DNA-fingerprint and RFLP analysis as tools to study genetic diversity in populations of fir, spruce and oak, pp. 95—107. In: Müller-Starck, G. and Ziehe, M. (Eds) Genetic Variation of Forest Tree Populations in Europe. Sauerländer-Verlag, Frankfurt/Main.

Kreitman, M. 1983. Nucleotide polymorphism at the alcohol dehydorogenase locus of *Drosophila melanogaster.* Nature 304: 412—417.

Kremer, A., Petit, R., Zanetto, A., Fougère, V., Ducousso, A., Wagner, D. and Chauvin, C. 1991. Nuclear and organelle gene diversity in *Q. robur* and *Q. petraea*, pp. 141—166. In:

Müller-Starck, G. and Ziehe, M. (Eds) Genetic Variation of Forest Tree Populations in Europe. Sauerländer-Verlag, Frankfurt/Main.

Ledig, F. T. and Conkle, M. T. 1983. Gene diversity and genetic structure in a narrow endemic, Torrey pine (*Pinus torreyana* Parry ex Carr.). Evolution 37: 79—85.

Leigh Brown, A. J. 1989. Population genetics at the DNA level: a review of the contribution of restriction enzyme studies. Oxford Surv. Ecol. Biol. 6: 207—242.

Li, T. (pers. comm.) Department of Forestry, University of Kentucky, Lexington, KY.

Libby, W. J., Stettler, R. F. and Seitz, F. N. 1969. Forest genetics and forest tree breeding. Ann. Rev. Genet. 3: 469—494.

Lidholm, J., Szmidt, A. E., Hällgren, J.-E. and Gustafsson, P. 1988. The chloroplast genomes of conifers lack one of the rRNA-encoding inverted repeats. Molec. Gen. Genet. 212: 6—10.

Lidholm, J., Szmidt, A. and Gustafsson, P. 1991. Duplication of the *psbA* gene in the chloroplast genome of two *Pinus* species. Molec. Gen. Genet. 226: 345—352.

Matos, J. A. and Schaal, B. A. (pers. comm.) Department of Biology, Washington University, St. Louis, MO 63130—4899.

McDaniel, R. G. 1984. Plant genetic engineering via organelle transfer. Plant Breeding Rev. 2: 283—302.

Mergen, F. and Thielges, B. A. 1967. Intraspecific variation in nuclear volume in four conifers. Evolution 21: 720—724.

Merkle, S. A., Chou, P. L. and Sommer, H. E. 1988. Stability of highly repeated sequences in the DNA of embryogenic cultures of yellow poplar, pp. 85—88. In: Cheliak, W. M. and Yapa, A. C. (Eds) Molecular Genetics of Forest Trees. Petawawa National Forestry Institute Inform. Rep. PI-X-80, Chalk River, Ontario, Canada.

Miksche, J. P. 1968. Quantitative study of intraspecific variation of DNA per cell in *Picea glauca* and *Pinus banksiana*. Can. J. Genet. Cytol. 10: 590—600.

Miksche, J. P. 1971. Intraspecific variation of DNA per cell between *Picea sitchensis* (Bong.) Carr. provenances. Chromosoma 32: 343—352.

Millar, C. I. and Westfall, R. D. 1992. Allozyme markers in forest genetic conservation. This issue (pp. 347—371).

Neale, D. B., Marshall, K. A. and Harry, D. E. 1991. Inheritance of chloroplast and mitochondrial DNA in incense-cedar (*Calocedrus decurrens*). Can. J. For. Res. 21: 717—720.

Neale, D. B., Marshall, K. A. and Sederoff, R. R. 1989. Chloroplast and mitochondrial DNA are paternally inherited in *Sequoia sempervirens* D. Don Endl. Proc. Natl. Acad. Sci. USA 86: 9347—9349.

Neale, D. B. and Sederoff, R. R. 1988. Inheritance and evolution of conifer organelle genomes, pp. 251—264. In: Hanover, J. W. and Keathley, D. E. (Eds) Genetic Manipulation of Woody Plants. Plenum Press, New York, New York.

Neale, D. B. and Sederoff, R. R. 1989. Paternal inheritance of chloroplast DNA and maternal inheritance of mitochondrial DNA in loblolly pine. Theor. Appl. Genet. 77: 212—216.

Neale, D. B., Wheeler, N. C. and Allard, R. W. 1986. Paternal inheritance of chloroplast DNA in Douglas-fir. Can. J. For. Res. 16: 1152—1154.

Neale, D. B. and Williams, C. G. 1991. Restriction fragment length polymorphism mapping in conifers and applications to forest genetics and tree improvement. Can. J. For. Res. 21: 545—554.

Nybom, H. 1990. DNA fingerprints in sports of 'Red Delicious' apples. Hortscience 25: 1641—1642.

Nybom, H. and Hall, H. K. 1991. Minisatellite DNA "fingerprints" can distinguish *Rubus* cultivars and estimate their degree of relatedness. Euphytica 53: 107—114.

Nybom, H. and Rogstad, S. H. 1990. DNA "fingerprints" detect genetic variation in *Acer negundo* (*Aceraceae*). Plant Syst. Evol. 173: 49—56.

Nybom, H., Rogstad, S. H. and Schaal, B. A. 1990. Genetic variation detected by use of the M13 "DNA fingerprint" probe in *Malus*, *Prunus*, and *Rubus* (Rosaceae). Theor. Appl. Genet. 79: 153—156.

Nybom, H. and Schaal, B. A. 1990a. DNA "fingerprints" reveal genotypic distributions in natural populations of blackberries and raspberries (Rubus, Rosaceae). Amer. J. Bot. 77: 883—888.

Nybom, H. and Schaal, B. A. 1990b. DNA "fingerprints" applied to paternity analysis in apples (*Malus* x *domestica*). Theor. Appl. Genet. 79: 763—768.

Nybom, H., Schaal, B.A. and Rogstad, S. H. 1989. DNA "fingerprints" can distinguish cultivars of blackberries and raspberries. Acta Hort. 262: 305—310.

Ohba, K., Iwakawa, M., Okada, Y. and Murai, M. 1971. Paternal transmission of a plastid anomaly in some reciprocal crosses of Sugi, *Cryptomeria japonica* D. Don. Silvae Genet. 20: 101—107.

Owens, J. N. and Morris, S. J. 1990. Cytological basis for cytoplasmic inheritance in *Pseudotsuga menziesii*. I. Pollen tube and archegonial development. Amer. J. Bot. 77: 433—445.

Owens, J. N. and Morris, S. J. 1992. Cytological basis for cytoplasmic inheritance in *Pseudotsuga menziessii*. II. Fertilization and proembryo development. Amer. J. Bot. 78: 1515—1527.

Painter, R. H. 1966. Lessons to be learned from past experience in breeding plants for insect resistance, pp. 349—355. In: Gerhold, H. D., Schreiner, E. J., McDermott, R. E. and Winieski, J. A. (Eds) Breeding Pest Resistant Trees. Pergamon Press, Oxford, UK.

Palmer, J. D., Jansen, R. K., Michaels, H. J., Chase, M. W. and Manhart, J. R. 1988. Chloroplast DNA variation and plant phylogeny. Ann. Missouri Bot. Gard. 75: 1180—1206.

Parks, C. R. and Wendel, J. F. 1990. Molecular divergence between Asian and North American species of *Liriodendron* (Magnoliaceae) with implications for interpretation of fossil floras. Amer. J. Bot. 77: 1243—1256.

Paterson, A. H., Damon, S., Hewitt, J. D., Zamir, D., Rabinowitch, H. D., Lincoln, S. E., Lander, E. S. and Tanksley, S. D. 1991. Mendelian factors underlying quantitative traits in tomato: comparison across species, generations and environments. Genetics 127: 181—197.

Perry, T. O. 1976. Maternal effects on the early performance of tree progenies, pp. 473—481. In: Cannell. M. G. R. and Last, F. T. (Eds) Tree Physiology and Yield Improvement. Academic Press, London, UK.

Pring, D. R. and Lonsdale, D. M. 1985. Molecular biology of higher plant mitochondrial DNA. Internat. Rev. Cytol. 97: 1—46.

Righter, F. I. and Stockwell, P. 1949. The fertile species hybrid, *Pinus murraybanksiana*. Madroño 10: 65—69.

Rogstad, S. H., Nybom, H. and Schaal, B. A. 1991. The tetrapod DNA fingerprinting M13 repeat probe reveals genetic diversity and clonal growth in quaking aspen (*Populus tremuloides*, Salicaceae). Plant Syst. Evol. 175: 115—123.

Rogstad, S. H., Patton, J. C. and Schaal, B. A. 1988a. A human minisatellite probe reveals RFLPs among individuals of two angiosperms. Nucleic Acids Res. 16: 11378.

Rogstad, S. H., Patton, J. C. II and Schaal, B. A. 1988b. M13 repeat probe detects DNA minisatellite-like sequences in gymnosperms and angiosperms. Proc. Natl. Acad. Sci. USA 85: 9176—9178.

Ryskov, A. P., Jincharadze, A. G., Prosnyak, M. I., Ivanov, P. L. and Limborska, S. A. 1988. M13 phage DNA as a universal marker for DNA fingerprinting of animals, plants and microorganisms. FEBS Lett. 233: 388—392.

Saiki, R. K., Gelfand, D. H., Stoffel, S., Scharf, S. J., Higuchi, R., Horn, G. T., Mullis, K. B. and Erlich, H. A. 1988. Primer-directed enzymatic amplification of DNA with a thermostable DNA polymerase. Science 239: 487—491.

Saylor, L. C. and Smith, B. W. 1966. Meiotic irregularity in species and interspecific hybrids of *Pinus*. Amer. J. Bot. 53: 453—468.

Schnabel, A. and Asmussen, M. A. 1989. Definition and properties of disequilibria within nuclear-mitochondrial-chloroplast and other nuclear-dicytoplasmic systems. Genetics 123: 199—215.

Sears, B. B. 1980. The elimination of plastids during spermatogenesis and fertilization in the plant kingdom. Plasmid 4: 233—255.

Sigurgeirsson, A., Szmidt, A. E. and Karpinska, B. 1991. Alaskan *Picea sitchensis* populations infiltrated with *Picea glauca* genes: a study using DNA markers, pp. 197—207. In: Fineschi, S., Malvolti, M. E., Cannata, F. and Hattemer, H. H. (Eds) Biochemical Markers in the Population Genetics of Forest Trees. SPB Academic Publishing bv, The Hague, The Netherlands.

Sitbon, F. and Gustafsson, P. 1988. Ribosomal DNA polymorphism in *Pinus sylvestris*, pp. 485—486. In: Hanover, J. W. and Keathley, D. E. (Eds) Genetic Manipulation of Woody Plants. Plenum Press, New York, New York.

Smith, R. L. and Sytsma, K. J. 1990. Evolution of *Populus nigra* (Sect. *Aigeiros*): introgressive hybridization and the chloroplast contribution of *Populus alba* (Sect. *Populus*). Amer. J. Bot. 77: 1176—1187.

Soliman, K., Fedak, G. and Allard, R. W. 1987. Inheritance of organelle DNA in barley and *Hordeum* x *Secale* intergeneric hybrids. Genome 29: 867—872.

Stine, M. (pers. comm.) School of Forestry, Wildlife, and Fisheries, Louisiana State University, Baton Rouge, LA.

Stine, M. and Keathley, D. E. 1990. Paternal inheritance of plastids in Engelmann spruce x blue spruce hybrids. J. Hered. 81: 443—446.

Stine, M., Sears, B. B. and Keathley, D. E. 1989. Inheritance of plastids in interspecific hybrids of blue spruce and white spruce. Theor. Appl. Genet. 78: 768—774.

Strauss, S. H., Bousquet, J., Hong, Y.-P. and Hipkins, V. D. 1992. Biochemical and molecular genetic markers in biosystematic studies of forest trees. New Forests. This issue (pp. 125—158).

Strauss, S. H., Hong, Y.-P. and Hipkins, V. (pers. comm.) Department of Forest Science, Oregon State University, Corvallis, OR.

Strauss, S. H. and Howe, G. T. 1990. An investigation of somatic variability for ribosomal RNA gene number in old-growth Sitka spruce. Can. J. For. Res. 20: 853—856.

Strauss, S. H., Neale, D. B. and Wagner, D. B. 1989. Genetics of the chloroplast in conifers: biotechnology research reveals some surprises. J. Forestry 87: 11—17.

Strauss, S. H., Palmer, J. D., Howe, G. T. and Doerksen, A. H. 1988. Chloroplast genomes of two conifers lack a large inverted repeat and are extensively rearranged. Proc. Natl. Acad. Sci. USA 85: 3898—3902.

Strauss, S. H. and Tsai, C.-H. 1988. Ribosomal gene number variability in Douglas-fir. J. Hered. 79: 453—458.

Sutton, B. C. S., Flanagan, D. J. and El-Kassaby, Y. A. 1991a. A simple and rapid method for estimating representation of species in spruce seedlots using chloroplast DNA restriction fragment length polymorphism. Silvae Genet. 40: 119—123.

Sutton, B. C. S., Flanagan, D. J., Gawley, J. R., Newton, C. H., Lester, D. T. and El-Kassaby, Y. A. 1991b. Inheritance of chloroplast and mitochondrial DNA in *Picea* and composition of hybrids from introgression zones. Theor. Appl. Genet. 82: 242—248.

Szmidt, A. E. (pers. comm.) Department of Forest Genetics and Plant Physiology, Swedish University of Agricultural Sciences, Umeå, Sweden.

Szmidt, A. E., Aldèn, T. and Hällgren, J.-E. 1987. Paternal inheritance of chloroplast DNA in *Larix*. Plant Molec. Biol. 9: 59—64.

Szmidt, A. E., El-Kassaby, Y. A., Sigurgeirsson, A., Aldèn, T., Lindgren, D. and Hällgren, J.-E. 1988. Classifying seedlots of *Picea sitchensis* and *P. glauca* in zones of introgression using restriction analysis of chloroplast DNA. Theor. Appl. Genet. 76: 841—845.

Szmidt, A. E. and Wang, X.-R. 1991. DNA markers in forest genetics. In: Müller-Starck, G. and Ziehe, M. (Eds) Genetic Variation of Forest Tree Populations in Europe. Sauerländer-Verlag, Frankfurt/Main (in press).

Tsai, C.-H. and Strauss, S. H. 1989. Dispersed repetitive sequences in the chloroplast genome of Douglas-fir. Curr. Genet. 16: 211—218.

van Damme, J. M. M. 1986. Gynodioecy in *Plantago lanceolata* L. V. Frequencies and spatial distribution of nuclear and cytoplasmic genes. Heredity 56: 355—364.

Wagner, D. B., Dong, J., Carlson, M. R. and Yanchuk, A. D. 1991c. Paternal leakage of mitochondrial DNA in *Pinus*. Theor. Appl. Genet. 82: 510—514.

Wagner, D. B., Furnier, G. R., Saghai-Maroof, M. A., Williams, S. M., Dancik, B. P. and Allard, R. W. 1987. Chloroplast DNA polymorphisms in lodgepole and jack pines and their hybrids. Proc. Natl. Acad. Sci. USA 84: 2097—2100.

Wagner, D. B. and Govindaraju, D. R. 1988. Associations of chloroplast DNA and cone morphology in the sympatric region of *Pinus banksiana* and *Pinus contorta*, pp. 492—493. In: Hanover, J. W. and Keathley, D. E. (Eds) Genetic Manipulation of Woody Plants. Plenum Press, New York, New York.

Wagner, D. B., Govindaraju, D. R. and Dancik, B. P. 1989a. Tests of cytonuclear population-genetic structure involving the chloroplast genome in a sympatric region of *Pinus banksiana* and *P. contorta*. Genetics 122: s42.

Wagner, D. B., Govindaraju, D. R., Yeatman, C. W. and Pitel, J. A. 1989b. Paternal chloroplast DNA inheritance in a diallel cross of jack pine (*Pinus banksiana* Lamb.). J. Hered. 80: 483—485.

Wagner, D. B., Nance, W. L., Nelson, C. D., Li, T., Patel, R. N. and Govindaraju, D. R. 1992. Taxonomic patterns and inheritance of chloroplast DNA variation in a survey of *Pinus echinata* Mill., *P. elliottii* Engelm., *P. palustris* Mill., and *P. taeda* L. Can. J. For. Res. 22: 683—689.

Wagner, D. B., Sun, Z.-X., Govindaraju, D. R. and Dancik, B. P. 1991b. Spatial patterns of chloroplast DNA and cone morphology variation within populations of a *Pinus banksiana — Pinus contorta* sympatric region. Am. Nat. 138: 156—170.

Wang, C. 1990. A DNA rearrangement found in chloroplast genome from eastern hemlock (*Tsuga canadensis*). M.S. Thesis, University of Louisville, Louisville, KY, 64 pp.

Wang, X.-R. and Szmidt, A. E. 1990. Evolutionary analysis of *Pinus densata* (Masters), a putative Tertiary hybrid. 2. A study using species-specific chloroplast DNA markers. Theor. Appl. Genet. 80: 641—647.

White, E. E. 1990a. Chloroplast DNA in *Pinus monticola*. 1. Physical map. Theor. Appl. Genet. 79: 119—124.

White, E. E. 1990b. Chloroplast DNA in *Pinus monticola*. 2. Survey of within-species variability and detection of heteroplasmic individuals. Theor. Appl. Genet. 79: 251—255.

Whittemore, A. T. and Schaal, B. A. 1991. Interspecific gene flow in sympatric oaks. Proc. Natl. Acad. Sci. USA 88: 2540—2544.

Wilcox, M. D. 1983. Reciprocal cross effects in *Pinus radiata*. N. Z. J. For. Sci. 13: 37—45.

Williams, J. G. K., Kubelik, A. R., Livak, K. J., Rafalski, J. A. and Tingey, S. V. 1990. DNA polymorphisms amplified by arbitrary primers are useful as genetic markers. Nucleic Acids Res. 18: 6531—6535.

Zimmerman, P. A., Lang-Unnasch, N. and Cullis, C. A. 1989. Polymorphic regions in plant genomes detected by an M13 probe. Genome 32: 824—828.

New Forests **6**: 391—407, 1992.

Review paper

Use of DNA markers in forest tree improvement research

D. B. NEALE, M. E. DEVEY, K. D. JERMSTAD, M. R. AHUJA, M. C. ALOSI and K. A. MARSHALL

Institute of Forest Genetics, Pacific Southwest Research Station, Box 245, Berkeley, CA 94701, USA

Received 15 December 1990; accepted 3 March 1992

Key words: DNA markers, genetic maps, RFLPs

Application. Development of DNA markers will provide abundant new genetic markers for forest tree improvement research. DNA markers will be most useful for estimating genetic diversity in breeding populations and for germplasm identification. Eventually, high-density maps may be used to identify quantitative trait loci and to practice marker-assisted selection.

Abstract. DNA markers are rapidly being developed for forest trees. The most important markers are restriction fragment length polymorphisms (RFLPs), polymerase chain reaction- (PCR) based markers such as random amplified polymorphic DNA (RAPD), and fingerprinting markers. DNA markers can supplement isozyme markers for monitoring tree improvement activities such as; estimating genetic diversity in breeding populations, germplasm identification, verifying controlled crosses, and estimating seed orchard efficiencies. Because the number of DNA markers is potentially limitless, it should be possible to map individual quantitative trait loci (QTL) by linkage analysis with high-density maps. Finally, if such associations can be found, it may also be possible to design marker-assisted breeding strategies for forest trees.

Introduction

Genetic markers are important tools for forest tree improvement. Isozyme markers have been applied extensively during the past 15 years and have contributed significantly to tree breeding programs (Adams 1981a, 1983; Adams et al. 1988; Cheliak et al. 1987; Friedman and Neale 1992; Miller et al. 1989; Wheeler and Jech 1992). Isozymes generally provide ample genetic information and are relatively inexpensive, rapid, and technically easy to apply, thus they should continue to play an important role in forest tree improvement. In recent years, DNA-based genetic markers have been developed, most notably restriction fragment length polymorphisms (RFLPs) and polymerase chain reaction- (PCR) based markers. DNA-based markers have the potential to overcome some of the limitations of isozymes for tree improvement applications. In this paper, we will describe some of the new DNA-based genetic markers and attempt to show where they might best be applied in forest tree improvement. The application of DNA markers in forest genetics research in general has been discussed in several recent papers (Cheliak and Rogers 1990; Friedman and Neale 1992; Gianfranceschi et al. 1991; Nance and Nelson 1989; Neale et al. 1989; Neale and Williams 1991; Wagner 1992).

DNA-based genetic markers

Genetic markers are of two general types; Mendelian and nonMendelian. Mendelian markers segregate as alleles at a locus, whereas allelism is not established for nonMendelian markers. NonMendelian markers are often called fingerprints. There are also two major assay approaches, RFLPs and PCR, both of which are used with Mendelian and nonMendelian markers. We will describe markers of both types and assay approaches.

Restriction fragment length polymorphisms

The concept of RFLP mapping of complex genomes was first described by Botstein et al. (1980) and has been described by many authors (Beckmann and Soller 1983, 1986a, 1986b; Landry and Michelmore 1987; Soller and Beckmann 1983; Tanksley et al. 1989). RFLPs are simple Mendelian genetic markers which result from various types of mutations and rearrangements of the DNA. The first step in RFLP detection is to isolate DNA from the organism of study and cleave the DNA with one or more restriction endonucleases. Restriction endonucleases are bacterial enzymes which cleave double-stranded DNA at unique palindromic recognition sequences, usually 4—8 nucleotides in length. The restriction endonuclease *Eco*RI, for example, cleaves the 6-base sequence GAATTC between the G and A (Fig. 1A). Insertions and deletions of small segments of DNA or the gain or loss of a restriction site are two types of RFLPs which are easily detected (Figs. 1B and 1C). Eukaryotic genomes, however, are very large so that when genomic DNA is cut with a restriction enzyme, a nearly continuous distribution of restriction fragment sizes results. It is thus impossible to visualize individual DNA fragments as shown in Fig. 1C.

The problem of visualizing individual DNA fragments is overcome by a technique called Southern blotting and probe hybridization. Following digestion with restriction enzymes, DNAs are fractionated electrophoreti-

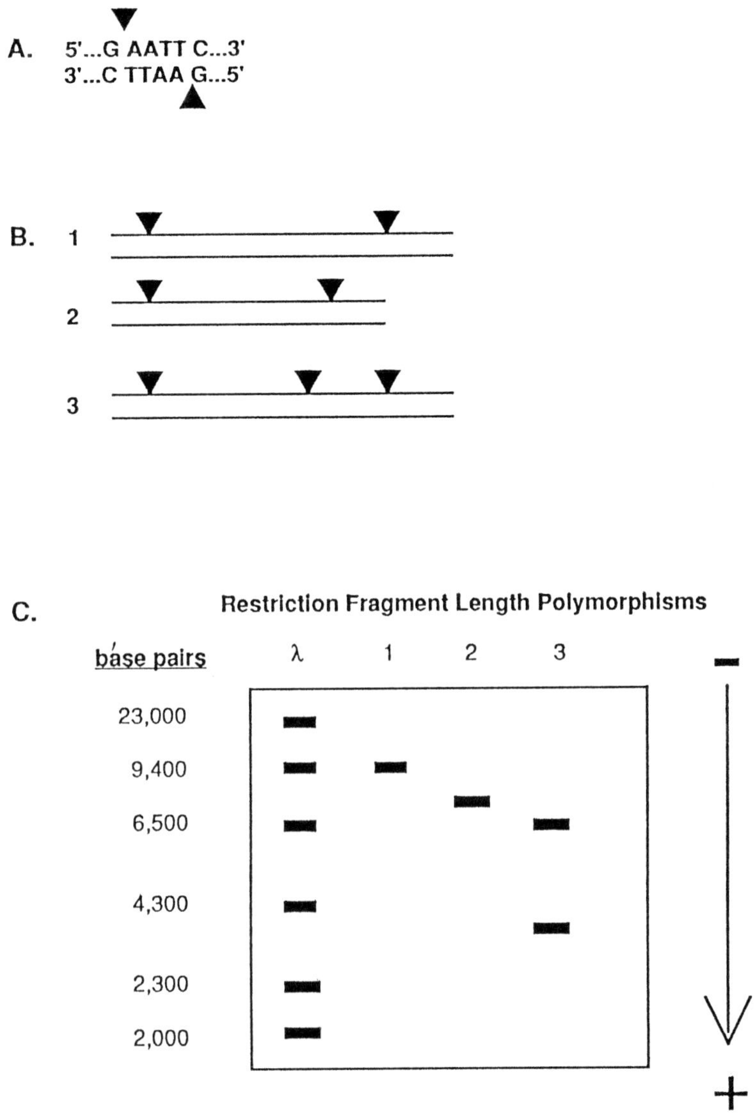

Fig. 1. Restriction fragment length polymorphisms. A. The recognition sequence for the restriction endonuclease *Eco*RI is 5′..GAATTC..3′. *Eco*RI cuts double-stranded DNA between the G and A within the palindromic 6-base sequence. B. DNA fragments of three trees cut with *Eco*RI. Tree 2 has a deletion of a small amount of DNA sequence between the two *Eco*RI sites, whereas tree 3 has gained an *Eco*RI site relative to trees 1 and 2. C. Gel separation of *Eco*RI-cut DNA fragments. Lanes 1, 2, and 3 show the separation of DNA fragments from the three trees. Molecular weight standards are shown in the first lane.

cally on agarose gels. The DNA is then denatured and single-stranded DNA molecules are transferred and covalently linked to nylon membranes (blots). Small DNA fragments, called probes, are then radiolabelled and allowed to hybridize to their complementary DNA sequences bound to the blots. The radioactive DNA hybrids on the blot are visualized by autoradiography (Fig. 2).

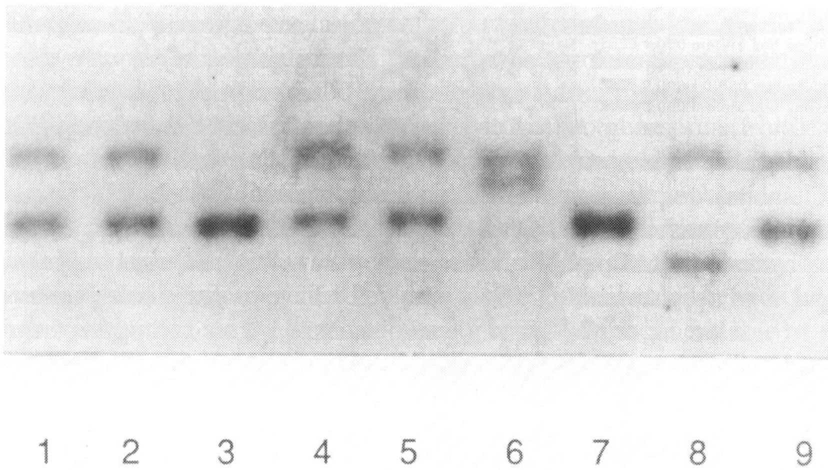

Fig. 2. Autoradiogram showing RFLPs among nine Douglas-fir seed orchard clones cut with *Eco*RI and hybridized with a cDNA probe from Douglas-fir.

Polymerase chain reaction-based markers

Amplification of specific DNA sequences by the polymerase chain reaction (Sakai et al. 1988) offers a new and powerful approach to developing genetic markers. Several approaches have recently been developed, however the most widely used method is the random amplified polymorphic DNA (RAPD) marker (Williams et al. 1990). Short oligonucleotide primers are used for PCR and resulting amplification products appear as visible bands on gels. If complementary sequences to the primer do not exist then no amplification product or visible band on gels results. Thus, these are dominant genetic markers because the homozygote for the amplification product can not be distinguished from the heterozygote. In conifers, however, the problem of dominance can be overcome if RAPD markers are assayed from the haploid megagametophyte tissue.

Fingerprinting markers

A common use of genetic markers is to establish whether two or more individuals are genetically the same or different. This application is often called "fingerprinting" from the traditional method of human forensic science. Genetic markers used for fingerprinting often reveal a large number of complex polymorphisms and are generally nonMendelian markers. The first well characterized DNA fingerprinting probes were the human minisatellite probes (Jefferies et al. 1985). Two of these probes, clones 33.15 and 33.6, have been shown to cross hybridize and reveal polymorphisms in rice (Dallas 1988) and in *Populus* (Rogstad et al. 1988). However, the lack of reports on the use of these probes with other plants suggests that their utility in plants is limited.

Another DNA fingerprinting probe is the bacteriophage M13 subrepeat probe. Vassart et al. (1987) showed that M13 would cross hybridize to mammalian minisatellite sequences and subsequently it was demonstrated that it would hybridize to plant DNA as well (Ryskov et al. 1988; Rogstad et al. 1988). Nybom and coworkers have used the M13 probe to fingerprint varieties of species with the Rosaceae (Nybom 1990; Nybom and Schaal 1990a, 1990b; Nybom et al. 1989, 1990). Rogstad et al. (1988) showed that M13 would cross hybridize to *Picea glauca, Pinus torreyana*, and *Populus* species and that polymorphisms could be detected. These preliminary studies, however, have not been followed by larger studies to evaluate the efficacy of using M13 to fingerprint in large populations of forest trees.

A PCR-based approach to revealing M13 polymorphisms has recently been developed (Welsh and McClelland 1990). The M13 DNA sequencing primers are used for PCR amplification of genomic DNA. This technique, called arbitrarily primed (AP-PCR), was used to fingerprint three varieties of rice (Welsh and McClelland 1990).

Finally, a new and promising marker called simple sequence length polymorphism (SSLP) might be used for fingerprinting in plants as well as for mapping individual genetic loci (Litt and Luty 1989; Tautz 1989; Weber and May 1989). Eukaryotic genomes contain stretches of simple sequences of tandem repititions of mono-, di-, tri-, and tetrameric motifs. Because these simple sequences vary in length and are dispersed throughout the genome, they are potentially useful markers for fingerprinting. Weising et al. (1991) have shown that such sequences exist in plants and can be used for fingerprinting. In addition, it may also be possible to design PCR primers from the unique sequence regions flanking the simple repeat sequences to reveal individual genetic loci (Tautz 1989). This approach is being used to map CA- repeat regions in humans (Litt and Luty 1989).

Advantages of molecular markers

DNA markers are true genetic markers which have several distinct advantages over other biochemical markers for some common applications in tree improvement. The first, and probably most important, advantage is that potentially an unlimited number of DNA markers can be detected. The number of biochemical markers, however, is limited to a small number of isozyme and monoterpene markers. A second advantage is that DNA markers do not vary among tissue types or developmental stages of the plant because the assays are based on the DNA itself and not the products of genes. There are clearly differences in the levels of expression of certain isozymes among tissue types commonly used in isozyme assays (megagametophytes, embryos, buds, needles). In Douglas-fir, there are only 11 isozyme loci which can be assayed from all four of the tissues listed above (Adams et al. 1990). DNA, however, can be isolated from almost any plant part, thus enabling DNA marker detection from many tissue types and at most developmental stages. A third advantage of DNA markers is that they are not affected by environmental variation. The presence or abundance of isozyme or biochemical marker products can be affected by environmental stimuli. A good example is the enzyme alcohol dehydrogenase (ADH). There are two isozyme loci on gels stained for ADH; however, one of the loci is anaerobically induced and is undetectable in noninduced tissue (Harry et al. 1988). Using DNA technology, both loci would be detected, whether the tissue were subjected to anaerobosis or not. There may also be allelic variation in response to environmental stimuli at certain isozyme loci.

RFLP and RAPD markers have some important advantages and disadvantages relative to one another. RAPD markers are generally simpler and faster to assay for than RFLPs. They require as little as 10ng of DNA per assay as opposed to RFLPs which require 10ng or more in conifers. PCR assays also do not require the use of radioisotopes which are used with RFLPs. The most severe limitation of RAPD markers is that they are dominant markers. Dominant markers are less informative than codominant markers for many of the applications in forest tree improvement. In addition, RAPD loci are diallelic; all that can be detected at a locus is the presence or absence of a PCR product. RFLPs, however, can detect multiple alleles at a locus. A PCR-based codominant genetic marker which would reveal multiple alleles would be an important development for genetic studies in forest trees and other plants.

Applications of molecular markers in tree improvement

Molecular markers can be used to monitor the success of numerous tree improvement activities. The application of isozyme markers to such problems has been well documented (Adams 1981a, 1983; Adams et al. 1988; Wheeler and Jech 1992). In this section, we will identify areas where DNA-based markers might be more powerful but also identify areas where isozymes would still be the marker of choice.

Estimation of genetic variability in domesticated populations

Isozymes have been used to assess the amounts of genetic variability in selected breeding populations versus wild stands from where selections were made, in seed orchard crops, and in commercial seed collections from wild stands (Adams 1981b). In Douglas-fir, these data do not suggest large losses in genetic variability due to domestication (Adams 1981b). In most cases, estimates from isozymes can be obtained rapidly and inexpensively, and a large number of loci can be assayed if megagametophytes are used. Thus, for rapid assessments of variability in domesticated populations and outputs from such populations (e.g., seed crops), isozymes are probably the best approach. However, in species in which levels of isozyme variability are low or in which estimates must be obtained from tissues where only a few loci can be resolved (e.g., seedlings in nursery crops), the application of DNA-based markers would be warranted. RFLPs would be a better marker than RAPDs for this application because multiple alleles at a locus can be detected.

Germplasm identification

Probably the most important application of genetic markers in tree improvement is the broad problem of germplasm identification. The array of identification problems ranges from simple to complex. For example, a simple identification problem would be to determine if two ramets are members of the same or different clone. This problem is akin to those in human forensics. Any reproducible genetic difference between the two ramets would be considered sufficient evidence to conclude that they were members of different clones. A more complex problem would be to assign seed source identification to a bulked seed collection. A problem of this type would generally involve gene frequency estimation. Whether a particular problem is simple or complex, the power to make an accurate identification is a function of genetic discrimination.

The power to genetically discriminate between individuals or groups of individuals is a function of the number of markers and the amount of genetic variability revealed by the markers. Forest trees, especially conifers, have high levels of isozyme variation, which makes these markers quite adequate for many germplasm identification problems. One limitation of isozymes is that a large number of loci can not be resolved from some tissues. This limitation could be overcome with DNA markers. RFLPs would be useful where large (>5g) amounts of tissue are available to extract the large amount of DNA needed. RAPD markers would be useful for some types of identification, but will have limited value because of dominance. The most important type of marker for germplasm identification are the fingerprinting markers such as M13, AP-PCR, SSLPs, or other hypervariable sequences. We recently used high-copy RFLP probes from loblolly pine to fingerprint 12 unlabelled longleaf pine seed orchard clones which we were unable to genetically differentiate using isozymes.

Controlled crosses

Controlled crossing is an important aspect of most tree breeding programs. Adams et al. (1988) have used isozymes to determine the accuracy of such crosses from operational programs. Their results showed that 30% or more of the Douglas-fir and loblolly pine crosses they studied were not correct and that these determinations could be made with as few as 6—10 isozyme loci and five seeds per cross. For these species, and possibly other forest trees, it would be unnecessary to use DNA markers. RFLPs would be useful for species with low isozyme variation, however, it would be necessary to grow the progeny to a size large enough to extract the large amount of DNA needed. RAPDs will be of little value for this application because of dominance.

Seed orchards

There are numerous applications of genetic markers in seed orchard management (Wheeler and Jech 1992). They include: (1) estimation of the extent of pollen contamination (Friedman and Adams 1985; Smith and Adams 1983), (2) estimating selfing rates and inbreeding (El-Kassaby et al. 1986), (3) estimating the success of supplemental mass pollination (Joly and Adams 1983; Wheeler and Jech 1986), (4) determining mating patterns within orchards (Erickson and Adams 1989), and (5) determining the effects of cultural practices, such as top pruning (Omi and Adams 1985). These applications are all centered around determining male parentage of seed produced in the orchard. By using isozymes, the genotype of the pollen gamete is inferred from the genotypes of the seed megagametophyte and embryo and then compared to male genotypes within and outside the orchard to infer parentage. The power of genetic discrimination among putative males is a function of the number of genetic markers, the amount of allelic variability at individual loci, the frequencies of alleles in the population, and the overall number of males in the population. Complete genetic discrimination among males by using isozymes is never possible in operational seed orchard situations and therefore statistical methods of inference must be employed. DNA markers could potentially increase genetic discrimination, but unfortunately, neither RFLPs or RAPDs are well suited for inferring the genotype of the male gamete from the embryo. A codominant PCR-based marker would be extremely useful for this purpose.

Quantitative trait mapping and marker-assisted selection

The uses of molecular and biochemical markers discussed previously do not require a full understanding of the linkage relationships among the markers. Isozyme linkage analyses have been conducted for numerous tree species (e.g., Adams and Joly 1980; Conkle 1981; Guries et al. 1978), however, the number of loci included in these studies was small, and the number of linkage blocks was generally less than the number of chromosomes. Such maps are useful for showing linkage relationships among isozyme loci, but restrict investigation to small segments of the genome. This limitation can be overcome by DNA markers because there is potentially an unlimited number that can be mapped. Once a saturated genetic map has been constructed it then becomes theoretically possible to map individual loci coding for quantitative traits (QTLs). After associations between markers and QTLs have been established it may be possible to use this information predictively in a marker-assisted selection scheme.

Quantitative trait locus mapping

The first demonstration of quantitative trait mapping using a saturated molecular marker map was in tomato (Paterson et al. 1988). Loci for three traits (fruit mass, soluble solids, and pH) were mapped in a backcross population of 237 plants derived from an interspecific cross. A subset of 68 (63 RFLP and 5 isozyme) evenly distributed markers were selected from a saturated map of over 300 markers. Interval mapping

(Lander and Botstein 1989) was used to identify chromosomal regions showing strong associations with quantitative traits. Each region includes one or more quantitative trait loci. Modifications of this approach could be applied to detecting QTLs of important traits in trees.

Stem volume is probably the most commercially important quantitative trait in forest trees. This trait is controlled by many genes and has a low heritability. Polygenic control and low heritability would make it difficult to map QTLs for stem volume. Nevertheless, if a small subset of genes having large effects could be detected, their identification would have large economic importance. Mapping of more highly heritable traits, controlled by a small number of genes (e.g., wood specific gravity, resistances to diseases, and phenological traits), would be easier and might be better choices for initial investigations.

Quantitative trait mapping in agricultural crops has been conducted using F_2 or backcross pedigrees. Such pedigrees generally do not exist for forest trees and are difficult to construct. QTL mapping in trees would have to be conducted using multigeneration outbred pedigrees. Unfortunately, the statistical theory for QTL mapping from outbred tree pedigrees is generally lacking. It is likely, however, that the interval mapping approach of Lander and Botstein (1989) can be extended to multigeneration outbred pedigrees.

Marker-assisted selection

Early genetic selection in forest tree breeding would have enormous practical importance because of the long generation times in trees. Early selection on height growth has been successfully practiced on seedlings at age two and older (Magnussen and Yeatman 1987; McKinley and Lowe 1986; Williams 1987) but has not been successful with one-year-old seedlings (Waxler and van Buijtenen 1981; Williams 1986). Indirect selection based on DNA markers could be practiced at a very early age if linkages to important traits are found. In addition, costs would be lower than progeny testing, selection intensities might be much higher, and selection may potentially be more efficient because of the higher heritabilities of the markers.

One cannot assume, however, that marker-assisted selection in forest trees can be practiced as effectively as it may be with highly inbred agricultural crops. Forest tree breeding populations are very large and are nearly in linkage equilibrium. Linkage equilibrium makes it difficult to achieve much genetic gain based on marker selection. If small elite breeding populations are employed, then it would be possible to practice within family selection, based on the markers, if the phases of parent tree marker genotypes were known. Further advancements in DNA mapping techniques and new statistical methods might enable marker-assisted selection under less restricted conditions. Forest geneticists should prepare for this possibility by developing high-density molecular maps and testing the efficiency of marker-assisted selection in experimental populations.

Mapping of DNA markers in conifers

We are constructing genetic maps in conifers using DNA markers. In this section, we will briefly describe the approaches we are using to construct these maps. Our initial project was to construct an RFLP map for loblolly pine (*Pinus taeda* L.) based on a segregating F_2 population of a 3-generation outbred pedigree. DNA is isolated from each of the four grandparents and two parent trees of the pedigree (Fig. 3). DNAs are digested with the restriction enzyme *Hin*dIII and Southern blots are prepared (Devey et al. 1991). Blots are then hybridized with ^{32}P-labelled random cDNA (complementary DNA) or genomic DNA probes (Fig. 4). Segregating RFLPs in this cross are identified from these hybridizations. We limit our screening for polymorphisms to just *Hin*dIII because nearly all

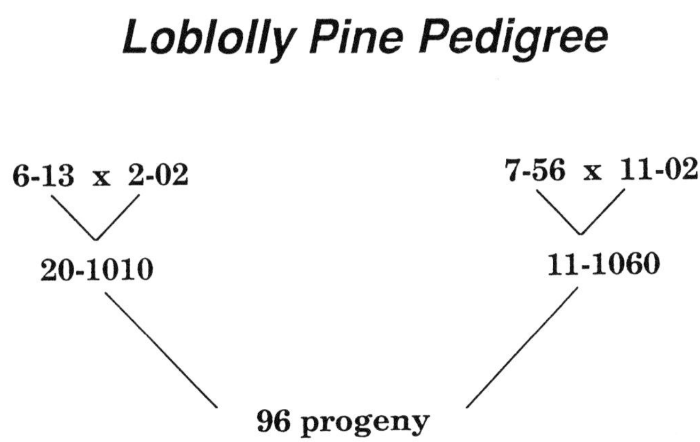

Fig. 3. Three-generation pedigree from loblolly pine to be used in construction of a high-density RFLP map.

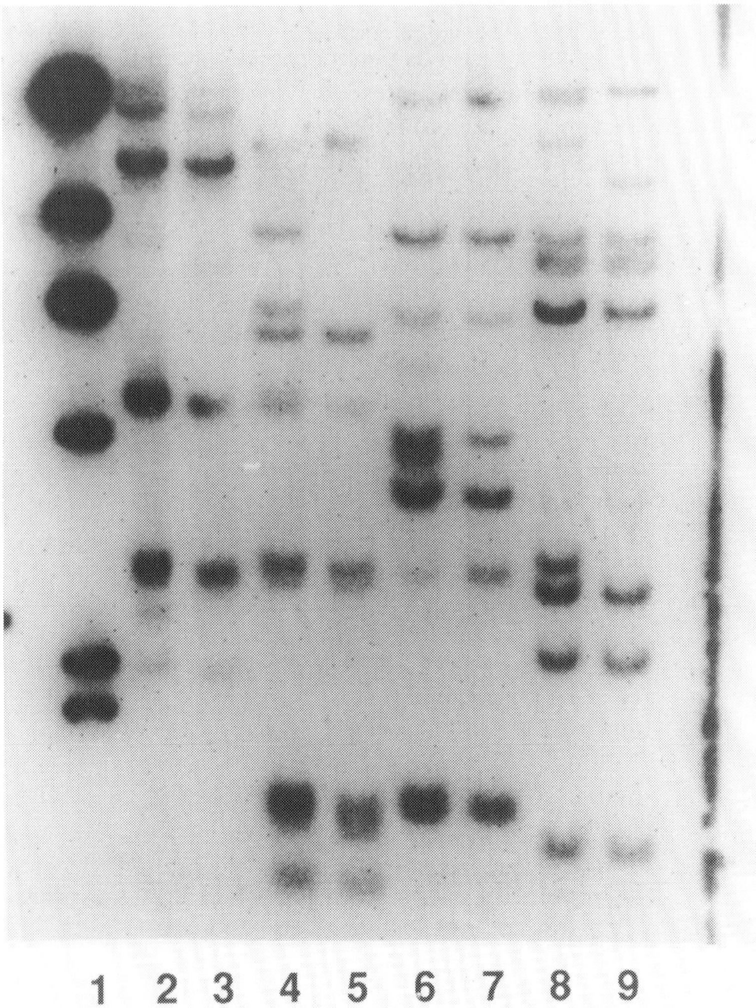

Fig. 4. Autoradiogram of two loblolly pine parent tree DNAs, 20-1010 and 11-1060, cut with four restriction enzymes and hybridized with loblolly pine cDNA probe pPtIFGc653. Lane 1, lambda*Hind*III molecular weight marker. Lanes 2—3, *Bam*HI; lanes 4—5, *Dra*I; lanes 6—7, *Eco*RI; and lanes 8—9, *Hin*dIII.

polymorphisms are length variants; other enzymes rarely reveal additional variation (Devey et al. 1991). Once a segregating RFLP is identified, the probe is relabelled and hybridized to a set of blots containing the 96 progeny DNA (Fig. 5). Segregating alleles are scored and tests for linkage

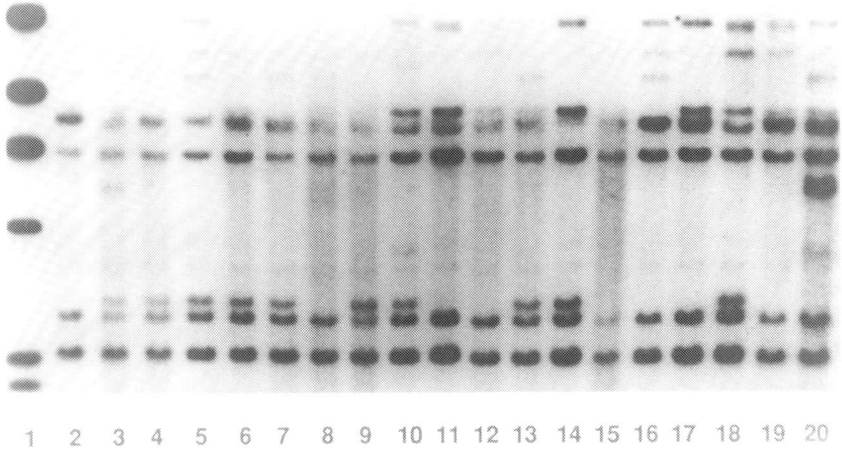

Fig. 5. Autoradiogram of segregating progeny of loblolly pine cross 20-1010 X 11-1060 cut with *Hin*dIII and hybridized with cDNA probe pPtIFGc653 (lanes 2-2-). Lane 1, lambda*Hin*dIII molecular weight marker.

are performed. We are following a nearly identical strategy for mapping RFLPs in a 3-generation Douglas-fir pedigree.

We have also begun to add RAPD markers to our loblolly pine and Douglas-fir RFLP maps. Segregating RAPDs in the pedigree are identified by screening random 10-mers (Operon Technologies) against DNA from four megagametophytes from both parent trees. Primers for segregating RAPDs are then used with DNAs from the F_2 progeny. RAPD markers can also be easily mapped using segregating megagametophytes from heterozygous mother trees.

Acknowledgments

We thank Bob Weir and Steve McKeand of North Carolina State University and Claire Williams and Nick Wheeler of the Weyerhaeuser Company for providing pedigrees and Dave Harry and Yousry El-Kassaby for reviews of the manuscript. This project was partially supported by U.S.D.A. Forest Biology Grant No. 88-33520-4074.

References

Adams, W. T. 1981a. Applying isozyme analyses in tree-breeding programs. pp. 60—64. In: Conkle, M. T. (Technical Coordinator) Proc. Symp. Isozymes North Am. Forest Trees and Forest Insects. USDA Forest Serv., Pacific SW Forest and Range Exp. Sta., Berkeley, California, Gen. Tech. Rep. PSW-48.

Adams, W. T. 1981b. Population genetics and gene conservation in pacific northwest conifers. pp. 401—415. Scudder, G. G. E. and Reveal, J. L. (Eds), Evolution Today, Proceedings of the Second International Congress of Systematic and Evolutionary Biology.

Adams, W. T. 1983. Application of isozymes in tree breeding. pp. 381—400. In: Tanksley, S. D. and Orton, T. J. (Eds), Isozymes in Plant Genetics and Breeding, Part A. Elsevier Science Publishers B. V. Amsterdam.

Adams, W. T. and Joly, R. J. 1980. Genetics of allozyme variants in loblolly pine. J. Hered. 71: 33—40.

Adams, W. T., Neale, D. B., Doerksen, A. H. and Smith, D. B. 1990. Inheritance and linkage of isozyme variants from seed and vegetative bud tissues in coastal Douglas-fir [*Pseudotsuga menziesii* var. *menziesii* (Mirb.) Franco]. Silvae Genet. 39: 153—167.

Adams, W. T., Neale, D. B. and Loopstra, C. A. 1988. Verifying controlled crosses in conifer tree-improvement programs. Silvae Genet. 37: 147—152.

Beckmann, J. S., and Soller, M. 1983. Restriction fragment length polymorphisms in genetic improvement: methodologies, mapping and costs. Theor. Appl. Genet. 67: 35—43.

Beckmann, J. S., and Soller, M. 1986a. Restriction fragment length polymorphisms and genetic improvement of agricultural species. Euphytica 35: 111—124.

Beckmann, J. S., and Soller, M. 1986b. Restriction fragment length polymorphisms in plant genetic improvement. Oxford Surv. Plant Mol. Cell Biol. 3: 196—249.

Botstein, D., White, R. L., Skolnick, M., and Davis, R. W. 1980. Construction of genetic linkage map in man using restiction fragment length polymorphisms. Am. J. Hum. Genet. 32: 641—656.

Cheliak, W. M. and Rogers, D. L. 1990. Integrating biotechnology into tree improvement programs. Can. J. For. Res. 20: 452—463.

Cheliak, W. M., Yeh, F. C. H. and Pitel, J. A. 1987. Use of electrophoresis in tree improvement programs. For. Chron. 63: 89—96.

Conkle, M. T. 1981. Isozyme variation and linkage in six conifer species. pp. 11—17. In: Conkle, M. T. (Technical Coordinator), Proc. Symp. Isozymes North Am. Forest Trees and Forest Insects. USDA Forest Serv., Pacific SW Forest and Range Exp. Sta., Berkeley, California, Gen. Tech. Rep. PSW-48.

Dallas, John F. 1988. Detection of DNA "fingerprints" of cultivated rice by hybridization with a human minisatellite DNA probe. Proc. Natl. Acad. Sci. USA. 85: 6831—6835.

Devey, M. E., Jermstad, K. D., Tauer, C. G. and Neale, D. B. 1991. Inheritance of RFLP loci in a loblolly pine three-generation pedigree. Theor. Appl. Genet. 83: 238—242.

El-Kassaby, Y. A., Yeh, F. C. and Sziklai, O. 1986. Clonal and seedling seed orchards: A comparison of outcrossing rates and heterozygosity in coastal Douglas-fir using allozyme markers. pp. 410—421. In: Hatcher, A. V. and Weir, R. J. (Eds). IUFRO Conference Proc., A Joint Meeting of Working Parties-S2.04.02, S2.04.03., and S2.03.03. Williamsburg, Virginia.

Erickson, V. J. and Adams, W. T. 1989. Mating success in a coastal Douglas-fir seed orchard as affected by distance and floral phenology. Can. J. For. Res. 19: 1248—1255.

Friedman, S. T. and Adams, W. T. 1985. Estimation of gene flow into two seed orchards of loblolly pine (*Pinus taeda* L.). Theoret. Appl. Genet. 69: 609—615.

Friedman, S. T. and Neale, D. B. 1992. Biochemical and molecular markers. In: Bramlett, D. (Ed), Pollen Management Handbook, Vol. II. SE Exp. Sta., Ashville, North Carolina, in press.

Gianfranceschi, L., Taramino, G., Binelli, G., Pe, M. E. and Ottaviano, E. 1991. RFLP markers in the genetic analysis of forest trees. In Effects of Pollution on the Structure of Forest Tree Populations, Consiglio Nazionale delle Ricerche, pp. 67—78.

Guries R. P., Friedman, S. T. and Ledig, F. T. 1978. A megagametophyte analysis of genetic linkage in pitch pine (*Pinus rigida* Mill.). Heredity 40: 309—314.

Harry, D. H., Kinlaw, C. S. and Serderoff, R. R. 1988. The anaerobic stress response and its use for studying gene expression in conifers. pp. 275—290. In: Hanover, J. W. and Keathely, D. E. (Eds), Genetic Manipulation of Woody Plants. Plenum Publishing Corporation, New York.

Jeffreys, A. J., Wilson, V. and Thein, S. L. 1985. Individual-specific "fingerprints" of human DNA. Nature 316: 76—79.

Joly, R. J. and Adams, W. T. 1983. Allozyme analysis of pitch x loblolly pine hybrids produced by supplemental mass-pollination. Forest Sci. 29: 423—432.

Lander, E. S. and Botstein, D. 1989. Mapping Mendelian factors underlying quantitative traits using RFLP linkage maps. Genetics 121: 185—199.

Landry, B. S. and Michelmore, R. W. 1987. Methods and applications of restriction fragment length polymorphism analysis to plants. In: Tailoring genes for crop improvement: an agricultural perspective. Ed. by: G. Bruening, J. Harada, and A. Hollaender. Plenum Press, New York. pp. 25—44.

Litt, M. and Luty, J. A. 1989. A hypervariable microsatellite revealed by in vitro amplification of a dinucleotide repeat within the cardiac muscle actin gene. Am. J. Hum. Genet. 44: 397—401.

Magnussen, S. and Yeatman, C. W. 1987. Four years' height growth of 25 jack pine (*Pinus banksiana* Lamb.) families in an accelerated nursery trial. pp. 47—64. In: Proc. 30th Northeastern For. Tree Improv. Conf., Orono, Maine.

McKinley, C. R. and Lowe, W. J. 1986. Juvenile-mature correlations. pp. 11-16. In: Proc. Workshop on Advanced Generation Breeding: Current Status and Research Needs. Baton Rouge, Louisiana. South. Coop. Series Bull. No. 309.

Miller, R. G., Conkle, M. T. and Friedman, S. T. 1989. The Forest Service laboratory for genetic analyses of trees. Three Planters' Notes 40: 25—29.

Nance, W. L. and Nelson, C. D. 1989. Restriction fragment length polymorphisms and their use in marker assisted selection in Southern pine improvement. pp. 50—59. In: Proceedings 20th Southern Forest Tree Improvement Conference. Charleston, South Carolina. The National Technical Information Service, Springfield, Virginia.

Neale, D. B., Tauer, C. G., Gorzo, D. M., and Jermstad, K. D. 1989. Restriction fragment length polymorphism mapping of loblolly pine: Methods, applications, and limitations. pp. 363—372. In: Proc. 20th Southern Forest Tree Imp. Conf., Charleston, South Carolina. The National Technical Information Service, Springfield, Virginia.

Neale, D. B. and Williams, C. G. 1991. Restriction fragment length polymorphism mapping in conifers and applications to forest genetics and tree improvement. Can. J. For. Res. 21: 545—554.

Nybom, H., Schaal, B. A. and Rogstad, S. H. 1989. DNA "fingerprints" can distinguish cultivars of blackberries and raspberries. Acta Horticulturae 262: 305—310.

Nybom, H. 1990. Genetic variation in ornamental apple trees and their seedlings (Malus, Rosaceae) revealed by DNA "fingerprinting" with the M13 repeat probe. Hereditas 113: 17—28.

Nybom, H. and Schaal, B. A. 1990a. DNA "fingerprints" applied to paternity analysis in apples (Malus x domestica). Theor. Appl. Genet. 79: 763—768.

Nybom, H. and Schaal, B. A. 1990b. DNA "fingerprints" reveal genotypic distributions in natural populations of blackberries and raspberries (rubus, rosaceae). Amer. J. Bot. 77: 883—888.

Nybom, H., Rogstad, S. H. and Schaal, B. A. 1990. Genetic variation detected by use of the M13 "DNA fingerprint" probe in Malus, Prunus, and Rubus (Rosaceae). Theor. Appl. Genet. 79: 153—156.

Omi, S. K. and Adams, W. T. 1985. Variation in seed set and proportions of outcrossed progeny with clones, crown position, and top pruning in a Douglas-fir seed orchard. Can. J. For. Res. 16: 502—507.

Paterson A. H., Lander, E. S., Hewitt, J. D., Paterson, S., Lincoln, S. E. and Tanksley, S. D. 1988. Resolution of quantitative traits into Mendelian factors by using a complete linkage map of restriction fragment length polymorphisms. Nature 335: 721—726.

Rogstad, S. H., Patton, J. C. and Schaal, B. A. 1988a. A human minisatellite probe reveals RFLPs among individuals of two angiosperms. Nucl. Acids Res. 16: 11378.

Rogstad, S. H., Patton, J. C. and Schaal, B. A. 1988b. M13 repeat probe detects DNA minisatellite-like sequences in gymnosperms and angiosperms. Proc. Natl. Aca. Sci. USA, 85: 9176—9178.

Ryskov, A. P., Jincharadze, A. G., Prosnyak, M. I., Ivanov, P. L. and Limborska, S. A. 1988. M13 phage DNA as a universal marker for DNA finger printing of animals, plants and microorganisms. FEBS LTRS 233: 388—392.

Sakai, R. K., Scharf, S., Faloona, F., Mullis, K. B., Horn, G. T., Erlich, H. A. and Arnheim, N. 1985. Enzymatic amplification of B-globin genomic sequences and restriction site analysis for diagnosis of sickle cell anemia. Science 230: 1350—1354.

Smith, D. B. and Adams, W. T. 1983. Measuring pollen contamination in clonal seed orchards with the aid of genetic markers. pp. 69—77. In: Proc. 17th South. Conf. on Forest Tree Improv.

Soller, M. and Beckmann, J. S. 1983. Genetic polymorphism in varietal identification and genetic improvement. Theor. Appl. Genet. 67: 25—33.

Tanksley, S. D., Young, N. D., Paterson, A. H. and Bonierbale, M. W. 1989. RFLP mapping in plant breeding: new tools for an old science. BioTechnology, 7: 257—264.

Tautz, D. 1989. Hypervariability of simple sequences as a general source for polymorphic DNA markers. Nucl. Acids Res. 17: 6463—6471.

Vassart, G., Georges, M., Monsieur, R., Brocos, H., Geguaere, A. S. and Christophe, D. 1987. A sequence in M13 phage detects hypervariable minisatellites in human and animal DNA. Science 235: 683—684.

Wagner, D. B. 1992. Nuclear, chloroplast, and mitochondrial DNA polymorphisms as biochemical markers in population genetic analyses of forest trees. New Forests 6: 373—390.

Waxler, M. S. and van Buijtenen, J. P. 1981. Early genetic evaluation of loblolly pine. Can. J. For. Res. 11: 351—355.

Weber, J. L. and May, P. E. 1991. Abundant class of human DNA polymorphisms which can be typed using the polymerase chain reaction. Am. J. Hum. Genet. 44: 388—396.

Weising, K., Beyermann, B., Ramser, J. and Kahl, G. 1991. Plant DNA fingerprinting with radioactive and digoxigenated oligonucleotide probes complementary to simple repetitive DNA sequences. Electrophoresis 12: 159—169.

Welsh, J. and McClelland, M. 1990. Fingerprinting genomes using PCR with arbitrary primers. Nucl. Acids Res. 18: 7213—7218.

Wheeler, N. C. and Jech, K. S. 1986. Estimating supplemental mass pollination (SMP) success electrophoretically. pp. 111—120. In: Proceedings Canadian Tree Improvement Association. Quebec City, Canada.

Wheeler, N. C. and Jech, K. S. 1992. The use of electrophoretic markers in seed orchard research. New Forests 6: 311—328.

Williams, C. G. 1986. Early genetic testing for loblolly pine growth traits. Ph. D. dissertation, North Carolina State University, Raleigh.

Williams, C. G. 1987. Influence of shoot ontogeny on juvenile-mature correlations in loblolly pine. For. Sci. 33: 411—422.

Williams, J. G. K., Kubelik, A. R., Livak, K. J., Rafalski, J. A. and Tingey, S. V. 1990. DNA polymorphisms amplified by arbitrary primers are useful as genetic markers. Nucl. Acids Res. 18: 6531—6535.

New Forests **6**: 409—420, 1992.

Review paper

A commentary on current approaches to forest population genetics

HANS-ROLF GREGORIUS[1] and PHILIPPE BARADAT[2]

[1] *Abteilung für Forstgenetik und Forstpflanzenzüchtung, Universität Göttingen, Büsgenweg 2, 3400 Göttingen, Germany;* [2] *INRA, Station de Recherches Forestières, B.P. 45 Gazinet, Pierroton 33610 Cestas, France*

Received 10 August 1991; accepted 18 May 1992

Key words: population genetics, forestry, genetic differentiation, mating systems, selfing, adaptation, biochemical and molecular markets, phylogeny

Abstract. Some of the most frequently chosen approaches to forest population genetics as reflected in the present volume are discussed, and some topics possibly deserving more attention are addressed. Among the first are studies of self-fertilization as a characteristic of forest tree mating systems, gene markers as indicators of adaptive or phylogenetic differentiation, and methodological aspects of the measurement of genetic differentiation. The latter mainly concern generalizable methods of analysis and the involvement of gene markers in the determination of mating systems, detection of mechanisms of frequency dependent selection, and the incorporation of genetic profiles (characteristics of frequency distributions of genetic types) into population genetic analysis.

Introduction

The topics and problems addressed by the contributors to this volume are so numerous and diverse that, despite the great interest each single one deserves, any attempt to summarize and evaluate the results and draw conclusions for future activities must remain unsatisfactory. Therefore, the present authors decided to select several areas of interest which, in their opinion, rank among the most frequently tackled in this volume and are probably worthy of critical discussion, and which determine to some extent current population genetic reasoning. Particularly this last point requires brief consideration of some aspects which might have deserved more emphasis or which are only implicitly addressed in this volume.

All reports referred to forest trees or shrubs, and many of the studies emphasized the following observations (see particularly the reviews of Loveless on tropical species, pp. 67—94, Moran on Australian, pp. 49—66, and Müller-Starck et al., pp. 23—47 on European species, as well as Strauss et al., pp. 125—158 on biochemical and molecular markers; all this issue):

(*i*) Heterozygosity increases in cohorts with increasing age.
(*ii*) In bisexual species the degree of outcrossing is quite large and may vary considerably between individuals.
(*iii*) The genetic variation within forest tree populations is larger than between populations. Moreover, the proportion of genetic variation within and between populations varies depending on taxonomic group, geographic distance, and class of genetic markers for which variation was measured.

Explanations of these observations stressed primarily the following mechanisms (the numbering corresponds to the above):

(*i*) For sessile organisms longevity requires the capacity to survive considerable temporal environmental heterogeneity. It is therefore most likely that such organisms show large genetic heterogeneity (heterozygosity) within individuals to ensure physiologically adaptive flexibility.
(*ii*) The need for individual heterozygosity explains the frequent observation of inbreeding depression, large degrees of which can be avoided by limiting self-fertilization to moderate amounts (though large ovule selfing proportions are occasionally reported; see e.g., Müller-Starck 1982, or Sorensen 1987). Self-fertility is nevertheless meaningful, since it allows for colonization by single or a few scattered individuals and since it helps reduce the genetic load by removing recessive sublethal genes.
(*iii*) Large amounts of gene flow and comparatively little adaptive differentiation of biochemical markers, at least among populations (of the same species) in a given geographic area, could be responsible for the low genetic differentiation. An alternative hypothesis was based on the possibility that populations are genetically similar as a consequence of common evolutionary history. It is noted that different types of genetic marker (proteins, secondary metabolites, DNA, etc.) refer to different time scales of divergence between populations depending on the degree of conservativeness of the marker applied.

In the following discussion these topics will be dealt with under more general headings for the purpose of enabling discussion in a broader framework.

Mating systems

Concerning the role played by mating systems in the analysis of heterozygosity and its adaptive significance or of gene flow, self-fertilization is found to attract most of the experimental interest in studies of forest tree populations (see Mitton, this issue, pp. 197—216). Two explanations of this situation are that dioecious species play a minor role in forest management, and a distinct characteristic of monoecious or hermaphroditic species is their potential to self-fertilize.

Most studies report proportions of self-fertilized ovules; other measures of self-fertilization such as the proportion of self-fertilizing pollen or the combined proportion of both gametic sexes are rarely considered, probably for reasons of experimental feasibility (for an exception see Gregorius et al. 1987). In single cases individual ovule selfing proportions are directly estimable for carriers of unique alleles at a marker locus. However, in the majority of studies, individual and population selfing proportions are estimated on a model-aided basis. The so-called "mixed mating model" is by far the most widely applied model, and the validity of its assumptions is a matter of frequent discussion (see Mitton, this issue, pp. 197—216).

While these estimates may give us an idea of the magnitude and distribution of self-fertilization, they provide per se no insight into the underlying mechanisms. For example, even if the effective population size is known, the observed selfing proportions may not be unambiguously attributable to random or non-random selfing, since limited pollen dispersal may define small neighbourhoods in which selfing may in fact occur at random. The resulting selfing proportions could then markedly exceed the reciprocal of the total population size and could therefore give rise to the misinterpretation that self-pollinating pollen have a fertilization advantage over cross-pollinating ones.

A major problem in experimental studies of plant mating systems is the identification of the individuals involved in a mating event, where a mating event is specified by the formation of a zygote. In practice, observations in early seed stages replace those of zygotes, and efficient methods of analysis mostly employ paternity exclusion and/or model-aided estimation procedures based on multiple locus gene markers (see Adams, this issue, pp. 217—240). Hypervariable gene markers are highly desirable for the identification or reconstruction of descent (parentage), and DNA techniques such as DNA-fingerprinting are particularly well suited for this purpose, provided they are developed to a degree allowing for simple genetic analysis and easy application to large numbers of individuals, as is presently the case with most enzyme markers.

It is also suspected that beyond reconstruction of descent, the usefulness of such hypervariable markers is probably quite limited in population genetics. The analysis of mechanisms of preferential mating, mating success, or other adaptive traits is more likely to benefit from genetic markers such as enzymes, terpenes (see Hanover's review, this issue, pp. 159—178), DNA gene probes, etc. whose biological functions are specifiable.

A negative effect of the concentration of research on self-fertilization is that most of our knowledge on tree mating systems refers to genealogical or spacial traits (see the part of Ellstrand's review in this volume which refers to gene flow within populations; also see Adams, this issue, pp. 217—240). Other traits, including the gene markers themselves, are very rarely considered with respect to their significance as determinants of mating (Müller-Starck and Gregorius 1988). This makes it difficult to judge the extent to which the consequences of self-fertilization affect adaptive potential. For example, notwithstanding a number of observations on incompatibility, little is known about the extent to which negative assortative mating for particular traits may reduce the potential for inbreeding depression. Clearly, such investgations require more generally applicable instruments of qualitative and quantitative analysis of mating systems including concepts of mating preference that do not change with the aspects or models considered (Gregorius 1989).

Distribution and maintenance of genetic variation

A strong motivation for studies on the spacial or ecological distribution of genetic variation stems from the desire to describe and understand the adaptational characteristics of populations and species. This, in turn, requires demonstration of mechanisms capable of maintaining the genetic polymorphisms required for the preservation of adaptability to unpredictably changing environments. The most frequently quoted (though not necessarily experimentally verified) mechanisms are (a) negatively frequency-dependent selection, (b) heterozygote advantage (overdominance) in fitness combined with low amounts of homotypic mating (selfing, phenotypically positive assortative mating, etc.), (c) differential selection and moderate gene flow among niches or demes, and (d) mutation.

Frequency-dependent selection — Negatively frequency-dependent selection is known as rare male mating advantage for a number of dioecious animals, and it was demonstrated in model analyses to arise in bisexual plants as a consequence of sexually asymmetrical variation in gamete production (Ross 1984). However, our knowledge about mechanisms conferring a fitness advantage solely because of rarity is still quite poor. The hypothesis of intraspecific competition for limited resources, for example, which predicts increasing mutual inhibition among individuals of similar type (utilizing the same resource spectrum) as their number increases, in appealing. (The methods of detecting spacial structure of genetic variation presented by Epperson in this volume may be useful for testing compatibility of the hypothesis with observations.) It also suggests wide applicability, since the definition of resources may range from space to nutritional components to the availability of mates, and a similar multitude of inhibition mechanisms (such as allelopathy in many tree species, see Harborne 1988, p. 277ff) is known or conceivable. Yet, in most real populations even the conditions for unambiguous detection of the phenomenon are difficult to realize, and the discovery of the true mechanisms generating this inhibition is more or less a matter of trial and error.

A typical example of this difficulty is provided by the interesting observation of Adams et al. (1990) that segregation distortion frequently occurs in favour of rare alleles. The selective advantage such alleles have in the haplophase must either vanish (i.e., the distortion must vanish) as they become more frequent, or it must be compensated by a disadvantage in the diplophase. In the second case the rare alleles are active in different stages and must have opposite effects on fitness (anatagonistic selection) such that stable equilibria are obtained at low frequencies. The reduction in population fitness implied by the segregation distortion is negligible as long as the total frequency of different alleles exhibiting distortion is small. Since the selection parameters in the haplo- and in the diplophase need not depend on frequencies, this would normally not be considered to be a case of frequency-dependent selection favouring rare alleles. Yet, the frequency dynamics are similar to those expected under the assumption of a rare type fitness advantage, so that techniques of analysis capable of telling the difference between the two forms of selection make high experimental demands.

Adaptive significance of gene markers — For forest trees it appears that, on the population level, no notable associations have been found, so far, between amounts of variation in isozymes and in basically adaptive morphological and physiological traits (Savolainen and Kärkkäinen, this issue, pp. 329—345). While this observation might suggest that enzyme gene markers have little adaptive significance, it must nevertheless be considered that, even in carefully designed experiments, it is difficult to study the possibility of direct selection at a marker locus.

Linkage or stochastic associations with the actually selected but unobserved loci may feign selection at the marker locus (see Bush and Smouse, this issue, pp. 179—196). To some extent this problem can be overcome by comparative studies based on offspring from controlled crosses between parents randomly sampled from a single population or based on observations in different populations, where in both cases variation of the genetic background is intended to reduce the effects of stochastic associations among loci. However, in partially self-fertilizing populations, apparent (associated) heterozygote advantage may arise at selectively neutral loci under different genetic backgrounds as long as heterozygotes are directly selectively advantageous at some other loci (Strobeck 1979). There is thus an urgent need for criteria which allow one to unambiguously distinguish associated from direct selection under manageable and realistic experimental conditions.

In view of the increasing evidence for selection (direct or by association) at marker loci (see Bush and Smouse, this issue, pp. 179—196), this distinction is essential in order to arrive at a realistic estimate of the number of directly selected loci: decline or growth of populations and thus changes in population fitness are affected by direct but not by associated selection. Moreover, the set of loci directly selected may be a different one at different ontogenetic stages or under different environmental conditions. Analogous considerations hold for a single locus, where different sets of alleles may be selectively equivalent under different conditions. More detailed knowledge about these ecological genetic aspects is quite likely to help put the selectionist-neutralist controversy on a more objective basis.

Enzymes provide a good starting position for such investigations in that their functions are quite well known and their activity is in many cases amenable to measurement. The potential for adaptive significance of these gene markers is thus evident, so that reliable information on adaptive or functional equivalence could be obtainable from studies of allozyme activity reaction norms based, for example, on temperature or substrate modfications (see e.g., Koehn et al. 1983). Differences between allozymes in their optimal activity on modified substrates may then serve as an explanation for heterozygote superiority if the enzyme regularly acts in intracellularly heterogeneous substrates. The reasoning is similar to that of spacial environmental heterogeneity, since the overall enzyme activity may be greater in an individual endowed with two allozymes differing in activity optima. Thus a heterozygote may utilize the intracellular substrate heterogeneity more efficiently than can either of the two corresponding homozygotes. Variation in other parameters affecting enzyme activity, such as temperature, may be used in a similar fashion for explaining heterozygote superiority in terms of temporally varying environments, as is well known.

Phylogeny and genetic differentiation — Selective or adaptive convergence may not provide a convincing causal hypothesis for the absence of significant differentiation for certain genetic traits between populations, because of the great number of incidences for which the populations must be assumed to concur. Leaving gene flow out of consideration for the time being (which, in any case, is relevant only for intraspecific comparisons), joint phylogenetic history might in this case offer a more satisfactory explanation (Strauss et al., this issue, pp. 125—158).

Of course, this hypothesis has to deal with the problem that the reconstruction of past events in anagenetic evolution is purely hypothetical or speculative when primarily based on presently observable genetic states. In particular, genetic similarity between populations as a result of common evolutionary (phylogenetic) descent is most likely to be expected if, for the traits considered, differential anagenetic evolution between the populations does not take place (unlikeliness of homoplasy). Concerning stochastic similarity in the sense of moderate divergence as the sole result of stochastic factors, our methods of analysis are mainly based on characteristics of genetic structures which are derived from stochastic dynamic models which are accepted as approximations to biologically reasonable situations. Yet, because these situations resist direct modeling and/or analysis, the quality of the approximations cannot be mathematically verified. Moreover, the non-selective stochastic effects of different modes of reproduction, systems of sexuality, and mating systems are not yet sufficiently explored, and it could also be difficult to unambiguously separate the stochastic consequences of selective neutrality from those of adaptation to specific forms of random variation in environment.

Therefore, genetic similarity as a result of common evolutionary descent can probably be expected predominantly for highly conservative genetic information, which is characterized by low rates of mutation yielding almost exclusively deleterious mutants, and which is thus fixed for a single genetic type. Among the various classes of gene markers supplying such information, extranuclear DNA markers are the most conservative even though, with increasing number of studies, extranuclear DNA previously thought to vary only between species is now found to even vary within individuals (review by Wagner, this issue, pp. 373—390). Since population genetic theory predicts that extranuclear variation is difficult to maintain in a population, it could easily be the case that this type of variation remains undetected. As a consequence, conspecific populations could be fixed for different extranuclear mutants, and this could lead to phylogenetic misclassification. On the other hand, for gene loci varying at the population level, similarity of patterns of variation among populations may at best be a consequence of *recent* common history of these populations but is very unlikely to be a primary consequence of common *evolutionary* descent.

As a causal hypothesis for genetic similarity at polymorphic loci between populations belonging to the same biological species, common descent is thus probably inferior to gene flow. Therefore, genetic markers varying within populations may not be ideal traits for phylogenetic classification. More complex genetic traits with little variation at the population or species level (such as integrated systems of regulatory genes governing development or biosynthesis of secondary metabolites) could perhaps provide more reliable indicators for common descent in connection with appropriate measures of homology. Above all, however, notwithstanding the intuitive appeal of various methods, there is no *conceptually* unique solution to the problem of phylogenetic inference (see e.g., the review of Felsenstein 1988). In many situations such inference has at best an auxiliary function, since the question to be answered is not directly one of descent. For example, it is not the origin or history of the demography and genetic structure of a population but rather the present state of both that determine this population's adaptive fate. Thus the detection of common descent of two populations may have little predictive power for possible differences in the evolution of future adaptations.

Genetic differentiation between and within populations — Studies of the distribution of genetic variation within and between populations probably enjoy greatest popularity among experimental population geneticists because of the large variety of basic problems that can be treated with such studies and the relative simplicity of experimental design. Speciation and phylogeny, adaptation, provenance research, migration, and conservation of genetic resourcs are typical areas of direct concern, and they will require different types of genetic traits (enzyme, terpenoids, DNA, etc.) for appropriate treatment. The almost exclusively applied method of analysis rests on the concept that the totality of genetic variation residing in a set of populations or subpopulations can be additively decomposed into individual genetic differences existing among members of the same collection and among members belonging to different collections. Thus, all of the individual genetic differences may be found within collections if they are all identical with respect to their frequency distribution of genetic types, and all differences exist between collections if no pair of collection shares a genetic type.

Accepting that reproductively isolated polymorphic populations can with high probability be expected to diverge genetically (compare the above remarks on common evolutionary descent), comparatively small amounts of genetic differentiation among populations suggest the existence of considerable gene flow as a homogenizing force. This seems to be the case for a remarkably wide range of tree species, and it accords nicely with the large degrees of cross-fertilization found in these species. Depending on the demographic stage at which differentiation is measured one expects different trends. Direct effects of gene flow show most reliably at the seed stage, yielding the lowest degrees of differentiation at this stage. In contrast, the opportunities of adaptive differentiation between populations increase with increasing age of the trees which therefore leads to larger measures of genetic differentiation.

The ubiquity of this finding (low differentiation with stage-specific trends) is meanwhile beginning to determine our views of adaptive strategies of tree species (see e.g., the compilation of Hamrick et al. in this issue, pp. 95—124), and it is therefore essential to guide one's attention to a serious weakness of the analytical basis of these views. As mentioned above, the quantification of the amounts of variation residing within and between populations assumes additive decomposability of the variation and is specified as a relative measurement by Wright's fixation index F_{ST} (which is mathematically identical to Nei's G_{ST}). As is inherent in Wright's derivations and wording, F_{ST} refers to the degree of genetic fixation among populations. This becomes quite evident if one considers the fact that $F_{ST} = 0$ only if all populations are genetically variable and show the same genetic frequency distributions, while the maximum value of 1 is attained only if all populations are genetically fixed. Thus, for example, according to F_{ST} a collection of populations is genetically completely "differentiated" if all of 1000 populations, say, are genetically fixed but 999 of them for the same allele. This is clearly a case of complete fixation but by no means of complete differentiation between populations. On the contrary, for all practical purposes, this case would rather serve as a demonstration of almost complete absence of genetic differentiation among populations (for a more detailed discussion see Gregorius and Roberds 1986).

Other measures of variation within and between populations, such as the information measure proposed by Lewontin (1972), are rarely applied, so that the statement "more than 50% of the variation is found within populations" may be true for one but not for the other index, and neither may satisfactorily reflect the actual data. One of the present authors even concluded that the current concept of "variation within and between populations" is arbitrary and should be replaced by more consistent measurements of differentiation (Gregorius 1988). In view of this ambiguity, generalizations on patterns of distribution of genetic variation in forest tree populations are probably premature. Particularly in the field of conservation of genetic resources, where knowledge of the distribution of genetic variation is of the utmost importance (Millar and Westfall, this issue, pp. 347—371), decisions based on ambiguous measurement may have grave and irreversible consequences.

Adaptability and genetic profiles — a suggestive area for future research

Conkle (this issue, pp. 5—22) pointed out that characteristics of genetic frequency distributions (genetic profiles) other than those currently most widely utilized (heterozygosity, diversity, etc.) may be important as indicators of more relevant adaptive or evolutionary processes or strategies. At times of drastic and globally active environmental changes accompanied by exorbitant rates of species extinction, this is a very substantial recommendation, since it addresses the genetic basis for preserving the adaptability of populations.

It therefore seems appropriate to conclude this commentary with the suggestion that more effort be put into studies of mechanisms supporting the preservation of adaptability under changing environments. As was argued earlier, phylogenetic studies are probably only of secondary significance in this context, but they may be useful in providing sampling strategies for populations when resources are limited. Moreover, at a purely descriptive level, multivariate techniques (see Westfall and Conkle, this issue, pp. 279—309) may help considerably in the stratification of genetic variability that considers information on linkage disequilibrium among loci. This also suggests as a promising field of research the characterization and analysis of genetic profiles of populations, however with more emphasis on their relations to currently (adaptively) latent and operating genetic potentials. Profiles such as the major and minor genetic polymorphisms discussed by Lewontin (1985) and the search for mechanisms producing them may then gain broader interest.

Minor polymorphisms, in particular, entail the existence of rare alleles, which received considerable attention in several of the contributions to this volume, even though their adaptive significance still largely remains in the dark. Rare or less frequent alleles may of course be viewed as an adaptive reserve (latent potential) which is maintained through constant gene flow from populations in which these alleles are frequent as a consequence of adaptive superiority. However, this hypothesis fails for allelic types which are rare in all populations. First attempts to relate extreme allele frequencies found in populations of the European silver fir to latent and operating genetic potentials were made by Bergmann et al. (1990).

Another aspect frequently addressed refers to genetic associations across gene loci as indicators of co-adaptation. The existence of such associations suggests complex adaptations, the relatively high specificity of which should have a pronounced effect on the adaptive flexibility of a population. Concerning heterozygosity, the analysis must focus on heterozygosity profiles and thus on characteristics of the distribution of the *degree* of heterozygosity (as defined for an individual by its number or proportion of heterozygous loci among a specified set of loci). Since the proportion of heterozygous loci in an average member of a population equals the average proportion of heterozygotes per locus, this widely used parameter cannot describe aspects of heterozygosity that are specifically due to multi-locus associations.

More relevant information can be expected, for example, from comparisons of observed heterozygosity profiles with the corresponding profiles resulting from stochastic independence among loci. In view of the various forces producing heterozygote advantage (temporal environmental heterogeneity, spacial heterogeneity combined with dominance of the locally adapted genetic variants, etc.) such comparisons could more efficiently be based on analytical methods detecting shifts of a heterozygosity profile to larger or lower degrees of heterozygosity, thus possibly revealing the existence of degrees which are characteristic of special mechanisms of adaptation and preservation of adaptability.

The above-mentioned aspects of adaptation should be understood as a more or less arbitrary selection of examples, demonstrating the potential use of genetic profiles in experimental population genetic analysis. Considering the frequently expressed desire for alternative methods of analysis of population genetic field data, the qualitative option inherent in the characterization of genetic profiles might give clues differing from those obtainable from current methods.

Acknowledgments

The authors wish to express their appreciation to three anonymous reviewers, whose comments helped to significantly improve the presentation of the main points of the paper.

References

Adams W. T. 1992. Gene dispersal in forest tree populations. This issue (pp. 217—240).

Adams W T., Neale, D. B., Doerksen, A. H. and Smith, D. B. 1990. Inheritance and linkage of isozyme variants from seed and vegetative bud tissues in coastal Douglas-fir (*Pseudotsuga menziesii* var. *menziesii* (Mirb.) Franco). Silvae Genetica 39(3—4): 153—167.

Bergmann F., Gregorius, H.-R. and Larsen, J. B. 1990. Levels of genetic variation in European silver fir *Abies alba*. Are they related to the species' decline? Genetica 82: 1—10.

Bush R. M. and Smouse, P. E. 1992. Evidence for the adaptive significance of allozymes in forest trees. This issue (pp. 179—196).

Conkle M. T. 1992. Genetic diversity — seeing the forest through the tree. This issue (pp. 5—22).

Ellstrand N. C. 1992. Gene flow among seed plant populations. This issue (pp. 241—256).

Epperson B. K. 1992. Spatial structure of genetic variation within populations of forest trees. This issue (pp. 257—278).

Felsenstein J. 1988. Phylogenies from molecular sequences: Inference and reliability. Annu. Rev. Genet. 22: 521—565.

Gregorius H.-R. 1988. The meaning of genetic variation within and between subpopulations. Theor. Appl. Genet. 76: 947—951.

Gregorius H.-R. 1989. Characterization and Analysis of Mating Systems. Ekopan Verlag, Witzenhausen.

Gregorius H.-R. and Roberds, J. H. 1986. Measurement of genetical differentiation among sub-populations. Theor. Appl. Genet. 71: 826—834.

Gregorius H.-R., Ziehe, M. and Ross, M. D. 1987. Selection caused by self-fertilization. I. Four measures of self-fertilization and their effects on fitness. Theor. Pop. Biol. 31: 91—115.

Hamrick J. L., Godt, M. J. W. and Sherman-Broyles, S. L. 1992. Factors influencing levels of genetic diversity in woody plant species. This issue (pp. 95—124).

Hanover J. W. 1992. Tree fitness as a function of terpene composition. This issue (pp. 159—178).

Harborne J. B. 1988. Introduction to Ecological Biochemistry. Academic Press.

Koehn R. K., Zera, A. J. and Hall, J. G. 1983. Enzyme polymorphism and natural selection. In: Nei, M. and Koehn, R. K. (Eds), Evolution of Genes and Proteins. Sinauer, Sunderland, Massachusetts, pp. 115—136.

Lewontin R. C. 1985. Population genetics. Ann. Rev. Genet. 19: 81—102.

Lewontin R. C. 1972. The apportionment of human diversity. Evolutionary Biology 6: 381—398.

Loveless M. D. 1992. Isozyme variation in tropical trees: Patterns of genetic organization and process influencing genetic structure. This issue (pp. 67—94).

Millar C. I. and Westfall, R. D. 1992. Allozyme markers in forest genetic conservation. This issue (pp. 347—371).

Mitton J. B. 1992. The dynamic mating systems of conifers. This issue (pp. 197—216).

Moran G. F. 1992. Patterns of genetic diversity in Australian tree species. This issue (pp. 49—66).

Müller-Starck G. 1982. Sexually asymmetric fertility selection and partial self-fertilization. 2. Clonal gametic contributions to the offspring of a Scots pine seed orchard. Silva Fennica 16: 99—106.

Müller-Starck G., Baradat, P. and Bergmann, F. 1992. Genetic variation within European tree species. This issue (pp. 23—47).

Müller-Starck G. and Gregorius, H.-R. 1988. Analysis of mating systems in forest trees. In: Weir, B. S. Eisen, E. J., Goodman, M. M., Namkoong, G. (Eds): Proceedings of the Second International Conference on Quantiative Genetics. Sinauer Associates, Sunderland, Massachusetts, pp. 573—595.

Ross M. D. 1984. Frequency-dependent selection in hermaphrodites: the rule rather than the exception. Biol. J. Linn. Soc. 23: 145—155.

Savolainen O. and Kärkkäinen, K. 1992. Effect of forest management on gene pools. This issue (pp.).

Sorensen F. C. 1987. Estimated frequency of natural selfing in Lodgepole Pine (*Pinus contorta* var. *murrayana*) from Central Oregon. Silvae Genetica 36: 215—216.

Strauss S. H., Bousquet, J., Hipkins, V. D. and Hong, Y.-P. 1992. Contributions of biochemical and molecular genetic markers to biosystematic studies of forest trees.

Strobeck C. 1979. Partial selfing and linkage: The effect of a heterotic locus on a neutral locus. Genetics 92: 305—315.

Wagner D. B. 1992. DNA as a biochemical marker in population genetic analyses of forest trees. This issue (pp. 373—390).

Westfall R. D. and Conkle, M. T. 1992. Allozyme markers in breeding zone designation. This issue (pp. 279—309).

FORESTRY SCIENCES

1. P. Baas (ed.): *New Perspectives in Wood Anatomy.* Published on the Occasion of the 50th Anniversary of the International Association of Wood Anatomists. 1982 ISBN 90-247-2526-7
2. C.F.L. Prins (ed.): *Production, Marketing and Use of Finger-Jointed Sawnwood.* Proceedings of an International Seminar Organized by the Timber Committee of the UNECE (Halmar, Norway, 1980). 1982 ISBN 90-247-2569-0
3. R.A.A. Oldeman (ed.): *Tropical Hardwood Utilization.* Practice and Prospects. 1982 ISBN 90-247-2581-X
4. P. den Ouden (in collaboration with B.K. Boom): *Manual of Cultivated Conifers.* Hardy in the Cold- and Warm-Temperate Zone. 3rd ed., 1982 ISBN Hb 90-247-2148-2; Pb 90-247-2644-1
5. J.M. Bonga and D.J. Durzan (eds.): *Tissue Culture in Forestry.* 1982 ISBN 90-247-2660-3
6. T. Satoo: *Forest Biomass.* Rev. ed. by H.A.I. Madgwick. 1982 ISBN 90-247-2710-3
7. Tran Van Nao (ed.): *Forest Fire Prevention and Control.* Proceedings of an International Seminar Organized by the Timber Committee of the UNECE (Warsaw, Poland, 1981). 1982 ISBN 90-247-3050-3
8. J.J. Douglas: *A Re-Appraisal of Forestry Development in Developing Countries.* 1983 ISBN 90-247-2830-4
9. J.C. Gordon and C.T. Wheeler (eds.): *Biological Nitrogen Fixation in Forest Ecosystems.* Foundations and Applications. 1983 ISBN 90-247-2849-5
10. M. Németh: *Virus, Mycoplasma and Rickettsia Diseases of Fruit Trees.* Rev. (English) ed., 1986 ISBN 90-247-2868-1
11. M.L. Duryea and T.D. Landis (eds.): *Forest Nursery Manual.* Production of Bareroot Seedlings. 1984; 2nd printing 1987 ISBN Hb 90-247-2913-0; Pb 90-247-2914-9
12. F.C. Hummel: *Forest Policy.* A Contribution to Resource Development. 1984 ISBN 90-247-2883-5
13. P.D. Manion (ed.): *Scleroderris Canker of Conifers.* Proceedings of an International Symposium on Scleroderris Canker of Conifers (Syracuse, USA, 1983). 1984 ISBN 90-247-2912-2
14. M.L. Duryea and G.N. Brown (eds.): *Seedling Physiology and Reforestation Success.* Proceedings of the Physiology Working Group, Technical Session, Society of American Foresters National Convention (Portland, Oregon, USA, 1983). 1984 ISBN 90-247-2949-1
15. K.A.G. Staaf and N.A. Wiksten (eds.): *Tree Harvesting Techniques.* 1984 ISBN 90-247-2994-7
16. J.D. Boyd: *Biophysical Control of Microfibril Orientation in Plant Cell Walls.* Aquatic and Terrestrial Plants Including Trees. 1985 ISBN 90-247-3101-1
17. W.P.K. Findlay (ed.): *Preservation of Timber in the Tropics.* 1985 ISBN 90-247-3112-7
18. I. Samset: *Winch and Cable Systems.* 1985 ISBN 90-247-3205-0

FORESTRY SCIENCES

19. R.A. Leary: *Interaction Theory in Forest Ecology and Management.* 1985
ISBN 90-247-3220-4
20. S.P. Gessel (ed.): *Forest Site and Productivity.* 1986 ISBN 90-247-3284-0
21. T.C. Hennessey, P.M. Dougherty, S.V. Kossuth and J.D. Johnson (eds.): *Stress Physiology and Forest Productivity.* Proceedings of the Physiology Working Group, Technical Session, Society of American Foresters National Convention (Fort Collins, Colorado, USA, 1985). 1986 ISBN 90-247-3359-6
22. K.R. Shepherd: *Plantation Silviculture.* 1986 ISBN 90-247-3379-0
23. S. Sohlberg and V.E. Sokolov (eds.): *Practical Application of Remote Sensing in Forestry.* Proceedings of a Seminar on the Practical Application of Remote Sensing in Forestry (Jönköping, Sweden, 1985). 1986 ISBN 90-247-3392-8
24. J.M. Bonga and D.J. Durzan (eds.): *Cell and Tissue Culure in Forestry.* Volume 1: General Principles and Biotechnology. 1987 ISBN 90-247-3430-4
25. J.M. Bonga and D.J. Durzan (eds.): *Cell and Tissue Culure in Forestry.* Volume 2: Specific Principles and Methods: Growth and Development. 1987
ISBN 90-247-3431-2
26. J.M. Bonga and D.J. Durzan (eds.): *Cell and Tissue Culure in Forestry.* Volume 3: Case Histories: Gymnosperms, Angiosperms and Palms. 1987 ISBN 90-247-3432-0
Set ISBN (Volumes 24-26) 90-247-3433-9
27. E.G. Richards (ed.): *Forestry and the Forest Industries: Past and Future.* Major Developments in the Forest and Forest Industries Sector Since 1947 in Europe, the USSR and North America. In Commemoration of the 40th Anniversary of the Timber Committee of the UNECE. 1987 ISBN 90-247-3592-0
28. S.V. Kossuth and S.D. Ross (eds.): *Hormonal Control of Tree Growth.* Proceedings of the Physiology Working Group, Technical Session, Society of American Foresters National Convention (Birmingham, Alabama, USA, 1986). 1987 ISBN 90-247-3621-8
29. U. Sundberg and C.R. Silversides: *Operational Efficiency in Forestry.*
Vol. 1: Analysis. 1988 ISBN 90-247-3683-8
30. M.R. Ahuja (ed.): *Somatic Cell Genetics of Woody Plants.* Proceedings of the IUFRO Working Party S2.04-07 Somatic Cell Genetics (Grosshansdorf, Germany, 1987). 1988 . ISBN 90-247-3728-1
31. P.K.R. Nair (ed.): *Agroforestry Systems in the Tropics.* 1989 ISBN 90-247-3790-7
32. C.R. Silversides and U. Sundberg: *Operational Efficiency in Forestry.*
Vol. 2: Practice. 1989 ISBN 0-7923-0063-7
Set ISBN (Volumes 29 and 32) 90-247-3684-6
33. T.L. White and G.R. Hodge (eds.): *Predicting Breeding Values with Applications in Forest Tree Improvement.* 1989 ISBN 0-7923-0460-8
34. H.J. Welch: *The Conifer Manual.* Volume 1. 1991 ISBN 0-7923-0616-3
35. P.K.R. Nair, H.L. Gholz, M.L. Duryea (eds.): *Agroforestry Education and Training.* Present and Future. 1990 ISBN 0-7923-0864-6
36. M.L. Duryea and P.M. Dougherty (eds.): *Forest Regeneration Manual.* 1991
ISBN 0-7923-0960-X

FORESTRY SCIENCES

37. J.J.A. Janssen: *Mechanical Properties of Bamboo.* 1991 ISBN 0-7923-1260-0
38. J.M. Bonga and P. Von Aderkas: *In Vitro Culture of Trees.* 1992 ISBN 0-7923-1540-5
39. L. Fins, S.T. Friedman and J.V. Brotschol (eds.): *Handbook of Quantitative Forest Genetics.* 1992 ISBN 0-7923-1568-5
40. M.J. Kelty, B.C. Larson and C.D. Oliver (eds.): *The Ecology and Silviculture of Mixed-Species Forests.* A Festschrift for David M. Smith. 1992 ISBN 0-7923-1643-6
41. M.R. Ahuja (ed.): *Micropropagation of Woody Plants.* 1992 ISBN 0-7923-1807-2
42. W.T. Adams, S.H. Strauss, D.L. Copes and A.R. Griffin (eds.): *Population Genetics of Forest Trees* (Corvallis, Oregon, USA, 1990) 1992 ISBN 0-7923-1857-9

KLUWER ACADEMIC PUBLISHERS – DORDRECHT / BOSTON / LONDON